Springer Series in
MATERIALS SCIENCE

44

Springer
Berlin
Heidelberg
New York
Barcelona
Hong Kong
London
Milan
Paris
Tokyo

Physics and Astronomy ONLINE LIBRARY

http://www.springer.de/phys/

Springer Series in
MATERIALS SCIENCE

Editors: R. Hull   R. M. Osgood, Jr.   J. Parisi   H. Sakaki   A. Zunger

The Springer Series in Materials Science covers the complete spectrum of materials physics, including fundamental principles, physical properties, materials theory and design. Recognizing the increasing importance of materials science in future device technologies, the book titles in this series reflect the state-of-the-art in understanding and controlling the structure and properties of all important classes of materials.

Series homepage – http://www.springer.de/phys/books/ssms/

Volumes 1–26 are listed at the end of the book.

Jacob Greenberg

Thermodynamic Basis of Crystal Growth

P–T–X Phase Equilibrium
and Non-Stoichiometry

With 126 Figures and 28 Tables

 Springer

Prof. Jacob Greenberg
Department of Inorganic and Analytical Chemistry
The Hebrew University of Jerusalem
Jerusalem 91904, Israel
jacgreen@vms.huji.ac.il

Series Editors:

Prof. Alex Zunger
NREL
National Renewable Energy Laboratory
1617 Cole Boulevard
Golden Colorado 80401-3393, USA

Prof. R. M. Osgood, Jr.
Microelectronics Science Laboratory
Department of Electrical Engineering
Columbia University
Seeley W. Mudd Building
New York, NY 10027, USA

Prof. Jürgen Parisi
Universität Oldenburg, Fachbereich Physik
Abt. Energie- und Halbleiterforschung
Carl-von-Ossietzky-Strasse 9-11
26129 Oldenburg, Germany

Prof. Robert Hull
University of Virginia
Dept. of Materials Science and Engineering
Thornton Hall
Charlottesville, VA 22903-2442, USA

Prof. H. Sakaki
Institute of Industrial Science
University of Tokyo
7-22-1 Roppongi, Minato-ku
Tokyo 106, Japan

Library of Congress Cataloging-in-Publication Data

Greenberg, Jacob H., 1938-
    Thermodynamic basis of crystal growth : P-T-X phase equilibrium and
nonstoichiometry / Jacob H. Greenberg.
        p. cm. -- (Springer series in materials science, ISSN 0933-033X ; 44)
    Includes bibliographical references and index.
    ISBN 3540412468 (alk. paper)
    1. Crystal growth. 2. Thermodynamics. I. Title. II. Series.

    QD921 .G73 2002
    548'.5--dc21
                                                                    00-049714

ISSN 0933-033X
ISBN 3-540-41246-8 Springer-Verlag Berlin Heidelberg New York

Springer-Verlag Berlin Heidelberg New York
a member of BertelsmannSpringer Science+Business Media GmbH

http://www.springer.de

Typesetting by the author
Data conversion by perform GmbH, Heidelberg
Cover concept: eStudio Calamar Steinen
Cover production: *design & production* GmbH, Heidelberg

Printed on acid-free paper   SPIN: 10778037   57/3141/mf   5 4 3 2 1 0

Preface

It is particularly symptomatic that a volume concerning P–T–X phase equilibrium should appear in the Materials Science Series. Entering the 21st century, progress in modern electronics is increasingly becoming associated with devices based not only on silicon but also on chemical compounds. These include both semiconductors and, in the last 15 years, multinary oxides with high-T_c superconductor properties. The critical role of chemical processes in the technologies of these materials is quite evident, and in recent years has stimulated vigorous research activity in the physical chemistry of materials, resulting in a renaissance of this field. The leading role in these efforts belongs to thermodynamics, in particular, computer modeling of chemical processes, phase equilibrium, and controlled synthesis of inorganic materials with preliminary fixed stoichiometric composition. Especially important contributions have been made regarding non-stoichiometry and our understanding of the crucial relationship between composition and properties of the materials since the development of the vapor pressure scanning approach to the phenomenon of non-stoichiometry. This method of the in situ investigation of the crystal composition directly at high temperatures proved to be of an unparalleled precision of 10^{-3}–10^{-4} at. % and made it possible to obtain in an analytical form functional dependences of the crystal composition on temperature, pressure, and composition of the crystallizing matrix for crystals with sub-0.1 at. % range of existence. It is believed that this approach has enormous potential, not least because of the tremendous possibilities for creating proprietary products and improved profit margins. Attesting to the growth of interest in phase equilibrium studies are the expanding research activities in this area in laboratories associated with industries devoted to the production of inorganic materials.

The present book is intended to provide, in a single small volume, an outline of the basic concepts of phase equilibrium in the temperature–pressure–composition P–T–X phase space and the utilization of these concepts in inorganic materials science. This is not a textbook on chemical thermodynamics. Nor does it attempt to provide an encyclopedic coverage of the literature on P–T–X phase equilibrium in inorganic materials. An endeavor has been made to give clear-cut, readily assimilated information about the selected factual matter of topical interest and to present the theoretical foundations in a rigorous albeit concise form.

This is not a reference book, but rather a book for reading, although admittedly not always an easy one. It consists of three interrelated chapters describing the theoretical elements of the geometrical thermodynamics, basic experimental methods of investigation of P–T–X phase equilibrium, and a compendium of experimental data on P–T–X phase equilibrium of selected semiconductor and oxide systems. Key research papers are discussed, in some cases even ones that were published 20, 30 or more years ago, but which are still relevant and sometimes unique. An attempt had been made to present a balanced synopsis of the tremendous effort made by the many scientists who have contributed to research on P–T–X phase equilibrium. Inevitably, when writing such a review one cannot ensure that all the material deserving discussion is included and properly referenced. Moreover, it is often said that an author of a review tends to emphasize his own work. To the extent that I am guilty of both of these sins, I apologize in advance both to the reader and to the researchers whose work may not have received a proper coverage. My hope is that in the attempts to keep the book within reasonable size I have not unwittingly introduced any debilitating misconceptions.

Books of this kind are seldom read from cover to cover. I can only hope that different parts of it will be found not only stimulating, but useful to various readers: practicing materials engineers working in industry, materials scientists at research centers and in academia, and graduate students at physics and materials engineering university departments.

It is a pleasure to express my gratitude to Prof. P. Rudolph of the Institute of Crystal Growth, Berlin, who actually initiated my writing of this book. I am grateful to Prof. U. El-Hanany of the IMARAD Imaging Systems, Ltd., for our long-standing cooperation in implementing some of my vapor pressure scanning ideas in the CdTe crystal growth technology. My special thanks go to Prof. L. Ben-Dor of the Hebrew University of Jerusalem for furnishing me with the independence that I needed to write this book, for her individual kindness and for concealing any dismay at my dilatory progress. I am indebted to Dr. V.N. Guskov and Dr. G.D. Nipan of the Institute of General and Inorganic Chemistry, Russian Academy of Sciences, Moscow, for our collaboration. I am also grateful to Prof. S. Yariv of the Hebrew University of Jerusalem for introducing me to the staff of the Department of Inorganic and Analytical Chemistry of this University.

Jerusalem, October 2001   *J. H. Greenberg*

Contents

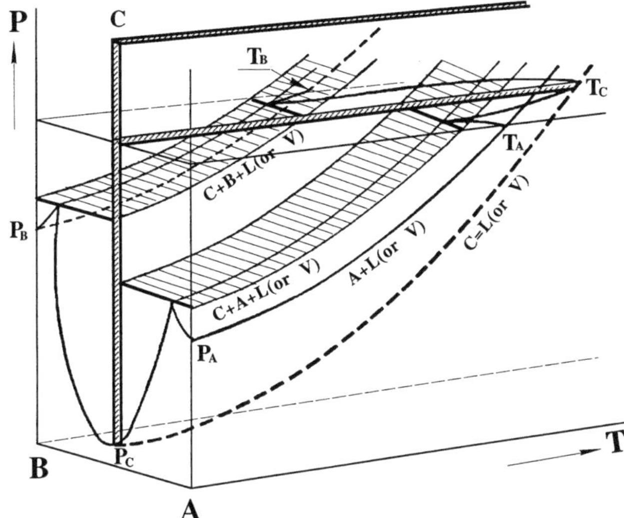

P–T–X diagram of a system with a binary non-stoichiometric compound. Congruent transition

Introduction

One of the fundamental tenets of chemistry is the law of definite proportions. It states that the ratio of the elements which make up a compound is an invariable simple fraction. This proportion was called *stoichiometry*. The law of definite proportions was universally accepted after the famous debate between Proust and Berthollet, which was unequivocally won (as it seemed at that time) by Proust. This law became a theoretical basis of Dalton's atomistic hypothesis and was a stimulus for rapid progress in chemistry, especially organic. More than 100 years passed, and at the beginning of the twentieth century Kurnakov showed [1] that intermetallic compounds do not necessarily have definite stoichiometric compositions. After that, deviation from stoichiometry was experimentally observed in various classes of crystalline solids (oxides, sulfides, etc.).

In 1930, Wagner and Schottky [2] established relationships between non-stoichiometry and lattice defects by statistical thermodynamic methods. At temperatures higher than absolute zero, crystals accumulate a certain number of imperfections. As a result, the entropy S of such a crystal is greater than that of a perfect crystal. At elevated temperatures the entropy term in the free energy G may dominate the endothermic process of point defect formation, and the minimum free energy $G = H-TS$ will occur at a non-stoichiometric composition. Analysis of the phase equilibrium in heterogeneous systems [3, 4], along with kinetic considerations [5], showed that pure components in a binary eutectic system are thermodynamically unstable, since a pure substance at constant pressure has a definite melting point T_m, whereas in a eutectic system it should melt over a temperature range T_E-T_m between the eutectic temperature T_E and T_m. This gradual change in the melting temperature was ascribed to the melting of the solid solution rather than a pure component and is known as the so-called *phase rule argument for the universality of a solid solution.*

Detailed investigations of solid–vapor equilibria show that as a rule these are bivariant processes, i.e. the composition of the crystal depends on two parameters (temperature and pressure), and consequently is not constant. In many applications of the phase equilibrium concepts (e.g. metallurgy) this dependence is of little importance. In these cases, information on the condensed *phase* equilibrium is sufficient. It is contained in the condensed phase diagram, which is a projection of the liquidus and solidus curves onto the $T–X$ (temperature vs. composition) plane. These curves represent the maximum solubility of the components in condensed phases.

Properties of some inorganic materials (semiconductors, in particular) depend strongly on the composition of the crystal, whereas the homogeneity range, or the maximum non-stoichiometry, is very often less than the precision of conventional analytical methods (roughly 0.1 at. %). Consequently, new and more sensitive methods must be developed for investigating non-stoichiometry. For many types of materials, such a technique can be vapor pressure measurement. By this method, composition X at the measured temperature T and pressure P can be directly obtained in many cases with an accuracy as high as 10^{-3}–10^{-4} at. %.

Presentation of P–T–X diagrams of compounds with narrow homogeneity ranges might constitute a graphical problem: the thickness of the line in the diagram might exceed the range of existence of the crystalline phase. Therefore, the composition variable is sometimes presented in various ways: as the logarithm of deviation from stoichiometry, $\ln \delta$; a subscript x in the formula AB_x of the compound, etc. However, since the investigation of non-stoichiometry is neither a geometrical problem nor an exercise in graphical construction, it seems reasonable to show the homogeneity range on an arbitrary scale in the diagram, while giving the actual experimental compositions at different T and P in separate tables. This is the main procedure adopted throughout the following chapters.

The exposition is in three parts. The first is a step-by-step approach to P–T–X phase equilibrium. The types of diagrams most frequently encountered in materials science are discussed. The composition of crystals grown from various matrices is presented in conjunction with P–T–X diagrams. In the second part, the principal experimental methods of investigation of P–T–X equilibrium are briefly described. In the third part, experimental P–T–X diagrams for a large variety of systems are presented. It would be highly advisable to read Chap. 1 and to practise in constructing the isothermal and isobaric sections of simple binary systems before going on to Chap. 3. More complex systems with polymorphism of the components and compounds, as well as metastable states, could be left for more in-depth studies.

Throughout the text emphasis is placed on the phase rule argument of universal solubility. This is where our approach differs from that taken in the encyclopedic book by Ricci [6], a generally recognized source for everybody who deals with P–T–X diagrams. In the first part, extensive use is made of geometrical and analytical investigations of phase equilibrium presented in [6–30]. The reader is referred to these publications as general texts.

1 Thermodynamic Fundamentals

1.1 Definitions

Throughout the exposition some fundamental thermodynamic terms will be in constant use; therefore it seems worthwhile to introduce some basic definitions.

A *thermodynamic system* is an isolated body to be investigated. The complexity of the system depends on the specific problems of the investigation. In thermodynamics the number of particles that constitute the system should be sufficient to be described by statistical laws and the concept of the state of aggregation. A set of experimentally measured properties characterizes *the state of the system.* Two kinds of properties are to be distinguished. If the value of the property for the whole system is equal to the sum of those of its separate parts, then it is called an *extensive quantity.* Volume and mass are extensive properties. Non-additive properties are called *intensive quantities.* These become uniform throughout the system when equilibrium is attained. Temperature and pressure are examples of intensive properties.

The state, that the system spontaneously attains when isolated from the physical world, is called *equilibrium.* In equilibrium, the properties of the system, which are called *parameters*, are independent of time. A *state function*, or thermodynamic potential, is a quantity for which the differential is an exact differential over the parameters of state. The following state functions are known

Gibbs energy $G = G\,(T,\,P,\,n_1,\,n_2,\,...\,,\,n_k)$;
Helmholtz energy $F = F\,(T,\,V,\,n_1,\,n_2,\,...\,,\,n_k)$;
Enthalpy $H = H\,(S,\,P,\,n_1,\,n_2,\,...\,,\,n_k)$;
Internal energy $U = U\,(S,\,V,\,n_1,\,n_2,\,...\,,\,n_k)$.

Here $n_1,\,n_2,\,...\,,\,n_k$ are the mole numbers of the components. The *components* are the minimal necessary constituents of the system capable of varying independently in concentration. The total of species, which make up the system, is greater that the number of the components if the concentrations of the species are correlated. This may be a consequence of chemical reactions between the species. Then the number of components n is the difference between the total number of species N and the number of restricting equations R: $n = N - R$. For example,

$$CaCO_3(s) = CaO(s) + CO_2(g)$$

can be considered a two-component system, since here $N = 3$ and $R = 1$.

A set of homogeneous portions of the system with identical and continuous thermodynamic properties is called a *phase*. Phases are separated by phase surfaces at which the properties change discontinuously. An *equation of state* is an analytical form, which relates the thermodynamic potential to its appropriate parameters. Any form of the *Gibbs fundamental equation*

$$dU = TdS - PdV + \Sigma\mu_i dn_i , \tag{1}$$
$$dF = - SdT - PdV + \Sigma\mu_i dn_i , \tag{2}$$
$$dH = TdS + VdP + \Sigma\mu_i dn_i , \tag{3}$$
$$dG = - SdT + VdP + \Sigma\mu_i dn_i \tag{4}$$

can be an equation of state. In Eqs. (1)–(4), μ_i is the *chemical potential* of the ith component which is the partial derivative of the corresponding characteristic function with respect to the number of moles n_i. For example, in Eq. (4)

$$\mu_i = (\partial G/\partial n_i)_{T,P,nj} . \tag{5}$$

A *thermodynamic function* is said to be *characteristic* if all the thermodynamic properties of the system can be expressed in terms of this chosen function together with its derivatives with respect to the corresponding parameters. In practical applications, a form of the fundamental equation

$$SdT - VdP + \Sigma n_i d\mu_i = 0 . \tag{6}$$

is frequently used. It is known as the *Gibbs–Duhem equation*, and at constant T and P it reduces to

$$\Sigma\, n_i d\mu_i = 0 . \tag{7}$$

The choice of a particular form of the fundamental equation depends on the experimental conditions. If the experimental parameters to be measured are pressure P, temperature T, and composition X (or mole numbers n_i), as is the case with P–T–X investigations, then Eq. (4) is the appropriate equation of state.

The *concept of equilibrium condition* is very important in thermodynamics. According to Gibbs, the necessary and sufficient condition for the system to be in equilibrium is either

$$(\delta S)_{U,V,ni,...,nk} \leq 0 , \tag{8}$$

or

$$(\delta U)_{S,V,ni,...,nk} \geq 0 , \tag{9}$$

where δ is a virtual displacement. Depending on the experiment, other forms of the equilibrium conditions can be used, which are equivalent to Eqs. (8,9). In the following discussion, it will be convenient to use the form

$$(\delta G)_{T,P,ni,\dots,nk} \geq 0 \,, \tag{10}$$

which states that the system is in equilibrium if the Gibbs free energy is at a minimum with respect to every infinitesimal isothermal-isobaric process.

The number of parameters, which are arbitrarily variable without changing the phase state of the system, is called the *number of degrees of freedom*, or *variance* of the system. The variance determines the number of parameters, which must be fixed to define fully the state of the system. Phase equilibria are classified according to the number of coexisting phases (single-phase, two-phase, etc.) and by the number of degrees of freedom (invariant, univariant, bivariant, etc.).

The equation, which defines the variance F of a system, made up of n components and φ phases,

$$F = n - \varphi + 2 \tag{11}$$

is known as the *Gibbs phase rule*. If the parameters are correlated in any way (by chemical reactions, conditions of equality of the composition, invariability), then the variance is reduced by the number of restricting equations R, and the general form of the phase rule is

$$F = n - \varphi + 2 - R \,. \tag{12}$$

For example, in a binary system, the bivariant liquid–vapor equilibrium ($F = 2$) becomes univariant on the azeotropic line where $X_L = X_V$ ($F = 1$, because $R = 1$).

The compositional dependences of the vapor pressure of the system at $T =$ const or of the boiling temperature at $P =$ const are given by the Gibbs–Konovalov equations:

$$(\partial P/\partial X)_T > 0 \text{ when } X_V > X_L, \text{ or } (\partial P/\partial X)_T < 0 \text{ when } X_V < X_L \,; \tag{13}$$
$$(\partial T/\partial X)_P > 0 \text{ when } X_V < X_L, \text{ or } (\partial T/\partial X)_P < 0 \text{ when } X_V > X_L \,. \tag{14}$$
$$\text{If } (\partial T/\partial X)_P = 0, \text{ then } X_V = X_L \,. \tag{15}$$

The vapor is enriched in the component that increases the vapor pressure (at $T =$ const) or decreases the boiling temperature (at $P =$ const) when added to the mixture. If the temperature and pressure pass through an extremum, then the compositions of the coexisting phases coincide.

1.2 Geometrical representation of phase equilibrium

As we have seen, the fundamental equation is an analytical description of the thermodynamic state of the system. For example, in the form of Eq. (4), it is the function $\psi(G, P, T, X) = 0$, which relates the Gibbs free energy to the temperature, pressure and the composition parameter X. Besides the analytical form of Eqs. (1)–(4), the state of the system, or the function ψ, can also be represented geometrically. Such a representation is called the *phase diagram*. Experimentally the most conveniently measured parameters are pressure, temperature, and composition. Therefore, the most common form of graphical representation of phase equilibrium is the P–T–X diagram, which is the projection of the G–P–T–X diagram, or the $\psi(G, P, T, X) = 0$ function, onto the phase space P–T–X.

1.2.1 One-component systems

1.2.1.1 *P–T* phase diagram

In a one-component system, the equation of state, Eq. (4) is reduced to

$$dG = -SdT + VdP ,\tag{16}$$

i.e., the ψ-function becomes $\psi(G, P, T) = 0$, and the state of the system can be described by the three-dimensional G–P–T diagram. Since the chemical potential, or the molar Gibbs energy, for the one-component system is the same in all coexisting phases, the state of the system is defined by two parameters, P and T. Consequently, a geometrical representation of a one-component system is a two-dimensional P–T phase diagram. This diagram (Fig. 1 [21]) comprises the fields of existence of the solid S, liquid L, and vapor V phases divided by two-phase curves, which converge in a triple point. The univariant two-phase equilibria SV (sublimation), SL (melting), and LV (vaporization) are described by Eq. (17),

$$V_{ij}dP = S_{ij}dT ,\tag{17}$$

where $S_{ij} = S_j - S_i$ and $V_{ij} = V_j - V_i$ are the changes in molar entropy and volume for the phase transition $i \to j$. Because $S_{ij} = H_{ij}/T$ (H_{ij} is the enthalpy change), it follows from Eq. (17) that

$$dP/dT = H_{ij}/V_{ij}T .\tag{18}$$

This relation is known as the Clausius–Clapeyron equation. The sign and the value of the dP/dT slopes for the SV, LV, and SL curves in the P–T diagram are determined by the molar volumes of the coexisting phases, because the entropy and enthalpy of sublimation, vaporization, and melting are positive.

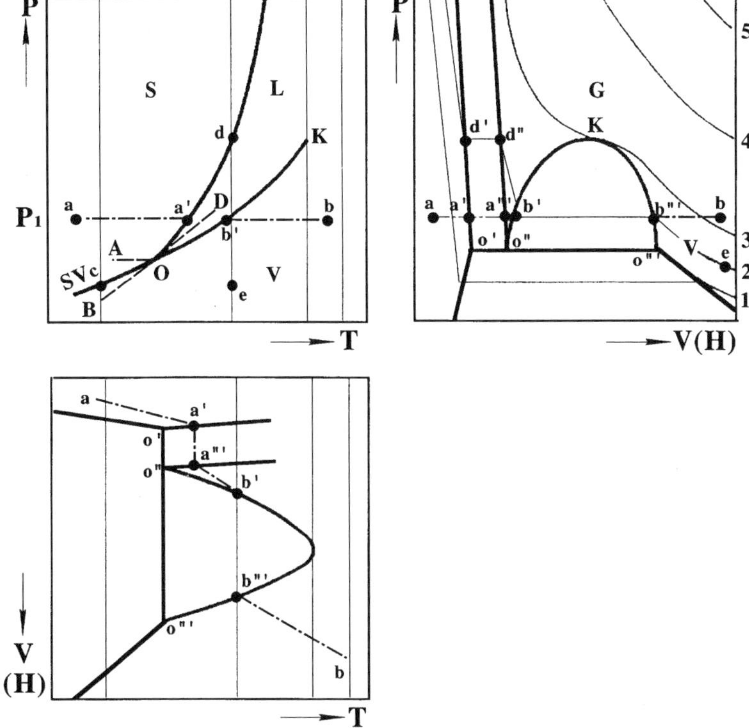

Fig. 1. Schematic $P–T$, $P–V$, and $V–T$ projections of the $P–V–T$ phase diagram of CO_2

The *sublimation curve* SV originates at the ($P=0$, $T=0$) point, has a positive slope $dP/dT > 0$ (the molar volume of the vapor is much greater than that of the solid), and ends up at the invariant point O. The *vaporization curve* LV originates at the invariant point O and ends at the critical point K, where L and V become identical. The slope of LV is also positive, because $H_{LV} > 0$ and $V_{LV} > 0$. The molar sublimation enthalpy is a sum of the melting (H_{SL}) and vaporization (H_{LV}) enthalpies. Since the change in volume for sublimation and vaporization is roughly the same, the slope of the sublimation curve is greater than that of the vaporization curve

$$(dP/dT)^{SV} > (dP/dT)^{LV} ; \tag{19}$$

this means that the sublimation curve is steeper than the vaporization curve. The *melting curve* SL starts at the invariant point O and describes the pressure dependence of the melting temperature. Melting is an endothermic process ($H_{SL} > 0$); consequently, the sign of dP/dT in Eq. (18), or the arrangement of the melting curve on the $P–T$ plane, is determined by the molar volumes of the solid and liquid phases. If $V_{SL} = V_L - V_S > 0$ (which is the case in Fig. 1), then the melting temperature rises with increasing pressure. The melting curve SL is considerably steeper than both SV and LV, because V_{SL} is much smaller than V_{SV} and V_{LV}.

The invariant *triple point* O (Fig. 1) is the intersection of the three univariant curves, SV, LV, and SL. All three phases (S+L+V) of carbon dioxide coexist at this point ($t = -56.6°C$, $P = 5.2$ atm). Dashed lines AO, BO, and OD in Fig. 1 correspond to the metastable extensions of the corresponding equilibria. AO describes the vapor pressure over a supercooled liquid, OD is for an overheated solid, and BO describes a supercooled state of the mixture (liquid + solid). The Gibbs energies of these states are higher than those of equilibrium; as a result, the metastable phase disappears spontaneously. For example, the vapor pressure over the supercooled liquid (AO in Fig. 1) is higher than that over the equilibrium solid phase (SV), and the liquid → solid transition is a spontaneous process.

1.2.1.2 Three-dimensional *P–V–T* diagram

A two-dimensional *P–T* diagram (Fig. 1) defines the system, i.e. describes the number and nature of phases at fixed P and T values. However, this diagram contains no information on the properties of the phases and the relative quantities of the phases that comprise the equilibrium state. The properties of the phases (molar volume, entropy, chemical potential, etc.) are functions of the temperature and pressure. They can be calculated from the equation of state, Eq. (16),

$$V = (\partial G/\partial P)_T; \quad S = -(\partial G/\partial T)_P. \tag{20}$$

Otherwise, they can be measured experimentally at a chosen P and T. The most common technique is the *P–V–T* experiment that consists of measurements of all three parameters and constructing the *P–V–T* diagram (Fig. 2). The state of the system here is completely defined by three parameters. All possible (P,V,T) values for a certain phase in three-dimensional space describe a surface that is called the *field of existence* of this phase. A combination of these fields makes up the three-dimensional *P–V–T* diagram.

Figure 2 is the *P–V–T* diagram of carbon dioxide [21]. It consists of the solidus S, liquidus L, and vaporus V surfaces, that describe the fields of existence of the solid, liquid, and vapor phases. At temperatures above the critical point K ($t_K = 31.1°C$, $P_K = 72.947$ atm for CO_2), the L and V surfaces converge into G. When the aggregation state of the system changes (phase transition), extensive properties (molar volume V or enthalpy H in Fig. 2, in particular) change discontinuously. As a result, the phase boundaries are shifted in the *P–V–T* space relative to one another, and the points outside these boundaries do not describe the equilibrium state of the system. For example, if a solid consumes heat in an isobaric, isothermal process, it would partially melt on reaching the phase boundary. On further heating at $T = $ const, $P = $ const, the amount of the liquid in the two-phase mixture (S+L) gradually increases at the expense of the solid. Point a″ (Fig. 2) corresponds to an

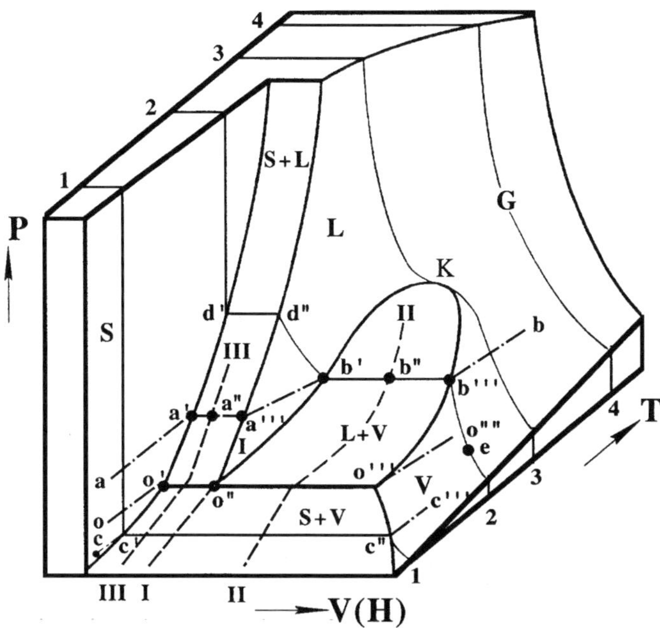

Fig. 2. Schematic P–V–T phase diagram of CO_2

overall specific volume of such a mixture and is a weighted average of the specific volumes a′ and a‴ of the individual phases, S and L in this case.

An important concept for describing phase diagrams is that of conjugated points, conjugated curves, and conodes. *Conjugated points* define the properties (V or H in Fig. 2) of individual coexisting phases in equilibrium at T = const, P = const. In Fig. 2 these are a′ and a‴, b′ and b‴, c′ and c‴ for the solid–liquid, liquid–vapor, and solid–vapor equilibria, respectively. Curves made up of the conjugated points are called *conjugated curves* (o′a′ and o″a‴, ko″ and ko‴in Fig. 2). The tie-lines between the conjugated points are called *conodes*; a′a‴ and b′b‴ in Fig. 2 are conodes. When P and T change continuously, the conodes generate three two-phase surfaces, S+L, S+V, and L+V. For example, the b′b‴ conode generates the (L+V) surface when the temperature rises from the triple point O to the critical point K. If the state of the system is defined by the point b″, then the masses of the liquid and vapor are inversely proportional to the segments of the conode:

$$(m''/m''') = (b''b'''/b'b'').$$ (21)

Equation (21) is known as the *lever rule*. P–T, P–V, and T–V projections of the three-dimensional diagram (Fig. 2) are shown in Fig. 1. A specific feature of the projections is that the points on the two-phase SV, LV, and SL curves do not necessarily correspond to two phases. For example, point a$'$ on the P–T projection of the SL equilibrium may correspond to two-phase state a$''$ (Fig. 2) as well as to individual solid a$'$ or liquid a$'''$. The triple point O in Fig. 1 is the projection of three points: o$'$ for the solid, o$''$ for the liquid, and o$'''$ for the vapor (Fig. 2). Isotherms 1-5 in Figs. 1 and 2 are the sections of the three-dimensional figure cut by the T = const planes. It should be pointed out that Fig. 2 is a general type of the diagram pressure–temperature–extensive property.

1.2.1.3 Phase processes

Phase processes in a one-component system may occur at constant pressure, temperature, volume, or entropy. The sequence of phases depends on whether the fixed parameter is higher, lower, or equal to that at the triple point.

1.2.1.3.1 Isobaric processes

Consider a state of the system defined by the point a (Figs. 1 and 2). The pressure here is higher than at the triple point O. If this system is isobarically (P=const) heated (aa$'$), it remains a single-phase solid until point a$'$ on the two-phase SL curve is reached (Fig. 1). Here it gradually melts along the a$'$a$''$a$'''$ (Fig. 2) line; the mass ratio m_L/m_S continuously changes from 0 (at point a$'$) to 1 (at point a$'''$) according to the lever rule, Eq. (21). At point a$'''$ all of the solid is melted, and further isobaric heating proceeds along the a$'''$b$'$ line inside the existence field of the pure liquid L. On reaching the point b$'$, vaporization is observed, which proceeds along the b$'$b$''$b$'''$ conode in the (liquid–vapor) field up to the moment when the liquid disappears completely at point b$'''$. Subsequent isobaric heating of the vapor corresponds to the b$'''$b line (Figs. 1 and 2) with no further phase transitions.

If the initial pressure of the system is lower than that at the triple point, P(c) < P(O), then the two-phase liquid-vapor field is missing. Heating of the solid leads to formation of the vapor phase with no prior melting, and the sequence of the phase states for the isobaric heating (Fig. 2) is solid (cc$'$) $\rightarrow$ solid–vapor (c$'$c$''$) $\rightarrow$ vapor (c$''$c$'''$). An example of this type of process is heating of iodine or carbon dioxide at P = 1 atm. A special case is isobaric heating of the system at the pressure at the triple point P = P(O). The volume of the initial solid S changes along the oo$'$ line, and liquid L is formed at point o$'$. Melting (a two-phase state of the system) proceeds at T = const along the o$'$o$''$ line up to point o$''$ where the solid phase disappears. Subsequently the added heat is consumed at T = const to form the vapor phase along the o$''$o$'''$ line (two-phase liquid–vapor equilibrium). The liquid L completely vaporizes at o$'''$, and the volume of the resulting vapor V (single phase) changes along the o$'''$o$''''$ line. It should be stressed that for a system heated at the constant pressure of the invariant point the maximum number of coexisting phases is φ = 2, which follows from the phase rule, Eq. (12), with one restricting condition P = const (R=1).

1.2.1.3.2 Isothermal processes

Isothermal processes can be followed in the $T = $ const sections 1–5 of the system (Figs. 1 and 2). Isothermal compression of the vapor (point e, section 2) follows the line eb''' (Fig. 2) in the single-phase field up to the point b''' where condensation into liquid is observed. The total density of the two-phase (liquid–vapor) system then increases from b''' up to b', at which point the vapor condenses completely, and the resulting single-phase liquid L is further compressed up to point d'' where crystallization begins. Further compression of the two-phase (solid–liquid) system proceeds at $T = $ const, $P = $ const, until all of the liquid is crystallized (point d').

The difference in the molar volumes of the conjugated liquid and vapor phases in the (S + L) state (the b'b''' conode) decreases when the temperature (and pressure) rise from T(O). This b'b''' conode describes a two-phase liquid–vapor surface o''b'Kb'''o''' in the P–V–T space. Point K corresponds to a state of the system, where the molar volumes of the liquid and vapor converge. It is known as a *critical point* and is characterized by the critical temperature, critical pressure, and critical volume (density), which are physical constants of a substance. Above the critical point, the phase boundary, observed in the liquid–vapor state, disappears. As can be seen in Figs. 1 and 2, at temperatures above $T_3 = T_K$, no condensation can be observed in the system, no matter how high the pressure. This is the principal difference between the *vapor* V and *gas* G, which can be defined as a superheated vapor at a temperature higher than T_K. On the other hand, the vapor can be defined as a gas at $T < T_K$ which can be either in equilibrium with the liquid (*saturated vapor*) or in a single-phase state (*unsaturated vapor*). The saturated vapor is described in Fig. 2 by the curve Kb'''o'''c'', and the field of existence of the unsaturated vapor is at $T < T_3$ to the right of this curve.

Section 2 corresponds to the temperature $T_2 > T(O)$ that is higher than that at the invariant point O. When the system is compressed at $T_1 < T(O)$, the sequence of phases is V → (S+V) → S; the vapor condenses on the sublimation curve SV, and no phase transitions are observed when the volume goes below point c'. Phase processes at $T = T(O)$ are similar to those at $P = P(O)$.

1.2.1.3.3 Isochoric processes

Changes in the system under constant volume condition V=const are known as isochoric processes. These are univariant processes, and consequently they proceed along the two-phase curves SV, SL, and LV (Fig. 1). Depending on the initial mass ratio of the coexisting phases, these processes are seen in Fig. 2 as I–I, II–II or III–III trajectories. In I–I the initial solid/vapor ratio is equal to that at point o'', where liquid L is formed from the (S+V) mixture as a result of the phase reaction at the invariant point. As a consequence, both solid and vapor phases are totally consumed to form the liquid L, which is the only remaining phase. For the II–II trajectory, the S/V ratio is lower than that at point o''. The result of the phase reaction at the triple point is complete melting of the solid phase, and the surplus vapor forms a two-phase (L+V) state. If the S/V initial ratio is greater than that at point o'' (III–III), the vapor is com-

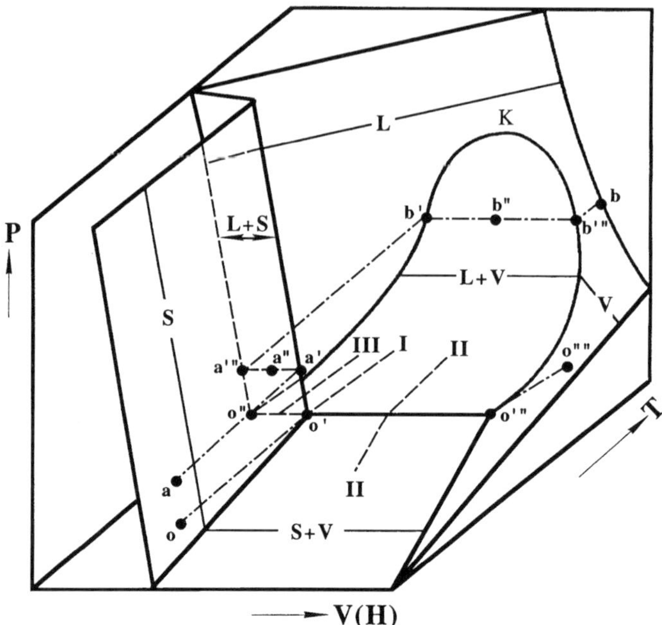

Fig. 3. Schematic P–V–T phase diagram of H_2O

pletely consumed (condensed) at the triple point, and the system proceeds along the SL two-phase curve.

In the system (Figs. 1 and 2) the molar volume of the liquid is greater than that of the solid, $V_L > V_S$, and the Clausius–Clapeyron equation, Eq. (18), requires that for such a system the slope of the SL curve is positive. On the contrary, if $V_L < V_S$, then the slope of the SL curve is negative, and the melting temperature decreases with rising pressure. An example of this type of system is shown in Fig. 3, which is the phase diagram of water [21]. Projections of this diagram are presented in Fig. 4. The diagram consists of the single-phase surfaces S, L, and V and the two-phase surfaces (S+V, S+L, and L+V) generated by the corresponding conodes. When heat is added at constant volume to the (S+V) state of this system, the pressure and temperature change along the SV sublimation curve (Fig. 4). If the S/V ratio is given by the II–II trajectory (Fig. 3), then at the invariant point, the solid is completely consumed, and the system proceeds into the (L+V) state. The cooling process of the (L+V) state in this system depends on the initial mass ratio L/V. In II–II, the liquid is completely crystallized, and the surplus vapor forms the (S+V) state below the invariant point O ($t_O = 0°C$, $P_O = 4.579$ mmHg). Cooling the mixture I results in formation of the single solid phase, whereas the composition III cools down to the (S+L) state because of complete condensation of the vapor into the solid phase. In other respects the system (Figs. 3 and 4) is believed to be self-explanatory.

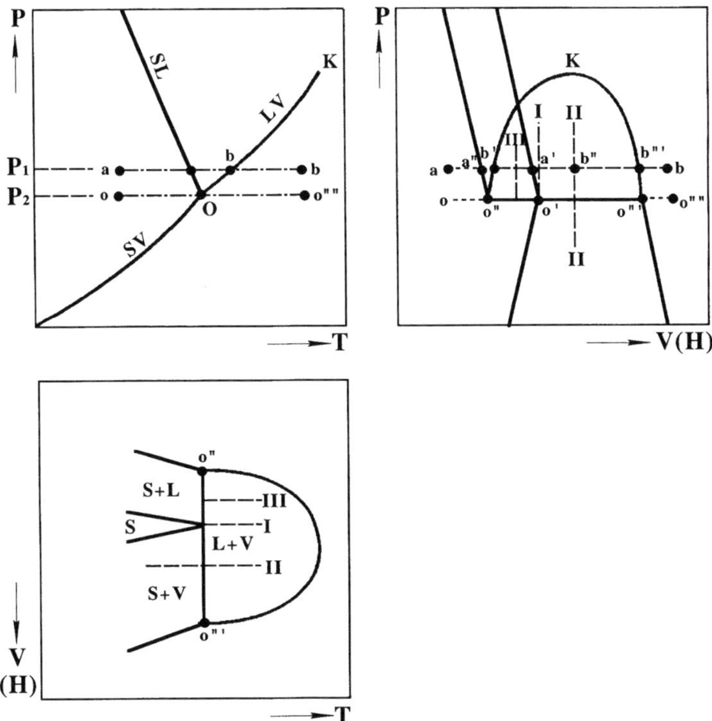

Fig. 4. Schematic P–T, P–V, and V–T projections of the P–V–T phase diagram of H_2O

1.2.1.4 Polymorphism and metastable states

Polymorphism is the ability of solids to crystallize in various structures in different temperature and pressure ranges. It is well known [31] that few chemical elements exist in only one single crystal structure, and some compounds form several (up to ten) polymorphs, particularly at high pressure [31,32]. Thus, polymorphism of solids may be considered a rule rather than an exception. An experimental phenomenon, also well-known, is the capacity of physico-chemical systems to retain *metastable states*. According to Gibbs [7], such states are stable with respect to all infinitesimal changes but are unstable with respect to the finite changes of parameters. In other words [14], the state of the system, which corresponds to the lowest (or the highest) of several possible extrema of the thermodynamic potentials (the global extremum), may be called the thermodynamically stable state, whereas all of the other local extrema describe metastable states. Phase diagrams with metastable states are only briefly mentioned by Ricci [6]. Systems with polymorphism, metastable states, and relationships between these phenomena were studied in detail in [33,34]. These systems are of the utmost interest, both basic and applied, because they are related to such problems as the preparation of synthetic diamonds, the stabilization of cubic

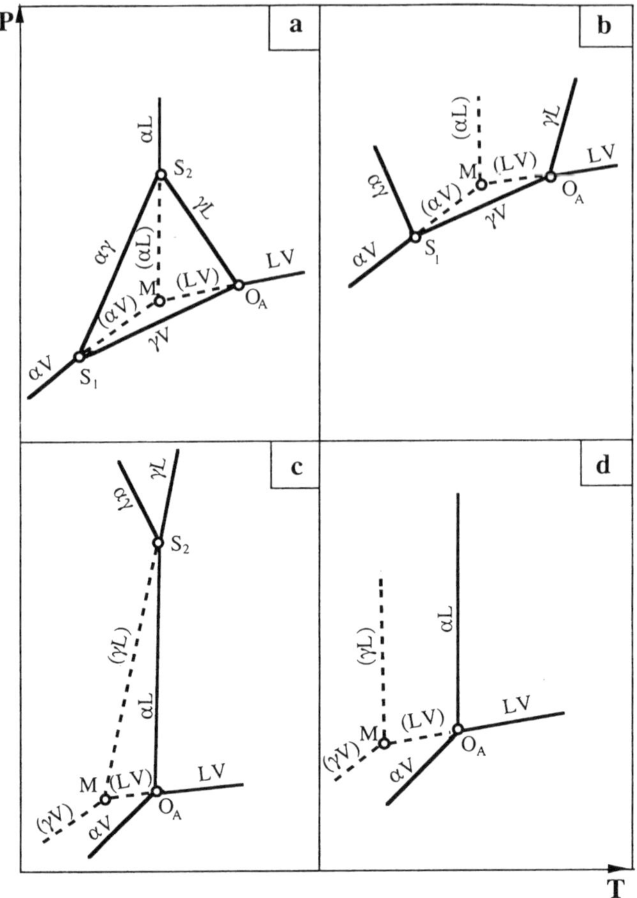

Fig. 5. *P–T–X* diagram of one-component system with four types of polymorphism

zirconia and, most recently, high temperature superconductivity, which is widely considered a property of metastable materials.

The subsequent discussion will be confined to systems with only two polymorphs of the crystalline solid. Formation of more than two polymorphs would complicate the visual representation considerably while adding nothing in principle to the description of the system. The following *symbolism* will be adopted throughout the discussion. The low-temperature polymorph of the solid will be labeled α, the high-temperature form γ, and stable univariant equilibria will be given by solid lines, whereas the metastable ones are denoted by dashed curves. On every line, the corresponding phase state of the system will be spelt out, and the inscriptions for metastable states will be in brackets. The invariant point of the three-phase equilibrium solid–liquid–vapor, αLV or γLV, will be denoted O_A. The solid–solid–vapor,

$\alpha\gamma V$, and solid–solid–liquid $\alpha\gamma L$ equilibria will be labeled S_1 and S_2, correspondingly, and the point of the metastable three-phase state, (αLV) or (γLV), will appear as M. For simplicity, the stable reversible equilibrium will be referred to just as an equilibrium.

If a crystalline substance exists in two forms, four types of P–T diagrams are possible (Fig. 5). In the type shown in Fig. 5a, the characteristic feature is the equilibrium sublimation and fusion of both polymorphs represented by the corresponding curves αV and γV for sublimation, and αL and γL for fusion. The $\alpha\gamma$ line describes the solid state phase transition equilibrium and shows the pressure dependence of the α–γ transition temperature. LV is the vaporization line. The γ-form is in three-phase equilibrium with the liquid and vapor at the triple point O_A, whereas the α-form gives a three-phase metastable state (αLV) with the liquid and vapor at point M. This state can be reached in three ways: by overheating the α-form from temperature T_{S1} to T_M along the (αV) curve, by supercooling the melt from T_{OA} down to T_M along the (LV) curve; or by going along the (αL) line from S_2 down to M. The equilibrium single-phase field of existence for the α polymorph is to the left of the αV, $\alpha\gamma$, and αL curves; that for the γ-form is within $S_1S_2O_A$; the liquid is to the right of the vaporization curve LV and the fusion lines γL and αL; and the vapor is below the sublimation curves αV, γV and vaporization LV. The single-phase fields for metastable phases are the following: for the liquid (L), it is between the lines γL, (LV) and (αL); for the vapor, between γV, (αV), and (LV); and for the α-form, between $\alpha\gamma$, (αL), and (αV). The γ-phase does not form metastable states. This type of polymorphism is frequently found in chemical elements, such as sulfur, iron (δ- and γ-forms, in particular), etc. [31].

In systems of the type shown in Fig. 5b, the invariant equilibrium $\alpha\gamma L$ (point S_2) is missing. As a result, the α-form melts only in a metastable process along the (αL) curve. This type of diagram is characteristic of Group II elements (Ca and Sr [31]).

In diagrams of the type shown in Fig. 5c, the γ-phase coexists in two-phase equilibria only with condensed phases α and L. The stable equilibrium $\alpha\gamma V$ (point S_1) does not appear, and as a consequence, the sublimation of the γ-form is a metastable process represented by the (γV) curve. The metastable single-phase region of γ is to the left of the curves (γV), (γL), and $\alpha\gamma$, and the α-form has no metastable field of existence. Examples of diagrams of this type are those of selenium and carbon [31]. According to [31], the triple point $\gamma\alpha L$ (diamond – graphite – melt) appears at $T > 4000$ K and $P > 10$ Gpa, and the triple point αLV (graphite – melt – vapor) temperature is about 4000 K. Sublimation of diamond is a non-equilibrium process; it may be in equilibrium either with graphite ($\alpha\gamma$ curve) or with the liquid (γL curve). All direct and catalytic syntheses of diamond are based on the phase transition graphite $\rightarrow$ diamond along the $\alpha\gamma$ line at temperatures above 3000 K and pressures above 10 GPa. Meanwhile, to cool the diamond down to low temperatures is apparently possible if the cooling process follows the metastable crystallization line (γL). The vapor-phase synthesis of diamond is evidently associated with non-equilibrium processes.

The characteristic feature of the diagram shown in Fig. 5d, is that the γ-form both melts and sublimes only along the metastable curves (γV) and (γL), meaning that the γ-phase appears only as a metastable state. This type of polymorphism has been observed for phosphorus [6] and benzophenone [18].

Thus, the following four types of phase diagrams are known for one-component systems with polymorphism:

- Fig. 5a — sublimation and fusion of both polymorphs are equilibrium processes; subsequently this type of polymorphism will be referred to as Type I.
- Fig. 5b — fusion of one of the forms is a metastable process; this diagram will be called Type II.
- Fig. 5c — sublimation of one of the polymorphs is a metastable process; this will be labeled Type III.
- Fig. 5d — both sublimation and fusion of one of the polymorphs are metastable processes, i.e. the region of single-phase existence of one of the polymorphs is metastable; we will call it Type IV.

Sometimes the diagrams in Fig. 5a and Fig. 5b are associated with the concept of enantiotropy, Type III with high-pressure polymorphism, and Type IV with the concept of monotropy [6].

The general feature of all types of diagrams is the possibility for one of the polymorphs to crystallize in a metastable process from the supercooled liquid. In Fig. 5, it means that the triple metastable point M may always be attained via the metastable state (LV) down from the LV equilibrium. Furthermore, in all four cases, only one three-phase metastable state M is possible. For example, three-phase metastable state ($\alpha\gamma$L) cannot be formed as an intersection of ($\alpha\gamma$), (αL), and (γL) lines (Fig. 6b) because in such a system it would be possible to crystallize the solid (γ) by heating the liquid (L) in isobaric conditions. The corresponding single-phase region of existence for the metastable liquid (L) is between the (αL) and (γL) lines (marked by two small arcs in Fig. 6b), and that for the solid (γ) is between ($\alpha\gamma$) and (γL) lines (a circular arc at the bottom of Fig. 6b). It is evident that such crystallization is inconceivable in a one-component system. Also impossible is the metastable state ($\alpha\gamma$V) [6] (Fig. 6c) because it is meaningless as well: the crystal (α) is not expected to be obtained by heating the vapor in isobaric conditions. The single-phase field of (α) in Fig. 6c is between (αV) and ($\alpha\gamma$), and that of the vapor (V) is between (γV) and (αV). In Fig. 6d the imaginary point is the intersection of (γV) and (αV) [6,24]. Here also it would be possible to precipitate a solid (α) by isobaric heating of the vapor.

Thus, a metastable state diagram can be formed only by those metastable lines, which on intersection result in a metastable triple point. It should be stressed that the metastable triple point cannot coincide with the equilibrium invariant point. For example, in the diagram of Fig. 6a, along with equilibrium sublimation AV, fusion AL, and vaporization LV, metastable processes (AV), (AL), and (LV) appear to be feasible. Considered separately (and shown in Fig. 6a by an arrow), these metastable curves constitute a system, in which it would be possible, by heating the vapor isobarically from the field bounded by the curves (LV) and (AV), to condense it to the

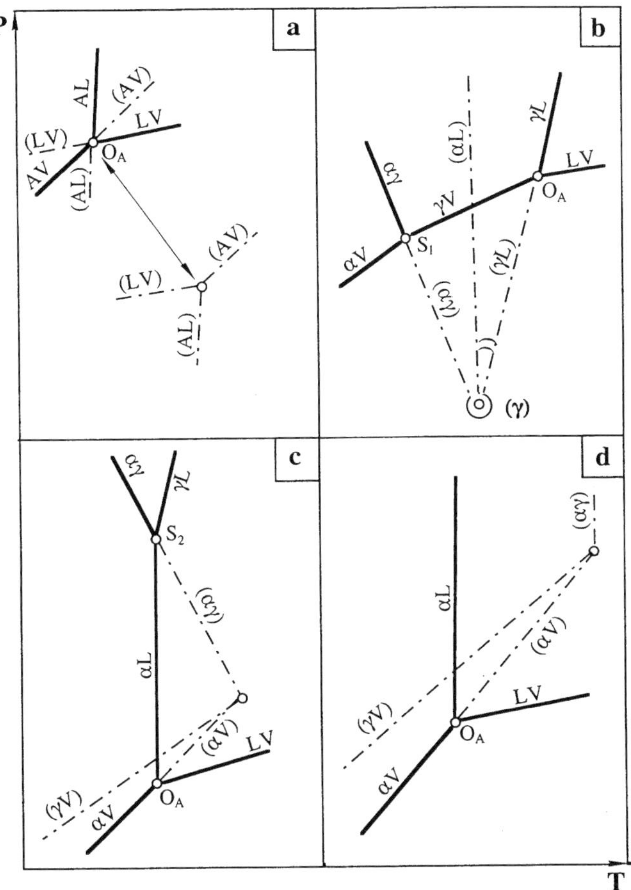

Fig. 6. *P–T–X* diagram of a one-component system with an imaginary triple point of metastable states

solid from the region between (AV) and (AL) or to the liquid bounded by (LV) and (AL) curves. Of course, such processes have no physical sense.

To summarize, metastable states in one-component systems are associated with polymorphism. In the process of metastable crystallization, one of the polymorphs is not formed, i.e., it exists only in an equilibrium state. The fields of existence of the three other phases are expanded as metastable extensions at its expense.

1.2.2 Binary systems

1.2.2.1 Evolution of *P–T–X* phase diagrams

In a binary system the composition parameter X in the $\psi\,(G, P, T, X) = 0$ function is usually expressed in terms of mole fractions or atomic percent. As a consequence, the P–T–X phase diagram of a binary system is three-dimensional. The phase surfaces in the P–T–X diagram are evolved from the Gibbs energies of the corresponding phases. As an example, we will show how the G–T–X diagram may be used to build up equilibrium curves of condensed phases in the T–X projection of the P–T–X phase diagram [10]. Figure 7 presents a system with complete miscibility in the liquid and a miscibility gap in the solid state. The Gibbs free energy surfaces for the liquid L and the solid solutions α and β are intersected in Fig. 7.1–7.7 by the isothermal planes T_1–T_7. The resultant G^L, G^α, and G^β curves are the composition dependence of the Gibbs energy G^i of phase i at $T = \mathrm{const}$. In Fig. 7 the melting points of the components are T_A and T_B, where $T_A < T_B$.

At $T_1 > T_B$ the free energy of the liquid is lower than that of both solids (Fig 7.1). As a result, at T_1 the liquid is the stable phase over the whole composition range (Fig. 7, T–X projection). When the temperature decreases, the free energies for the liquid and solid solutions are changed. For each phase, the temperature dependence of free energy is derived from Eq. (4):

$$(\partial G/\partial T)_{P,X} = -\,S. \tag{22}$$

Because the entropy of the liquid is greater than that of the solid, the change in temperature will affect the free energy of the liquid G^L more than those of the crystalline phases. As a consequence, the G^α and G^β curves will sink relative to G^L, and at $T_2 = T_B$, when the free energies of the liquid and solid B become equal ($G^L = G^B$), solidification of liquid B is observed. In the remaining composition range $X < 1$ (X is the mole fraction of B), the stable phase is the liquid solution (Fig. 7.2). At $T_3 = T_A$, because $G^L = G^A$, component A crystallizes. For compositions $X < l''_3$, $G^L < G^\alpha$ (and G^β), and therefore this is the region of the liquid. For $X > \beta_3$, on the contrary, $G^\beta < G^L$ (Fig. 7.3), and β is the stable phase (a solid solution of component A in component B). In the composition range $l''_3 < X < \beta_3$ the lowest free energy corresponds to a mixture of two phases, L and β. The criterion, Eq. (10), for equilibrium between two phases, L and β, is met, if a simultaneous tangent can be drawn to the G^L and G^β isotherms. The points of contact, l''_3 and β_3, indicate the compositions of the coexisting, or *conjugated phases* L and β at T_3 (Fig. 7, T–X projection). These points are said to be conjugated, and the isothermal tie-line between them is a *conode*. Thus, a conode determines the equilibrium compositions of the coexisting phases. It will be important for the subsequent sections to note that *the conjugated points do not correspond to the minima of the free energy curves.*

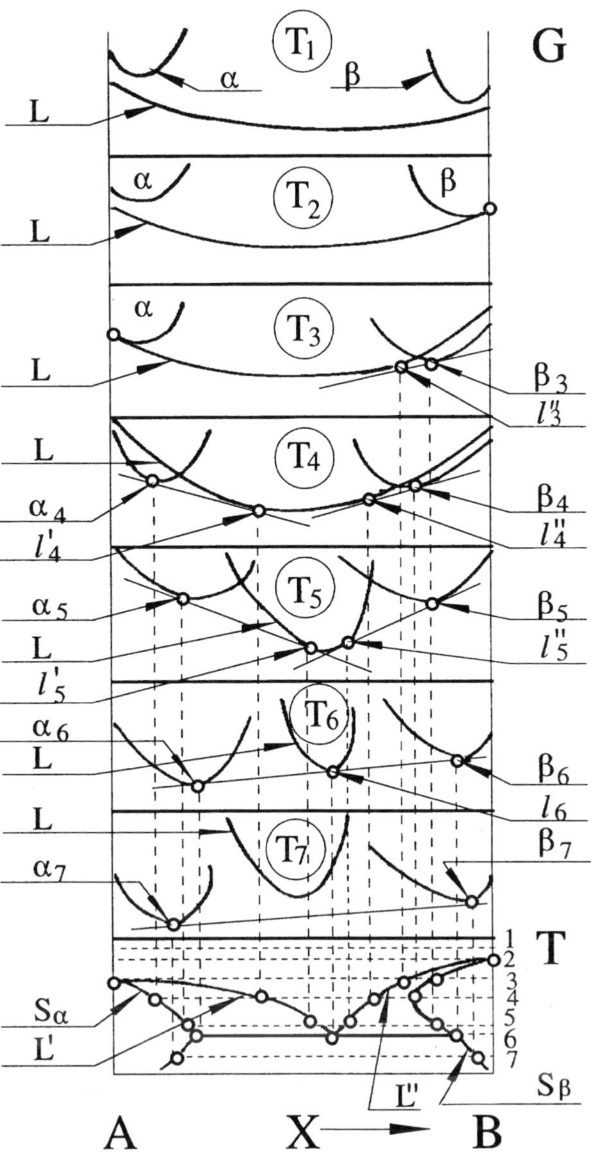

Fig. 7. Construction of the solidus and liquidus in a *T–X* projection from the free energy curves

On a further decrease of temperature (Figs. 7.4 and 7.5), the isothermal sections of the $G^{\alpha}(X,T)$, $G^{L}(X,T)$, and $G^{\beta}(X,T)$ surfaces give a continuous sequence of G curves that generate sequences of conjugated points. These are projected onto the T-X plane as *conjugated curves* S_{α} and L'; S_{β} and L" (Fig. 7, T–X projection), known as the *solidus* and *liquidus curves*. The solidus describes the maximum mutual solubility of the components in the crystalline phases over the whole interval of existence of the solid solutions α and β. The temperature of the equilibrium coexistence of the three phases is T_6, (Fig. 7.6), because at T_6 a simultaneous tangent can be drawn to all three G curves. This tangent shows that the chemical potentials of the components are the same in all three condensed phases. The coordinates of such a mixture (known as the *eutectic*) are $T_6 = T_e$, α_6, l_6, β_6 (Fig. 7, T–X projection). At $T_7 < T_e$ the free energy of the liquid for certain compositions is lower than that of the components A and B (Fig. 7.7). For these compositions the liquid is more stable than the pure components. But with respect to the solid solutions α and β, the liquid is unstable, because the tangent to the curves G^{α} and G^{β} lies below the G^{L} curve (Fig. 7.7). Therefore, at $T < T_6$ a mixture of two solid phases, α and β, is stable. Sometimes the solidus curves below the eutectic temperature are called the *solvus* curves.

As a rule, the shape of the solidus is such that the maximum solubility in the solid phases corresponds to the eutectic temperature (S^{α} curve in Fig. 7, T–X projection), and point α_6 in Fig. 7, T-X projection, is called the point of the saturation limit. However, the Van der Waals equation,

$$[(V'' - V') - (X'' - X')(\partial V/\partial X)_{P,T}]dP =$$
$$[(S'' - S') - (X'' - X')(\partial S/\partial X)_{P,T}]dT + (X'' - X')(\partial^2 G/\partial X^2)dX, \tag{23}$$

which expresses the equilibrium condition in a binary system, imposes no restrictions on the coexisting phases (in Eq. (23) primed symbols correspond to different phases). Therefore, the shape of the solidus, which is the temperature dependence of the composition of the solid and is determined by the sign of the derivative dX/dT, may be arbitrary. In particular, the common case is that of S_{α} in Fig. 7, T–X projection, at $T > T_e$ with $dX/dT < 0$, whereas S_{β} has a more complicated shape. At $T_e < T < T_4$, $dX/dT < 0$, and the solubility of A in the solid solution β increases with rising temperature. At temperatures $T_4 < T < T_B$, $dX/dT > 0$, and the solubility decreases. The maximum solubility corresponds to the extremum of the function $X(T)$, or to $dX/dT = 0$, i.e. to the composition β_4.

In this connection it would be appropriate to follow the cooling process of a sample with the composition $X = \beta_3$. At $T = T_3$, it crystallizes, and it is a single phase β at temperatures $T_3 < T < T_5$. On further cooling down to T_5, the sample partially fuses to form a liquid of the composition l''_5. The final solidification is observed at the eutectic temperature. This behavior is readily explained from the free energy standpoint. At temperatures $T_3 < T < T_5$, the free energy of the solid solution $\beta > \beta_4$ is lower than that of the liquid G^{L} (Fig. 7.4), and as a consequence, the stable phase is β. Meanwhile, at $T < T_5$ the G^β curve for the composition β_3 lies above the simultaneous tangent to the G^L and G^β curves (Fig. 7.6). Therefore, the stable state at these temperatures is the two-phase mixture ($L + \beta$). When the maximum solubility corresponds to a temperature that is intermediate between the eutectic T_e and the melting point T_B, the solidus shape is called *retrograde*.

This type of diagram is quite frequent in the materials science of semiconductors (e.g., doping of germanium, silicon, some of the III–V compounds).

1.2.2.2 The *P–T–X* space model

It has already been shown that the phase equilibrium in a binary system can be geometrically represented as a three-dimensional diagram with the orthogonal co-ordinates P, T, X. The composition X is usually expressed in terms of the number of moles N of the components A and B:

$$X = X_B = N_B/(N_A + N_B). \tag{24}$$

According to the phase rule, Eq. (11), a single-phase space in a binary system is trivariant: $F = 2 - 1 + 2 = 3$. Therefore the fields of existence of single phases in the $P–T–X$ space are volumes. Within these volumes, all three co-ordinates are independent continuous variables. These volumes are enclosed in curved surfaces, which are called *phase surfaces*. Outside of these are the volumes, in which two-phase equilibria are observed. Such equilibria in a binary system are bivariant ($F = 2$). Accordingly, an arbitrary choice (within certain limits) of T and P ($T =$ const, $P =$ const) fixes the compositions of the coexisting phases. The straight line $P =$ const, $T =$ const (isotherm-isobar) in the $P–T–X$ space is orthogonal to the $P–T$ plane, and its extremities lie on the corresponding phase surfaces. These extremities determine the compositions of the coexisting phases at given P and T. Hence, the isobar-isotherm of the two-phase equilibrium is a conode. When P and T vary arbitrarily, the two extremities of the conode describe a pair of continuous (in the limits of the two-phase equilibrium) curved surfaces, that are the bounding surfaces of the corresponding single-phase volumes.

If the isobar-isotherms of two two-phase equilibria with a common phase fall into one straight line, all three phases are in equilibrium. The intersection points of the conode and the three-phase surfaces define the compositions of the phases. Because three-phase equilibria are univariant in a binary system (Eq. 11), only one parameter can be changed arbitrarily without disturbing the phase state. When one variable changes continuously (e.g., P or T), the three intersection points describe three continuous curves in the $P–T–X$ space that define the compositions of the co-existing phases. Because the conodes are orthogonal to the $P–T$ plane, the imaginary surface generated by these three curves is also orthogonal to the $P–T$ plane. It may be called the *three-phase surface*. As a consequence, all three curves are projected onto the $P–T$ plane as one curve, whereas in $T–X$ and $P–X$ projections, they are seen as three individual curves, that describe the compositions of the phases in the three-phase equilibrium as a function of temperature (in $T–X$ projection) and pressure (in

P–X projection). For example, the three-phase equilibrium between solid S, liquid L, and vapor V appears as a single curved line SLV in *P–T* projection and as three lines (solidus, liquidus, and vaporus) in *T–X* and *P–X* projections.

Four-phase equilibrium in a binary system is invariant ($F = 0$). The compositions of the phases lie on a single tie-line P = const, T = const, and correspond to the intersection of this line with four phase surfaces. In the *P–T* projection, the four-phase equilibrium is seen as a single point, whereas in the *T–X* and *P–X* projections, four points are discerned. For example, the eutectic equilibrium in a binary system is represented by a single eutectic point (P_e, T_e) in the *P–T* projection and by four points in the *T–X* and *P–X* projections with the eutectic compositions of the condensed phases S_1, S_2, L and the vapor V at T_e and P_e.

The arrangement of the three-phase lines in the *P–T–X* space is determined by the thermodynamic properties of the coexisting phases. Thus, for example, if the volatilities of the components are not significantly different, then S_1S_2V or L_1L_2V curves (sublimation or vaporization) are directed so that $dP/dT > 0$, because the enthalpies of sublimation and vaporization are positive. The slopes of other three-phase lines in the *P–T* projection are arbitrary. Nevertheless, their disposition around the invariant point is specified by the *Schreinemakers rule* [23]: if phases P_1, P_2, P_3, P_4 are arranged according to increasing X; if the three-phase curves are labeled so that phase P_i does not participate in the equilibrium i; if the metastable extension of the equilibrium i through the invariant point is called i', then the sequence of the univariant curves around the invariant point in the *P–T* projection must be 1-2′-3-4′-1′-2-3′-4.

In general, single-phase volumes are separated by two-phase volumes. Therefore, as a rule phases of different compositions participate in two-phase equilibria. Such equilibria are called *incongruent*. According to Eq. (11), these equilibria are bivariant in binary systems. However, it might so happen that in a certain interval of the parameter values, the boundary surfaces of two single-phase volumes are internally tangent. Then in this interval the conodes degenerate into points, i.e., the compositions X_j and X_k of the coexisting phases j and k coincide. The locus of tangency of the surfaces of the phases j and k, for which $X_j = X_k$, is known as *the congruent phase transition* curve between phases j and k. Because of the restriction $X_j = X_k$, it follows from the phase rule, Eq. (12) with $R = 1$, that the congruent phase transition line in a binary system is univariant. For example, congruent vaporization (two-phase equilibrium liquid-vapor) is represented in the *P–T–X* space by a univariant curve known as the *azeotropic line*.

If the compositions of two phases coincide in a three-phase equilibrium, then on account of Eq. (12), such a state in a binary system is invariant ($F = 0$ with $R = 1$). For example, if the congruent melting curve S = L is tangent to the three-phase SLV line, the point of tangency is called the *congruent melting point*. Because this state of the system is invariant, the congruent melting point is the one with a fixed temperature, pressure, and composition of the vapor, which does not coincide with the composition of the liquid and solid. The compositions of all three phases (j, k, l) in a three-phase equilibrium cannot coincide in a binary system, because when $X_j = X_k = X_l$, there are two restricting equalities, i.e., in Eq. (12) $R = 2$ and consequently $F = -1$. Thus, if the composition of the saturated vapor X_V

were equal to X_L and X_S ($X_V = X_L = X_S$) at the congruent melting point, the variance of the system would be minus one, which, of course, is impossible.

Points of tangency of the three-phase line SLV and the congruent sublimation (S = V) and congruent vaporization (L = V) are called *congruent sublimation* and *congruent vaporization* (azeotropic) *points*.

1.2.2.3 Methods of graphical representation

The traditional and most popular method of investigating phase equilibrium is differential thermal analysis (DTA). If it is carried out in an open system in a flow of an inert gas, then, strictly speaking, the number of components is increased by unity. Because of this, along with thermodynamic complications, some kinetic effects might arise. If the inert gas proves to be neutral, the result of the thermal analysis is the isobaric T–X section of the P–T–X diagram at the pressure of the inert gas. Such studies are made, if the own vapor pressure of the system is lower than that of the inert gas. Because the boundaries of the condensed phase are expected to be only slightly affected by pressure, the resultant $(T$–$X)_P$ section is very similar to the T–X projection of the P–T–X diagram.

Thermal analysis is often carried out in evacuated and sealed tubes. Then the recorded temperatures correspond to phase transitions in condensed phases under the vapor pressure of the system. Neither the vapor pressure nor the composition of the vapor is measured in the DTA experiment. The resultant *condensed phase diagram* is the T–X projection of the P–T–X diagram without the vaporus line. In some applied sciences, e.g., the metallurgy of metals with high boiling points and relatively low melting points, condensed phase diagrams are quite sufficient for practical use. However, it is to be remembered that in the T–X projection the phase curves correspond to the maximum solubility of the components in the phases in three-phase equilibria. Therefore, when the equilibrium includes the vapor, the boundary compositions of the condensed phases depend on the pressure and composition of the vapor.

The complete graphical representation of phase equilibrium is given by three-dimensional P–T–X diagram. Because of the complexity of the shapes of the curved phase surfaces, it is more convenient to represent the three-dimensional models by their projections onto three orthogonal coordinate planes, P–T, T–X, and P–X. In these projections, univariant curves and invariant points are seen. In a binary system, these are three-phase equilibrium curves, two-phase congruent lines, two-phase equilibria of the pure components, triple points of pure components, quadruple points and points of congruent phase transitions. To elucidate the phase relations, isothermal and isobaric cross-sections of the P–T–X space diagram are usually made at certain representative temperatures (pressures).

In the subsequent discussion, phase equilibria will be presented by two projections, P–T and T–X, because two projections are sufficient to reproduce a three-dimensional construction. To facilitate the visual aspect of the discussion, either isothermal or isobaric cross-sections of the P–T–X diagrams will also be represented.

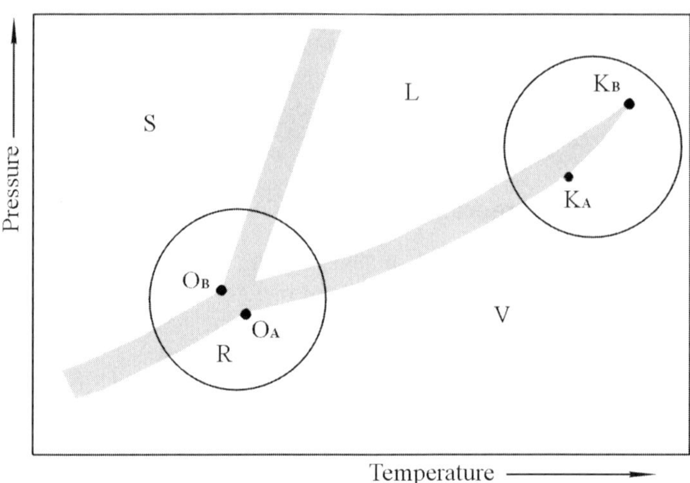

Fig. 8. *P–T–X* diagram of a binary system with complete solubility in all phases

1.2.2.4 Types of *P–T–X* phase diagrams of binary systems

In this section the types of *P–T–X* diagrams, which are most frequent in inorganic materials science and technology, will be considered. The approach will be from the simplest possible diagram to more complex systems, gradually introducing complications into phase behavior. Throughout the following exposition, the sequence of phases in all of the equilibrium labels will follow the increase in the content of component B in the equilibrium phases.

1.2.2.4.1 Complete miscibility in all phases. Ideal solutions

The phase space is restricted by two planes: $X = 0$ (pure component A) and $X = 1$ (pure component B). These planes contain two-phase equilibrium curves of the components: $S_A V_A$, $L_A V_A$, $S_A L_A$, and $S_B V_B$, $L_B V_B$, $S_B L_B$ (sublimation, vaporization and melting curves for A and B, respectively), which meet at the invariant triple points O_A and O_B (Fig. 8 [6]). In the $X = 0$ and $X = 1$ planes three single-phase planar $P–T$ regions are present for pure components: the vapors are below $S_i V_i$ and $L_i V_i$, the solids are above $S_i V_i$ and $S_i L_i$, and the liquid components *i* are between $S_i L_i$ and $L_i V_i$. When passing along the composition axis, the dimensionality of the system is increased by unity. As a result, each pair of $P–T$ curves for a certain two-phase equilibrium (e.g., $S_A V_A$ and $S_B V_B$) gives rise in the $P–T–X$ space to a pair of curved surfaces enclosing a volume, which represents two-phase equilibria between respective solutions (e.g., two-phase equilibrium SV between solid and vapor solutions). Pairs of single-phase planar $P–T$ fields for pure components give rise to volumes, within which arbitrary variation of the parameters P, T, and X corresponds to a single phase – a solid, liquid, or vapor binary solution. Thus, a point within a single-phase volume

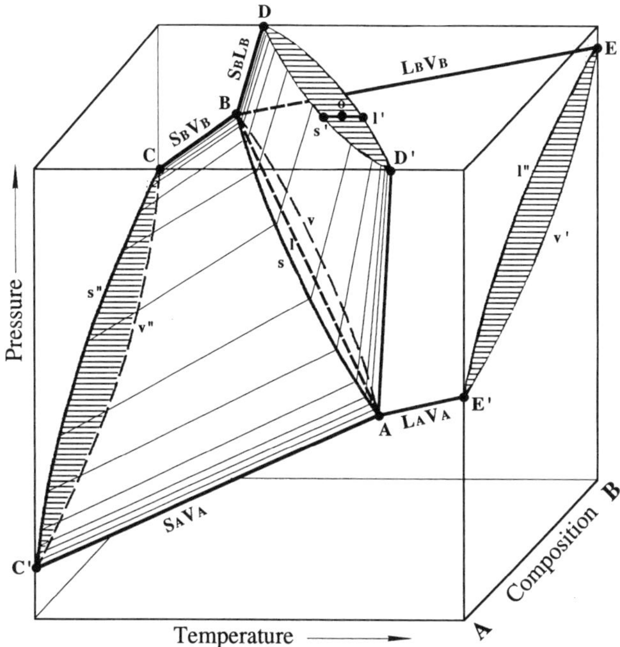

Fig. 9. *P–T–X* diagram of a binary system with complete ideal solubility in the solid, liquid and vapor phases

fixes the pressure, temperature, and composition of a particular phase. On the other hand, a point within a two-phase volume defines only an overall composition of a sample at a fixed pressure and temperature. This point lies on a conode, which ties the compositions of two conjugated phases. Therefore the two-phase volumes may be thought of as hollow spaces filled up with conodes (horizontal tie-lines in Fig. 9 [21]). In Fig. 8 they are seen as points within the three bands originating from the melting region R because the conodes are perpendicular to the *P–T* plane. The upper circle in Fig. 8 is the critical region.

In the melting region the triple points O_A and O_B are joined by three conjugated curves, *s, l,* and *v*, that represent the compositions of the solid, liquid, and vapor in the three-phase equilibrium SLV. Figure 9 presents these lines in a three-dimensional model of the melting region. The single-phase volume of an individual phase falls between the surfaces of this particular phase in two neighboring two-phase equilibria. Thus, the field of existence of the solid solution is between the solidus surfaces in SV and SL equilibria (A*s*BC*s″*C′A and AD′*s′*DB*s*A), the liquid is between the liquidus surfaces in LV and SL (AE′*l″*EB*l*A and AD′*l′*DB*l*A), and the vapor is between AC′*v″*Cb*v*A and AE′*v′*Eb*v*A. The compositions of the phases in the three-phase equilibrium SLV are defined by the curves *s, l,* and *v*, which are the intersecting lines of the corresponding surfaces: solidus surfaces A*s*BC*s″*C′A and AD′*s′*DB*s*A for A*s*B, liquidus AE$'l''$EBlA and AD$'l'$DblA for A/B and vaporus AC$'v''$CbvA and AE$'v'$EbvA for AvB. These three curves link the triple points of the pure components and lie on the three-phase "ruled surface", which is perpendicular to the P–T plane.

Three-dimensional phase diagrams are very complicated even in the simplest case of complete miscibility in all phases (Fig. 9). That is why, for practical purposes, projections of the space model onto the P–T and T–X planes are used. P–T and T–X projections of the diagram, Fig. 9, are shown on the left-hand side of Fig. 10. Here it is assumed that the melting temperatures of both components increase with rising pressure, $\partial P/\partial T > 0$ for both SL curves, and that the triple point O_A is higher in temperature and lower in pressure than O_B, i.e. B is the low-melting and more volatile component. On the P-T plane the univariant curves of pure components, which meet in the triple points O_A and O_B (thin lines in the P–T projection) are projected, as well as the three-phase SLV curve, which extends from O_A to O_B. On the T–X plane, the curves, which originate from the intersection of the phase surfaces, are projected with the same lettering, when three two-phase equilibria, SV, LV, and SL, meet in the melting region. In Fig. 10 these are the curves s, l, and v (the curve v is the thin one). Because Fig. 10 represents an ideal system, O_A and O_B are the only invariant points. The two-phase equilibrium SV (incongruent sublimation of the solid solution) is projected on the P–T plane as the region within the univariant "knife-edge" curves S_BV_B, SLV, S_AV_A, and the incongruent vaporization of the liquid (LV equilibrium) is within L_BV_B, SLV, and L_AV_A. For example, the heating process I–I for the sample X_1 at $T < T_1$ passes through the two-phase region SV. At $T = T_1$ the system undergoes the phase transition SV $\rightarrow$ SLV. The three-phase equilibrium comprises the solid solution of the composition $s(T_1)$, liquid $l(T_1)$, and vapor $v(T_1)$. In the melting region $T_1 < T < T_2$ the compositions of the solid, liquid, and vapor follow the curves s, l, and v (T–X projection, Fig. 10), and the vapor pressure curve is SLV. The P–T projection shows that the saturated vapor pressure decreases from $P(T_1)$ to $P(T_2)$ with the rising temperature. This is a consequence of condensation of the volatile component B. At T_2 a second phase transition occurs, SLV $\rightarrow$ LV; the solid disappears, and above T_2 vaporization is observed (two-phase equilibrium LV).

Points $s(T_1)$, $l(T_1)$ and $s(T_2)$, $l(T_2)$ determine the maximum solubility of the components in the solid and liquid solutions in the three-phase equilibrium SLV at the phase transition temperatures T_1 and T_2. These are of primary importance for materials science because they determine the maximum efficiency of purification or doping processes at the corresponding temperatures.

If the composition of the saturated vapor in the three-phase equilibrium is known, then the composition of the solid S(T_1) (in mole fractions of B) is readily calculated:

$$\mathrm{S}(T_1) = (N_B - n_B)/\left[(N_A + N_B) - (n_A + n_B)\right]. \tag{25}$$

Here N_A, N_B are the numbers of moles of A and B in the initial sample, and n_A, n_B are those in the vapor. Composition S(T_1) corresponds to the solid in SLV because the amount of the liquid is infinitesimal. A similar expression with S(T_1) substituted by L(T_2) gives the composition of the liquid in SLV at T_2 due to the infinitesimal amount of solid S(T_2).

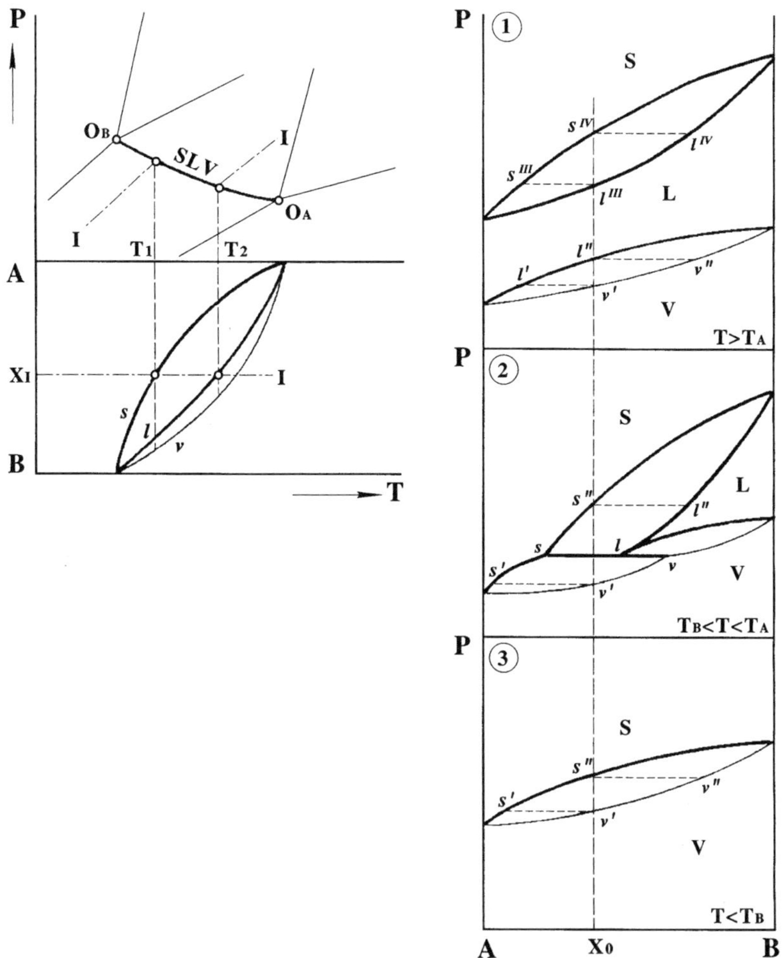

Fig. 10. Complete miscibility in all phases. P–T, T–X projections, and isothermal sections

All of the parameters of Eq. (25) may be measured in a vapor pressure experiment. Temperatures T_1 and T_2 are registered on the vapor pressure curve I–I as points of discontinuity of the function $P = P(T)$ at the phase transitions, and n_A, n_B are calculated from the ideal gas equation if the composition of the vapor at T_1 and T_2 is known. When it is unknown, additional investigation becomes necessary. The ways of probing the solidus surface in an incongruent sublimation process using the total vapor pressure measurements is discussed at a later stage.

A series of vapor pressure experiments with different N_A, N_B gives a number of curves I–I; each provides one point on the solidus line s and one point on the liquidus l. Thus, by vapor pressure measurements, s and l lines are obtained, i.e., the maximum solubility of the components in the solid and liquid.

When dealing with projections of the P–T–X diagram, it should be kept in mind that there is no one-to-one correspondence between the phase state of the system and the projection onto a single coordinate plane. A fixed phase state is described in projection by a certain geometrical image (e.g., a point in the P–T projection). However, the reverse is not true: a point in the projection may correspond to different phase states, which lie on the same perpendicular to the plane of this projection. This rather trivial concept of projective geometry is, however, sometimes neglected, when conclusions are drawn either from T–X or P–T projection alone. Complete information about the equilibrium state is obtained from simultaneously considering two projections of the diagram, and the compositions of the phases in two-phase equilibria are most conveniently represented by cross-sections of the space model.

Isothermal sections of the diagram with complete miscibility in all phases are shown on the right-hand side of Fig. 10 for three temperature intervals: above T_A (triple point of A), below T_B, and between them. In these sections (as well as in all others to appear in subsequent exposition), single-phase areas are lettered S, L, and V, and fields bounded by the limiting curves correspond to two-phase equilibria.

At $T > T_A$ (Fig. 10.1), the phase state of the sample with the composition X_0 is vapor at $P < P(v')$ (the single-phase region V in Fig. 10.1). When the vapor is compressed to the point $P(v')$, liquid appears whose composition is l'; that is why this point is called the *dew point*. Isothermal compression is accompanied by a change in the composition of the phases in the two-phase equilibrium LV along $l'l''$ (liquid) and $v'v''$ (vapor). The relative amounts of the phases m_v and m_l at a fixed pressure are given by the lever rule:

$$m_v/m_l = lX_0/X_0v,$$

where lX_0 and X_0v are the distances between the vertical $X_0 = \text{const}$ and the liquidus (and vaporus) at a fixed P, $P(v') < P < P(v'')$. At $P = P(l'')$, the amount of the vapor v'' is infinitesimal, and the phase transition LV $\rightarrow$ L is observed. $P(l'')$ is called the *boiling point* at the temperature of Fig. 10.1; when the pressure isothermally decreases along $X_0 = \text{const}$, the liquid boils upon reaching $P(l'')$.

In the pressure limits $P(l'') < P < P(l''')$, the $T = \text{const}$ plane cuts the single-phase space L. At $P(l''')$ the liquid solidifies to produce solid s'''. Upon further compression, the sample X_0 becomes two-phase SL; the compositions of the phases are represented by $s'''s^{IV}$ and $l'''l^{IV}$. At $P(s^{IV})$ the liquid is completely crystallized.

$(P$–$X)_T$ sections below T_B (Fig. 10.3) are different from Fig. 10.1 in one respect: the $T = \text{const}$ plane does not cross the liquidus space because T_B is the lowest melting temperature in the system. Accordingly, only two single-phase spaces, V and S, are seen in Fig. 10.3, which are separated by a two-phase region SV. Intersection of the solidus and vaporus surfaces with the $T = \text{const}$ plane results in two composition lines in Fig. 10.3, $s's''$ for solid and $v'v''$ for vapor, when the sample X_0 is compressed from $P(v')$ to $P(v'')$. The latter is called the *sublimation point* of sample X_0 at the temperature of Fig. 10.3.

Sections $T_B < T < T_A$ (Fig. 10.2) are somewhat more complicated. It is seen from the P–T projection, Fig. 10, that these isotherms cut three single-phase areas of component B (V_B, L_B, and S_B with rising pressure) and two of A (V_A and S_A). In addition, the three-phase equilibrium SLV is crossed; the composition of the phases lies on the solidus, liquidus, and vaporus curves of the T–X projection, Fig. 10. For example, in the section at $T = T_2$, Fig. 10, the compositions of the phases comprising the sample X_0 in the three-phase equilibrium are $s(T_2)$, $l(T_2)$, and $v(T_2)$ (see the T–X projection). When the temperature rises from T_B to T_A, the shape of the section, Fig. 10.2, changes, so that the points s, l, and v, originating from the triple point O_B and meeting in O_A, describe three lines, which are projected onto the T–X plane as s, l, and v curves and on P–T as a single SLV curve. In Fig. 10.2 it is seen that the change in the phase states of a sample, subjected to isothermal compression, depends on its composition. At $X < X_s$, the sequence of phase transitions is V $\rightarrow$ SV $\rightarrow$ S; at $X_s < X < X_l$ (e.g. X_0) it is V $\rightarrow$ SV $\rightarrow$ SLV $\rightarrow$ SL $\rightarrow$ S; at $X_l < X < X_v$ it is V $\rightarrow$ SV $\rightarrow$ SLV $\rightarrow$ LV $\rightarrow$ L $\rightarrow$ SL $\rightarrow$ S; and at $X > X_v$, V $\rightarrow$ LV $\rightarrow$ L $\rightarrow$ SL $\rightarrow$ S. In all of the two-phase equilibria the compositions of the conjugated phases lie on the extremities of the corresponding conodes (e.g., $s'v'$, $s''l''$); the amounts of phases are defined by the lever rule written for the corresponding equilibrium.

1.2.2.4.2 Complete miscibility in all phases. Non-ideal behavior

In a previous section it was pointed out that two single-phase volumes may be internally tangent in a certain interval of parameter values. The result is a curve of equal composition of the two phases. According to the Gibbs–Konovalov law, such two-phase equilibria result in extrema with respect to T and P on both phase surfaces. Extreme values of the parameters are the consequence of the difference in interaction energies between the same (A–A and B–B) and the opposite (A–B) kinds of atoms in solutions. If the A–B interaction is predominant, a negative deviation is to be expected with P_{min} and T_{max} in the two-phase equilibrium. Formation of a compound is the extreme case of this type. On the contrary, if the A–A type of interaction is predominant, the solution tends to a positive deviation from the ideal behavior. It is accompanied by P_{max} (at T = const) and T_{min} (at P = const). The extreme case of such a system is a miscibility gap in the solid (eutectic) or liquid. A three-dimensional P–T–X model of the liquid–vapor phase equilibrium for a system with a positive deviation from ideal behavior is shown in Fig. 11 [21]. The line mm of internal tangency between the liquidus and vaporus surfaces is the congruent vaporization curve, along which the compositions of the liquid and vapor coincide (*azeotropic mixture*). Because the azeotropic composition changes with temperature and pressure, the azeotropic mixture cannot be treated as a one-component state.

The tangent mm appears in the T–X and P–T projections as a congruent phase transition curve, which, according to the phase rule, Eq. (12), is univariant. Figure 12 is an example of a system with congruent processes in all three two-phase equilibria: S = L, S = V, and L = V. They are seen in projections as $s = l$, $s = v$, and $l = v$ curves. Figure 12 is a system with P_{max} and T_{min} in two-phase equilibria. The case of P_{min} will be referred to later in connection with compound formation.

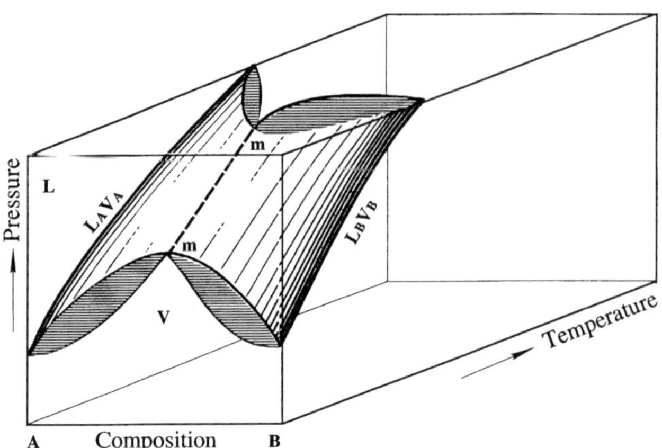

Fig. 11. Two-phase equilibrium liquid–vapor in a system with positive deviation from ideal behavior

On reaching the melting region, the congruent curves touch the three-phase surface at three invariant points with the following coordinates: (P_{cm}, T_{cm}, X_{cm}), (P_{cs}, T_{cs}, X_{cs}), (P_{cv}, T_{cv}, X_{cv}) for congruent melting, sublimation, and vaporization, respectively. Figure 12 is a diagram with T_{min} in SLV equilibrium. It is to be noted that the continuous curve SLV may pass thorough a pressure minimum, maximum, or no pressure extremum at all (as in Fig. 12).

Phase behavior in the system, Fig. 12, is more complicated than in Fig. 10, as is seen in the isothermal sections (Fig. 13) presented in a sequence of decreasing temperatures. $T > T_A$ planes cut all three single-phase volumes and congruent lines S = L and L = V. As a result, in Fig. 13.1, single-phase areas S, L, and V appear together with two pairs of phase lines with P_{max} in each. The pressure extrema are the points of tangency, or equal composition. With decreasing temperature, the liquidus surfaces touch at the invariant triple point O_A. On further cooling, the univariant state extends along the X-axes as the *slv* horizontal (Fig. 13.2), and the pure component A crystallizes from the vapor. At $T = T_B$, the second three-phase equilibrium appears in the section (with the pure component B), which also extends along X at $T < T_B$ (the upper horizontal *vls* in Fig. 13.3). It is seen in Fig. 12 that P decreases with T more rapidly in the L = V equilibrium than in the VLS. As a consequence, point $l = v$ approaches the *vls* horizontal and finally impinges on it at $T = T_{cv}$. The corresponding section, Fig. 13.4, crosses the VLS line at the point of congruent vaporization. T_{cv} is the lowest temperature of the congruent vaporization. At $T < T_{cv}$, the liquid evaporates incongruently (the two-phase area LV in Fig. 13.5).

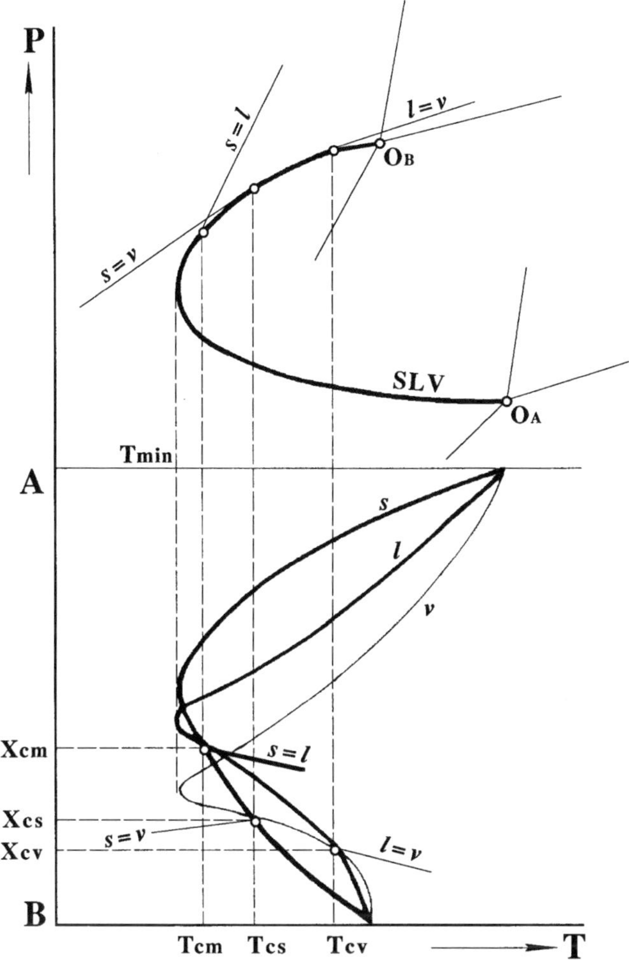

Fig. 12. *P–T* and *T–X* projections of the diagram with complete miscibility in all phases and congruent melting, vaporization and sublimation

At $T = T_{cv}$, the isothermal plane meets the congruent sublimation point, and $s = v$ appears on the *lvs* horizontal. $T < T_{cs}$ planes cross the S = V congruent line with P_{max} in SV equilibrium (Fig. 13.5). The lowest temperature of the congruent melting is T_{cm}. At $T_{min} < T < T_{cm}$, the melting of the solid solution is an incongruent process. T_{min} is the temperature minimum of the three-phase equilibrium SLV (Fig. 13.6). The composition of the phases at this temperature is determined by the minimum points of the solidus, liquidus and vaporus curves in the *T–X* projection (Fig. 12). $T < T_{min}$ planes cut two single-phase spaces, S and V, and the two-phase equilibrium SV with P_{max}.

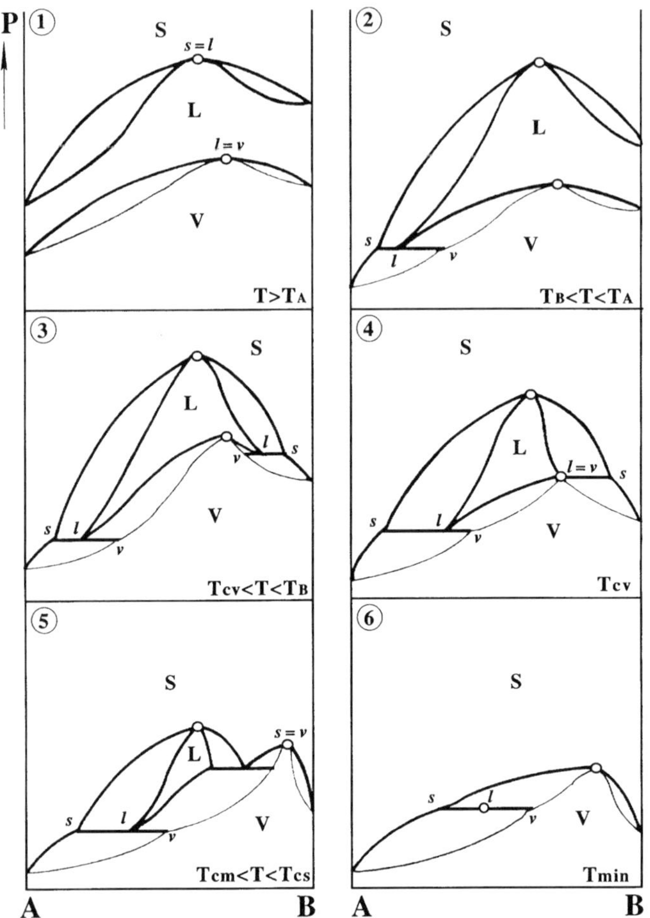

Fig. 13. Isothermal sections of the diagram of Fig. 12

It is worthwhile to emphasize here that, because the congruent melting temperature depends on pressure (increases with rising P in Fig. 12), the $s = l$ line in the P–T projection, Fig. 12, is not a vertical, and consequently, it touches the SLV curve at $T < T_{min}$. The characteristic feature of the congruent transition points is that the composition sequence of phases in the three-phase equilibria is changed at these points. For example, at T_{cm}, lsv (Fig. 13.5) is changed to slv (Fig. 13.6), when the phases are written according to the increase of component B in them.

1.2.2.5 Miscibility gap in the solid phase

A miscibility gap (MG) in the condensed solution appears as a result of considerable positive deviations from ideal behavior. The MG may be open (with respect to P and T) or closed; it may be entirely contained in a single-phase volume or cross, or touch two- or three-phase equilibrium spaces. If the miscibility gap is closed, it passes through temperature and pressure extrema (four, as a rule), which are called *critical points*. When the whole MG lies within a single-phase volume, it is a two-phase area. Unlike the other two-phase spaces, phase points within the MG represent mixtures of two different compositions of the same phase. If the MG cuts a two-phase space, three-phase equilibrium appears in certain limits of the parameters. For example, if a liquid–vapor equilibrium is crossed by a MG in the liquid, the result is the L_1L_2V equilibrium. Like any three-phase equilibrium, it is projected onto the P–T plane as a single univariant curve (L_1L_2V in this case), whereas three different curves appear in T–X and P–X projections (l_1, l_2, and v). If the two-phase equilibrium involves a congruent curve, and the MG crosses it, then an invariant congruent point appears in the three-phase equilibrium (e.g., $l = v$ in L_1L_2V).

When the MG crosses a three-phase equilibrium, the result is an invariant point of equilibrium of four phases. This is a *eutectic*, if the three-phase extrema are crossed by the MG; otherwise, the equilibrium is of a *peritectic* type.

In this section we discuss a miscibility gap in the solid solution ($S_1 + S_2$), which cuts the three-phase equilibrium SLV. A MG in the liquid ($L_1 + L_2$) will be considered later, in connection with the experimental data for a number of semiconductor systems.

1.2.2.5.1 Eutectic systems

Figure 14 is a three-dimensional phase diagram of a eutectic system [6]. Here B is the low-melting volatile component ($T_B < T_A$, $P_B > P_A$), and the vapor pressure at the eutectic point E is intermediate between that at the triple points of the components ($P_B > P_E > P_A$). There is no pressure or temperature maximum or minimum in the EO_A or EO_B lines, and the liquid–vapor equilibrium is of the simplest type, similar to that of Figs. 9 and 10. In Fig. 14, two-phase equilibria for pure components A and B are shown, which converge in the triple points O_A and O_B, as well as the three-phase surfaces AVB and ALB. Two series of the corresponding conodes are also shown in Fig. 14. Intersection of these three-phase surfaces with the liquid–vapor loop results in points l and v, which define the composition of the liquid and vapor phases in the eutectic equilibrium with pure solid components A and B at temperature T_E (to simplify this rather complicated three-dimensional figure, the mutual solubility of the components in the solid state is neglected). Analysis of the phase equilibrium is facilitated, if the three-dimensional diagram is projected onto the coordinate planes. P–T and T–X projections of the system, Fig. 14, are shown on the left-hand side of Fig. 15. Thin lines represent two-phase equilibria for pure components, originating in the triple points O_A and O_B (P–T projection), and the vaporus line (T–X projection). In the P–T projection E is the invariant eutectic point of the four-phase equilibrium ALBV. Four univariant curves meet at this point. Phases on these curves are labeled

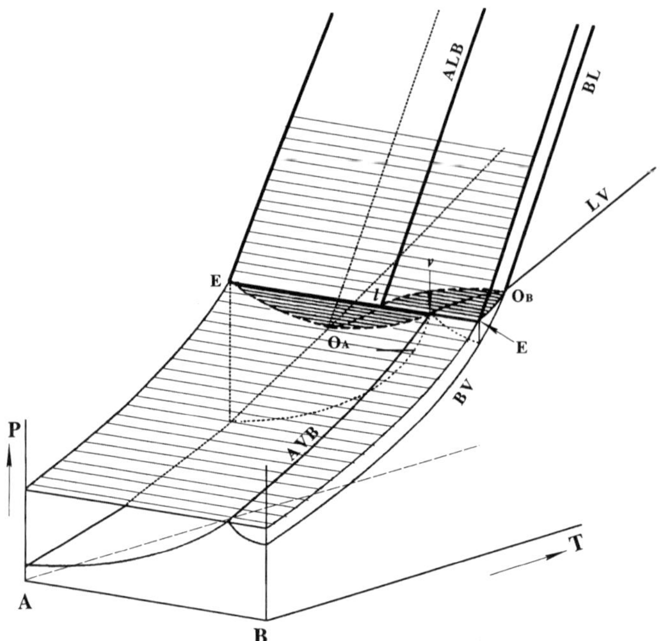

Fig. 14. The melting region for a binary system with complete immiscibility of solids

according to increasing X. LBV equilibrium is represented by the continuous curve EO_B, which rises (both in T and P) from (P_E, T_E) to O_B. The composition of phases in LBV is described by the corresponding branches of the liquidus l, solidus s, and vaporus v, originating in T_B in the T–X projection. ALV is represented by the line EO_A in the P–T projection and by the s, l, and v curves meeting at T_A in the T–X projection. The vapor pressure in this equilibrium goes down with increasing temperature due to condensation of the volatile component B.

In the AVB equilibrium, the composition of the vapor is given by the line v', and that of the solid solutions corresponds to the solidus at $T < T_E$ (T-X projection). ALB gives the pressure dependence of the eutectic temperature; the corresponding compositions lie on the curve l' (for the liquid), s' and s'' (for the solids). The two-phase equilibrium AV of the solid solution A and the saturated vapor V is projected onto the P–T plane as an area within the univariant curves ABV, ALV, and $S_A V_A$; BV is within $S_B V_B$, LBV, and ABV, and LV is within LBV, ALV, and $L_B V_B$.

Upon heating the sample X_1 in a closed tube, the vapor pressure is described by the curve I–I with three points of discontinuity: at T_1 (complete sublimation of the solid solution B), T_2 (appearance of the liquid), and T_3 (complete sublimation of the solid solution A).

If the composition of the vapor is known, then the composition of the solid solution can be determined from a single vapor pressure curve from Eq. (25) for the

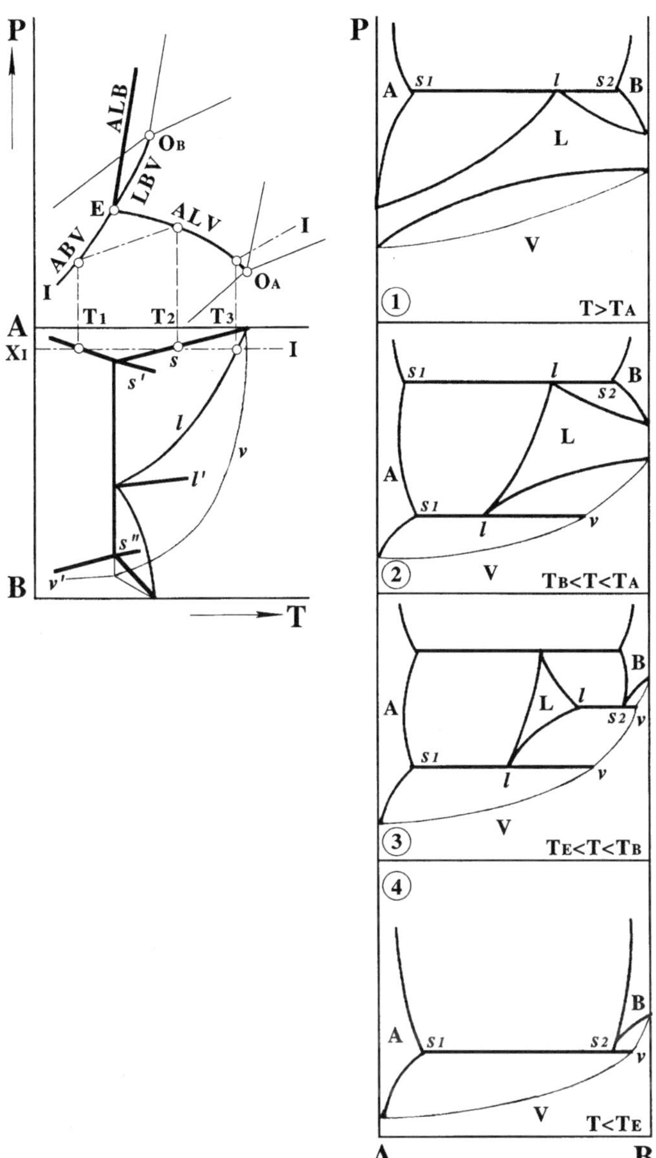

Fig. 15. Projections and isothermal sections of the diagram with complete miscibility in liquid and vapor and a miscibility gap in the solid (eutectic)

temperatures $T_1 \leq T \leq T_2$, as well as the composition of the liquid at $T \geq T_3$. By changing either the initial X_1 or the reduced mass of the sample (mass per cm^3), it is possible to determine the maximum solubility of the components in the crystalline and liquid phases at different temperatures, i.e., to construct the liquidus and solidus curves in the T–X projection from the vapor pressure measurements.

Phase equilibria are explicit in the isothermal sections of the system (the right-hand side of Fig. 15). The $T > T_A$ planes (Fig. 15.1) cut four single-phase spaces (A, B, L and V), the two-phase equilibria AB, AL, AB, and LV, and the three-phase equilibrium ALB. When the temperature is decreased to T_A, the second three-phase equilibrium appears in the section, i.e., point O_A. At still lower temperatures (e.g., T_2 or T_3 in projections, Fig. 15), the vapor may be condensed (depending on the composition) either into liquid (at $X_V > v$) or into solid solution A (with $X_V < v$) (Fig. 15.2). When $X_V = v$, isothermal condensation results in three-phase equilibrium s_1lv with fixed compositions of the phases. Isothermal compression of the liquid leads to formation of two solid solutions with compositions s_1 and s_2 at $P = P(s_1ls_2)$.

The isotherms $T_E < T < T_B$ cross a small part of the liquidus volume within ALV, LBV, and ALB (Fig.15.3). The planes $T < T_E$ are lower in temperature than the existence range of the liquid and therefore do not cross it (Fig. 15.4). Vapor with the composition $X_V < v$ condenses upon isothermal compression into the solid solution A; at $X_V > v$, solid B is formed, and from $X_V = v$, a mixture of A and B crystallizes with fixed compositions s_1 and s_2.

1.2.2.5.2 Peritectic systems

The system shown in Fig. 16 is different from that of Fig. 15 in two respects:

1. The peritectic point P is between O_A and O_B in temperature and higher than both of them in pressure, whereas the eutectic E is between O_A and O_B in pressure and below both of them in temperature.
2. LV equilibrium in Fig. 16 involves P_{max} (and T_{min}).

As a consequence, the phase behavior of the system, Fig. 16, is somewhat different from that of Fig. 15. In the liquid–vapor equilibrium in Fig. 16.1, an azeotropic point $l = v$ is observed, and the three-phase horizontal is lower in pressure than the boiling point of B at $T > T_A$. Furthermore, in equilibria s_1s_2l and BL the relation of the liquid and solid solution compositions is $X_L > X_B$, whereas in Fig. 15.1 it is $X_L < X_B$. At the temperatures of Fig. 16.2, isothermal compression of the vapor results in either solidification (when $X_V < v$) or liquefaction ($X_V > v$); the liquid is enriched in B when $X_V > X_{l=v}$ and the vapor contains more B than the liquid, when $v < X_V < X_{l=v}$. In the temperature interval $T_{cv} < T < T_A$, the univariant lines L = V and ALV draw together, and at $T = T_{cv}$, the azeotrope $l = v$ impinges on the s_1lv horizontal. This is the lowest temperature of congruent vaporization (the azeotropic point), and at $T < T_{cv}$ evaporation is an incongruent process (Fig. 16.3). At T_{cv}, the composition sequence of phases in the three-phase equilibrium is changed from ALV to AVL (Fig. 16.3). The isotherms in the interval $T_B < T < T_P$ cross the liquidus volume contained within the VBL equilibrium and the melting and vaporization

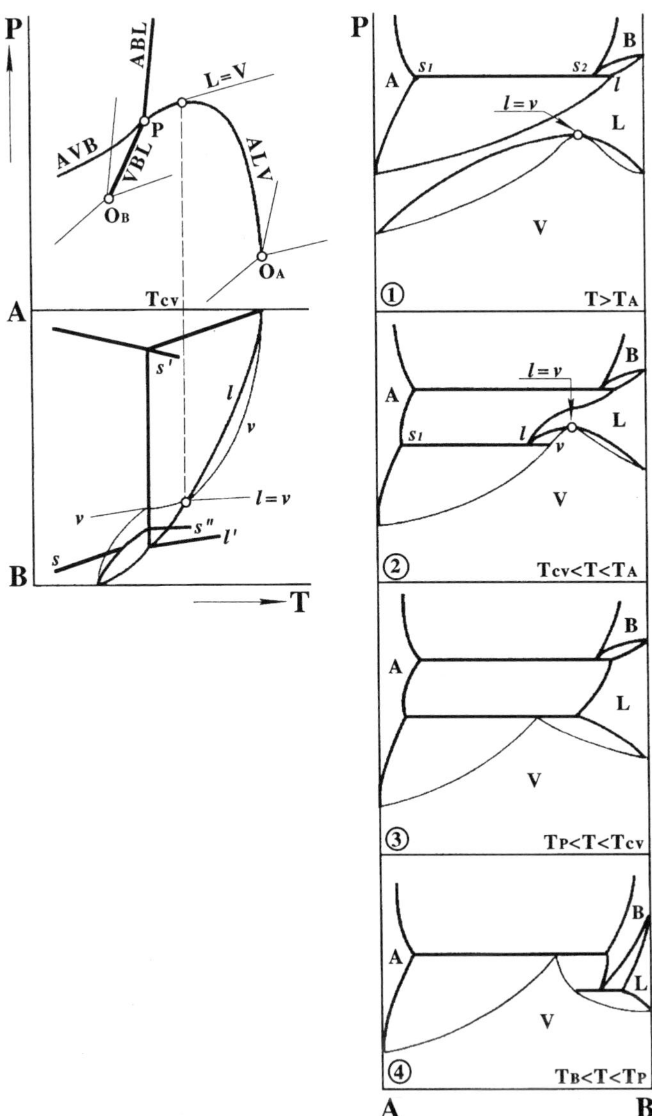

Fig. 16. Complete miscibility in the liquid and vapor (with azeotropic curve) and a miscibility gap in the solid (peritectic)

curves of pure component B. Crystallization of the liquid results in formation of solid solution B (Fig. 16.4).

1.2.2.6 Polymorphism of components

In subsequent exposition, the evolution of phase equilibria will be followed in two projections, P–T and T–X. To facilitate the visual perception, we will also provide isobaric T–X sections of the P–T–X model for crucial regions. The following *symbolism* will be adopted throughout the graphics. The low-temperature solid solution of the component A will be labeled α, the high-temperature one, γ, and the solid solution based on component B will be denoted β.

For *P–T projections*, bold lines will represent three-phase equilibria; the corresponding phases will be spelled out on the lines in the sequence following the increase of component B in the phases. Thin lines will represent univariant equilibria of pure components A and B. Triple points of the components will be labeled in the same way as that of the one-component systems: O_A will describe the solid – liquid – vapor equilibrium for component A; O_B, the corresponding equilibrium for B; $\alpha - \gamma$ – vapor will be S_1^A for A and S_1^B for B; $\alpha - \gamma$ – liquid will be S_2^A for A and S_2^B for B. Invariant quadruple points will be labeled N with the corresponding subscripts. All of the invariant points will be spelled out, framed, and shown separately. The sequence of phases in them will also follow the increase in component B.

For *T–X projections*, triple point temperatures will be given for pure components, as well as the curves describing the composition of the phases in three-phase equilibria. To facilitate the recognition of the relation between geometrical images in P–T and T–X projections, each T–X line will bear the corresponding label. The vaporus will be labeled v; the liquidus, l; and the solidus, α, β, or γ. In most cases, the vaporus will be a thin line; liquidus, medium bold; and the solidus, bold. Four-phase equilibria in T–X are four discrete points linked by a conode.

In *T–X isobaric sections*, single-phase fields, corresponding to homogeneous volumes, will be shown: V for vapor; L for liquid; α, β, and γ for the solids. They are separated by two-phase heterogeneous regions. If three conjugated heterogeneous volumes are crossed by the section plane, then three-phase equilibrium appears on the isobar. The compositions of the three participating phases will be connected by a tie-line which is a horizontal line on the isobar. On each section a pressure limit will be spelled out, for which this type of isobar is observed.

For every type of the diagram we will at first examine the phase equilibria with parallel polymorphism and then proceed to the non-parallel polymorphism of the components. All four types of polymorphism (Fig. 5) will be examined.

1.2.2.6.1 Component A with Type I polymorphism

Parallel polymorphism. Suppose that components A and B have the same number of polymorphs (two, to be precise) and the same type of P–T diagram. Suppose, furthermore, that complete miscibility in all aggregate states is observed. A P–T–X phase diagram of this type is presented in Fig. 17. P–T projection, along

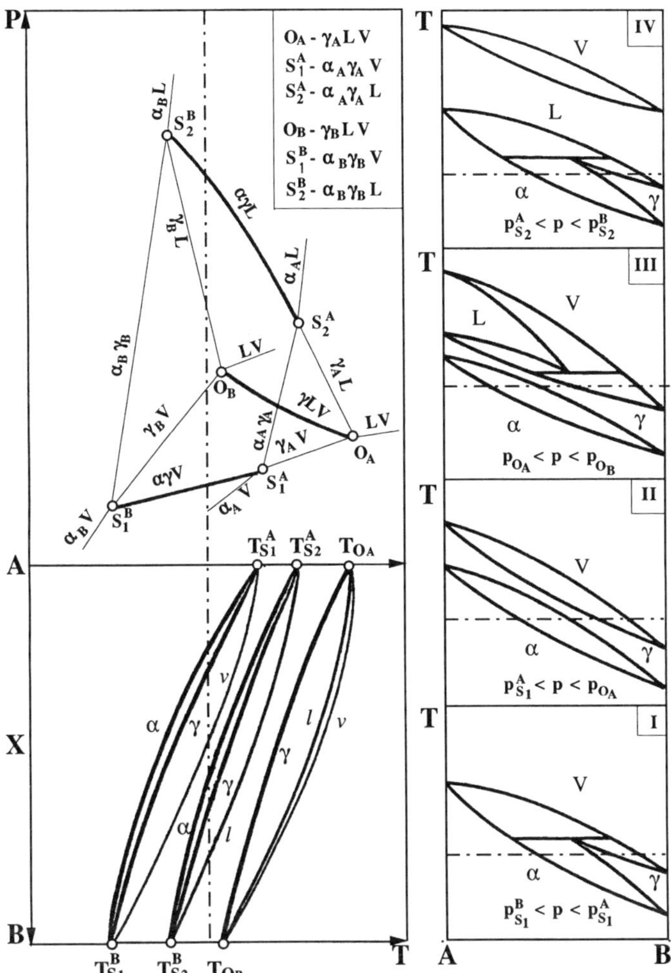

Fig. 17. *P–T–X* diagram of a binary system with Type I parallel polymorphism and complete miscibility of the components in all aggregate states

with unary equilibria (for pure components), shows three univariant curves running between the corresponding triple points: αγV is for the three-phase equilibrium solid solution α – solid solution γ – vapor; αγL is for solid α – solid γ – liquid; and γLV corresponds to the equilibrium between the solid solution γ, liquid, and vapor. On the *T–X* plane, each of these univariant equilibria is projected in the form of three curves, which determine the temperature dependence of the composition of the corresponding phases involved in these equilibria. It is clear that the *T–X* projection is not single-valued: it is impossible to distinguish on it between the homogeneous fields pertaining to single-phase volumes and heterogeneous regions representing two-phase volumes. The projections of these volumes onto the T–X plane overlap.

Phase relations in this system are conveniently followed in T–X isobaric sections. At pressures below $P_{S1}{}^{B}$ and low temperatures, the solid is single-phase α over the whole compositional range. On heating up to a certain temperature, sublimation is observed when the figurative point reaches the heterogeneous loop αV. Further isobaric heating will result in complete sublimation of the solid when the single-phase field V is attained; subsequent heating will not change the aggregate state of the system. At pressures falling within the $P_{S1}{}^{B} - P_{S1}{}^{A}$ range (Fig. 17,I), along with single-phase volumes α and V, solid solution γ appears, as well as the three-phase surface $\alpha\gamma$V. As a consequence, besides the heterogeneous region αV, two more heterogeneous equilibria are observed in the section, $\alpha\gamma$ and γV. On intersection, these three heterogeneous fields result in three-phase equilibrium which appears in Fig. 17,I, as a tie-line running between the compositions of the three equilibrium phases, α, γ, and V. At pressures from the interval of Fig. 17,I, the solubility of component A in solid γ is restricted. At higher pressures, the single-phase field of γ expands, and in the interval shown in Fig. 17,II, it goes through the entire compositional range. At these pressures isobaric heating of the solid solution α of an arbitrary composition results at first in the α–γ phase transition in the two-phase loop $(\alpha+\gamma)$ then single-phase γ is formed which on further heating sublimes in the $(\gamma+V)$ field and finally becomes unsaturated vapor (above the vaporus line). At these pressures the α-form does not sublime. At still higher pressures (Fig. 17,III), the liquid appears with restricted solubility of component B. Two-phase regions (fusion γL and vaporization LV) appear in the section, as well as the three-phase equilibrium γLV. A further increase of pressure leads to expansion of the liquid-phase area toward component B and, finally, complete miscibility at P_{OB}. The pressure range of $P_{OB} < P < P_{S2}{}^{A}$ involves neither sublimation of the γ-form in γV nor three-phase equilibrium γLV. When heated, the solid solution γ melts while passing through the γL loop; then the single-phase liquid L is observed, which vaporizes in the LV loop and finally reaches the unsaturated vapor V limit.

Further increasing pressure (Fig. 17,IV) results in restricted solubility of A in the γ-form. The melting region of α and the three-phase equilibrium $\alpha\gamma$L appear in the section. The isobars $P > P_{S2}{}^{B}$ do not cross the γ-region; the α-phase at these pressures fuses over the αL loop into liquid L, which subsequently vaporizes (the LV loop) and proceeds into the single-phase V field.

Identification of parallel polymorphism, which is the simplest case of polymorphism in binary systems, is sometimes inhibited by the fact that one or both of the triple points S_1 and S_2 of a component are outside the real temperature scale. Then the isotherm $T = 0$ K divides the P–T–X phase diagram into two parts, real and imaginary. Accordingly, P–T and T–X projections are also split by this isotherm (dot-and-dash line in Fig. 17) into real and imaginary parts. For example, in Fig. 17, points $S_1{}^{B}$ and $S_2{}^{B}$ could be imaginary. As a consequence, the sections of Fig. 17 are also restricted by the real temperature scale (dot-and-dash lines in the sections). As a result, at low temperatures the initial state appears as two-phase, e.g., $\alpha\gamma$ and γV in Fig. 17,I. These "clipped" sections are discussed in more detail by Ricci [6].

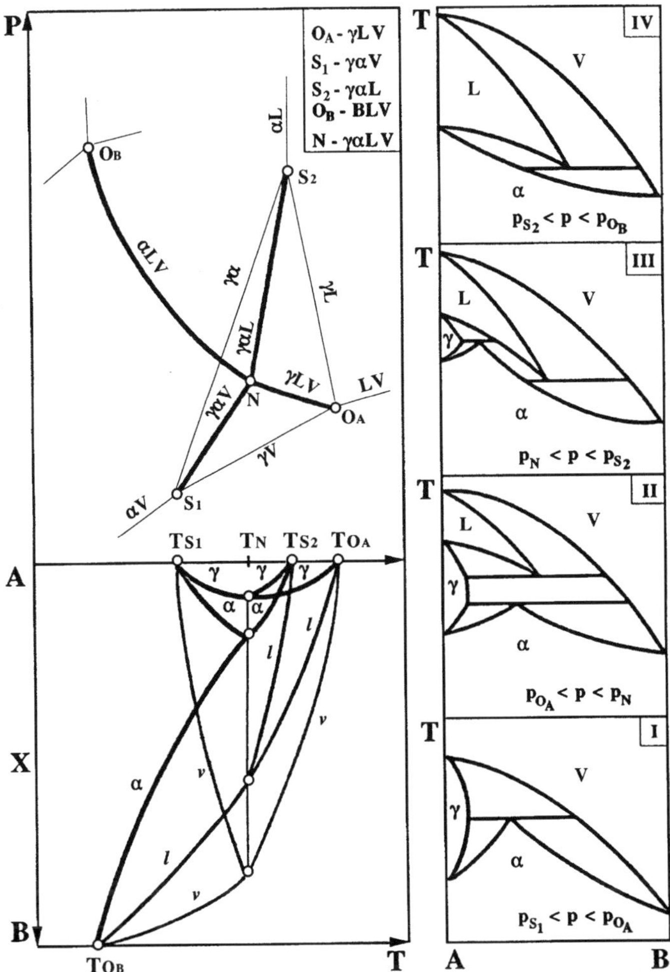

Fig. 18. *P–T–X* diagram of a binary system with Type I non-parallel polymorphism. The solubility of B is complete in the α-phase and restricted in the γ-phase

Non-parallel polymorphism. In this case, pure components A and B give rise to different types of *P–T* diagrams. We will consider cases of non-parallel polymorphism when component A forms two polymorphs, whereas component B exists in only one crystal structure.

(1) In the system shown in Fig. 18, the miscibility of component B in the α-phase is unrestricted, whereas that of the γ-phase is limited. Four-phase equilibrium $\gamma\alpha$LV is observed in this system. In the *P–T* projection, it is seen as point N, and in *T–X* projection four separate points appear on a tie-line, which correspond to the temperature T_N of this invariant. This quadruple point is formed by the intersection of four three-phase surfaces: γLV, $\gamma\alpha$V, $\gamma\alpha$L, and αLV. The lines of composition for each of the three phases, which form these surfaces, meet at the corresponding triple point. T–X sections in Fig. 18 clearly show that, whereas the α-form exists in the whole compositional and pressure ranges, the single-phase volume of γ is localized in the P–T–X phase space: it exists within the pressure range from P_{S1} to P_{S2} and dissolves component B only up to a certain limit (q.v. the single-phase field of γ on the sections). It is also seen in Fig. 18 that in all of the equilibria involving both α and γ, the α-polymorph is enriched in component B compared to γ. It is relevant to point out here that the reverse case, viz., complete miscibility of γ and B and restricted miscibility for α and B, is essentially that of the type of Fig. 17 at temperatures above the dot-and-dash line.

(2) The solubility of B in α is greater than that in γ (Fig. 19). An additional quadruple point appears in this system for the four-phase equilibrium αLβV (point N_2 in the P–T projection). In the T–X projection the second conode is seen at temperature T_{N2}. In contrast with Fig. 18, the three-phase curve αLV in the P–T projection is terminated not at the unary triple point O_B but rather at the quadruple point N_2. This three-phase surface is projected onto the T–X plane as three individual lines, α, l, and v, running between two conodes, whereas at the triple point at T_{OB} three lines (l, β, and v) meet; the corresponding P–T projection is the univariant LβV curve. Two more univariant curves meet at point N_2: $\alpha\beta$V, rising from low temperature, and αLβ, going up to high pressures. The compositions of phases in these equilibria are seen in the T–X projection. As in Fig. 18, the single-phase volume of γ is localized in the P–T–X phase space (the γ field in the sections, Fig. 19). Both α and β fields are restricted in composition, i.e., α exhibits limited solubility of B, and β is partially miscible with A.

Now we will follow the changes in the phase state of the system for isobaric heating of a two-phase (α+β) sample (Fig. 19,II). As the sample is heated, it passes the $\alpha\beta$V three-phase equilibrium, at which vapor appears. On complete sublimation of the volatile β-form, the sample proceeds to the sublimation area αV. When the temperature of the three-phase $\gamma\alpha$V equilibrium is reached, the γ-polymorph appears; its sublimation field γV corresponds to the temperatures between the $\gamma\alpha$V and γLV tie-lines. When the temperature of γLV is reached, the sample fuses. Subsequent phase transitions depend on the composition of the sample. Substances enriched in component B are heated to the vaporization field LV and finally to the unsaturated vapor V. For A-rich samples, complete condensation of the vapor is observed at the temperature of the γLV equilibrium. Further on, as the liquidus surface in γL is attained, the γ-phase fuses completely, and the samples get into the single-phase L field. On further heating, the liquidus in LV is reached, and vaporization is observed through the LV loop, finally coming to the vaporus surface in the LV loop and the single-phase V field.

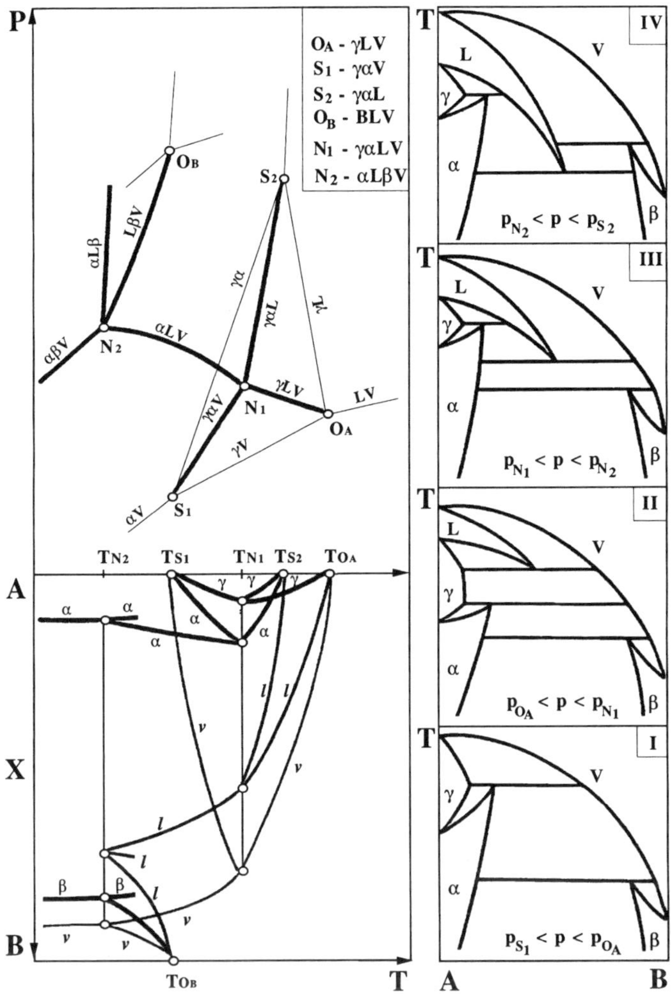

Fig. 19. *P–T–X* diagram of a binary system with Type I non-parallel polymorphism and restricted solid state miscibility. The solubility of B in α is greater then that in γ

(3) The solubility of component B in the γ-phase is greater than that in the α-form (Fig. 20). The difference between this type of system and that of Fig. 17 is that the univariant curves γLV, αγL, and αγV on the *P–T* projection are terminated not at the unary triple points of component B, but rather at the invariant quadruple points of four-phase equilibria γLβV at N_1, αγLβ at N_2, and αγβV at N_3. Three corresponding conodes are observed in the *T–X* projection. The invariant equilibria are linked via three three-phase surfaces projected onto the *P–T* plane as curves γLβ, αγβ, and γβV and onto the *T–X* plane as the corresponding

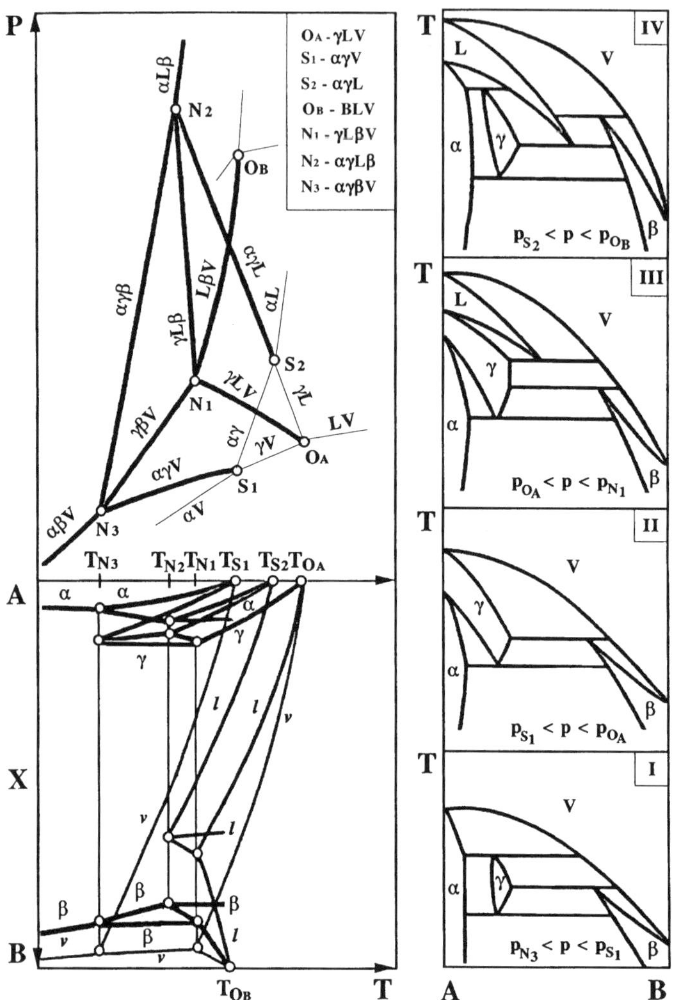

Fig. 20. *P–T–X* diagram of a binary system with Type I non-parallel polymorphism and restricted solid state miscibility. The solubility of B in γ is greater then that in α

curves γ, l, and β (between T_{N1} and T_{N2}); α, γ, and β (between T_{N2} and T_{N3}) and γ, β, and v (between T_{N1} and T_{N3}). The single-phase volume of the γ-phase is confined in the *P–T–X* phase space (q.v. the sections in Fig. 20), and in a specific interval of pressure and temperature the γ-polymorph shows restricted miscibility with both components (Figs. 20,I and 20,IV).

1.2.2.6.2 Component A with Type II polymorphism

Parallel polymorphism for this type does not involve the univariant three-phase condensed equilibrium between solid solutions α, γ, and the liquid L. The difference between this type of system and that of Fig. 17 is that in $P\text{--}T$ projection the $\alpha\gamma$L curve is not observed, and in $T\text{--}X$ projection the corresponding α, γ, and l curves are missing. The single-phase volume of γ is open to high pressures; as a consequence, at $P > P_{OB}$ a sequence of single-phase fields α, γ, L, and V appears in the isobaric sections, which span the unary planes of A and B and are separated by the heterogeneous fields $\alpha\gamma$, γL and LV. In the real temperature scale point S_1^B might not appear. In such a case, as well as that of Fig. 17, the $T = 0$ K isotherm will divide the fields α, $\alpha\gamma$, γ, γL, and L into real and imaginary parts.

Non-parallel polymorphism
(1) If the solubility of component B in the α-form is complete and that in γ is restricted, then the triple point S_2 of Fig. 18 is no longer observed in the $P\text{--}T$ projection, and the curves $\gamma\alpha$, γL, and $\gamma\alpha$L do not cross. Accordingly, in the $T\text{--}X$ projection, unlike that of Fig. 18, the composition curves of the γ, α, and l phases do not meet at temperature T_{S2}. As a result, the single-phase volume of γ becomes unrestricted in pressure; therefore, section Fig. 18,III continues up to P_{OB} from P_N. At $P > P_{OB}$ (see Fig. 18,III) the three-phase equilibrium αLV will disappear together with two-phase loop αV and the single-phase liquidus L will extend up to component B.

    (2) If the solubility of component B in both polymorphs is restricted and the miscibility for the α-form is greater than that for γ, then such a system will be different from the diagram in Fig. 19 only in that that point S_2 and the melting curve αL will no longer be there, and curves $\gamma\alpha$, αL, and $\gamma\alpha$L will not intersect. If the three-phase curves αLβ and $\gamma\alpha$L do not intersect either, then the γ-form is not localized in the $P\text{--}T\text{--}X$ phase space, and sections of the type of Fig. 19,IV fall within the pressure interval $P_{N2} < P < P_{OB}$. But if αLβ and $\gamma\alpha$L do intersect, then we have an additional invariant four-phase equilibrium, $\gamma\alpha$Lβ (Fig. 21). In $P\text{--}T$ projection it appears as a quadruple point N_2 and the N_3 invariant is for the αLβV equilibrium. In $T\text{--}X$ projection a new tie-line appears at T_{N2}. The composition curves for phases on the three-phase surface, which is bounded by the points N_2 and N_1, are projected onto the $T\text{--}X$ plane as γ, α, and l lines, while those for the αLβ equilibrium, running from N_2 to N_3, are α, l and β. The phases participating in the N_2 invariant give rise to two more three-phase equilibria, $\gamma\alpha\beta$ at $T < T_{N2}$ and $\gamma l\beta$ going up in temperature, $T > T_{N2}$. The γ-field is not localized in the $P\text{--}T\text{--}X$ phase space.

    The characteristic feature of this type of system is that at $P > P_{N2}$ only solid solution γ melts isobarically: no melting region αL is observed in Fig. 21,IV. The diagrams of the Fig. 21 type are characteristic of Group IV metal nitrides TiN, ZrN, and ThN [24].

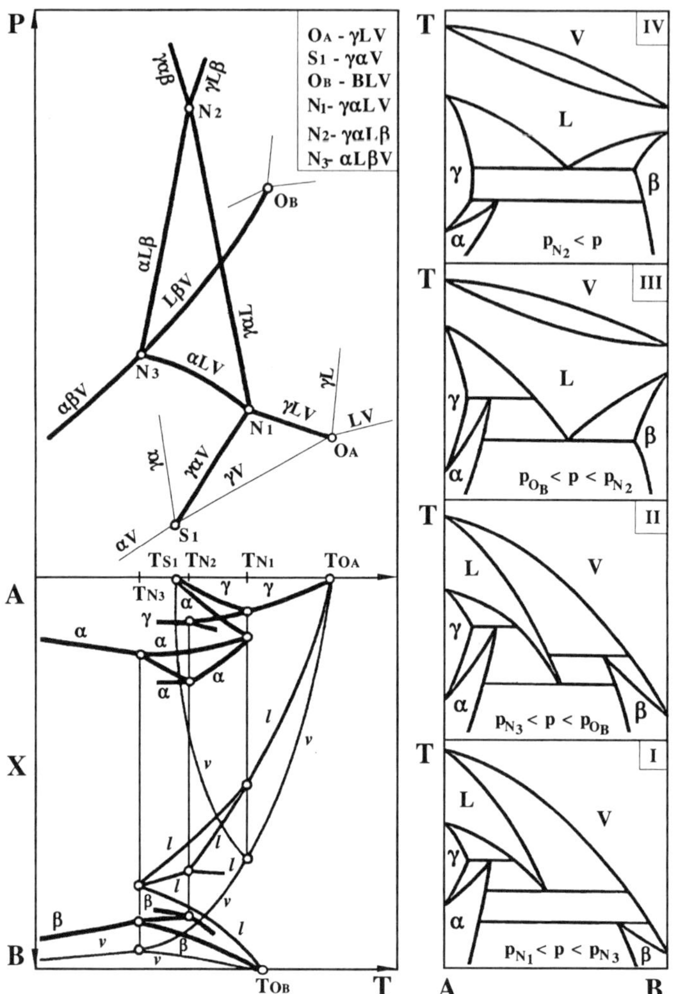

Fig. 21. P–T–X diagram of a binary system with Type II non-parallel polymorphism and restricted solid state miscibility. The solubility of B in α is greater than that in γ

(3) In the case of limited solubility of component B in both polymorphs, that in the γ-form being greater than in α, neither triple point S_2 nor the melting curve αL are observed (unlike the diagram of Fig. 20). If the αγβ and γLβ surfaces do not cross, the invariant quadruple point for αγLβ and univariant equilibria αγL and αLβ are also missing. But if the αγβ and γLβ surfaces intersect, then the αγL curve, originating from the point N_2 in the P–T projection (Fig. 22), should run at $T > T_{N2}$ so that it would not meet the phase transition αγ and melting γL curves because such an intersection is possible only for Type I (Fig. 5,a) polymorphism. The γ-phase field in Fig. 22 is not localized in the P–T–X phase space, and the

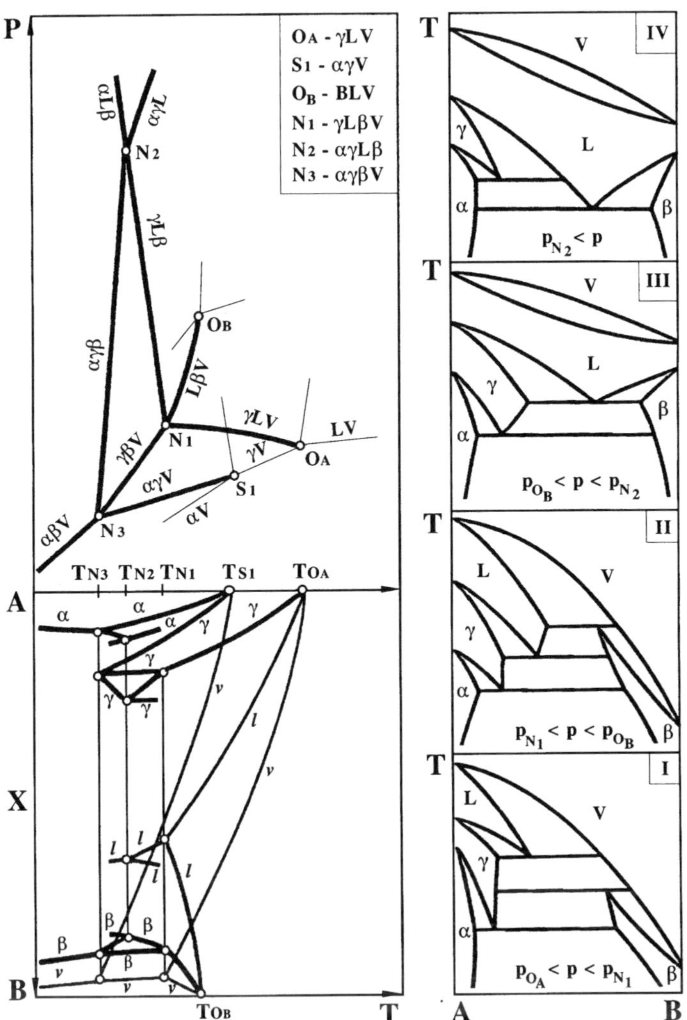

Fig. 22. *P–T–X* diagram of a binary system with Type II non-parallel polymorphism and restricted solid state miscibility. The solubility of B in γ is greater than that in α

shape of the melting region depends on whether the quadruple point N_2 is observed in the system. In the former case all of the isobars at $P > P_{OB}$ are of the type of Fig. 22,III., i.e., only the γ-phase can melt, whereas in the latter case, the isobars at $P > P_{N2}$ are of the type of Fig. 22,IV, and both polymorphs melt when passing the corresponding αL and γL loops. Diagrams of the type of Fig. 22 are known for many transition metal hydrides, UH_3 in particular [24].

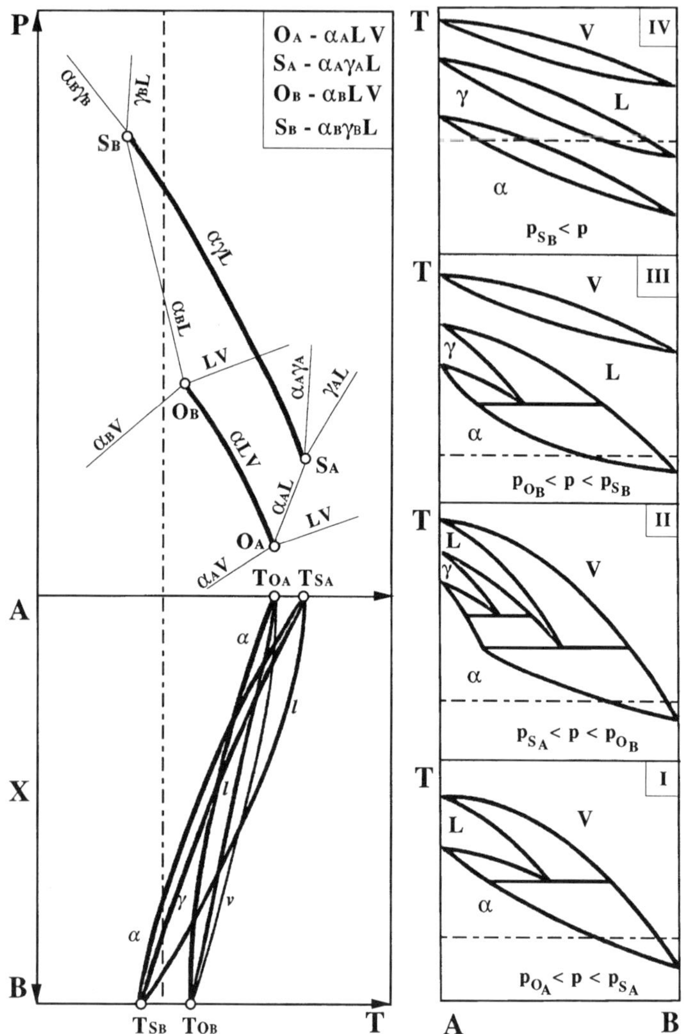

Fig. 23. *P–T–X* diagram of a binary system with Type III parallel polymorphism and complete miscibility in all aggregate states

1.2.2.6.3 Component A with Type III polymorphism

Parallel polymorphism. When the polymorphism of A is paralleled to that of the second component, two three-phase surfaces are observed in the *P–T–X* phase space (Fig. 23). They are projected onto the *P–T* plane as two invariant curves, αLV and αγL, onto *T–X* as the α, *l,* and *v* lines at temperatures T_{OA} to T_{OB}, and as α, γ, and *l* lines at $T_{SA} < T < T_{SB}$. As can be seen in the sections, the α-form can be precipitated in isobaric conditions from both the liquid and the vapor, when cooled over the melting αL or sublimation αV range, whereas the γ-phase is crystallized only from the melt (no γV range is found on isobars, Fig. 23). Furthermore, at $P < P_{SB}$ the γ–form is miscible with the component B only to a certain extent. On a real temperature scale the triple point S_B might not be attainable. Then the $T = 0$ K plane will divide the P–T–X diagram into the real and imaginary parts (dot-and-dash lines in Fig. 23), and experimental T–X diagrams will show limited solubility of B in the α-phase.

Non-parallel polymorphism

(1) We shall at first consider the case, in which the γ-phase is completely miscible with component B, whereas the α-phase dissolves B only to a certain extent (Fig. 24). In this system a four-phase equilibrium $\alpha\gamma LV$ is observed which is a quadruple point N in the P–T projection, where four univariant curves meet: αLV, $\alpha\gamma L$, γLV, and $\alpha\gamma V$. The last one is of special interest. The γ-polymorph of pure component A cannot be crystallized from the vapor in equilibrium conditions because no sublimation curve γV is observed in the P-T diagram of A (Fig. 5, c). Meanwhile, the solid solution γ does form a sublimation area γV. Consequently, when the γ-form is doped with the second component B, it can be prepared by condensing the vapors at pressures within the interval of Fig. 24,I and Fig. 24,II. Moreover, if the pressure is within $P_N < P < P_{OB}$ (Fig. 24,II), then γ is the only polymorph in equilibrium with the vapor.

(2) Now we will refer to the case of limited miscibility of B with both polymorphs; the solubility in γ is greater than that in α (Fig. 25). Two four-phase equilibria are seen in this system, $\alpha L\beta V$ and $\alpha\gamma L\beta$. In the P–T projection the corresponding points are N_1 and N_2, and the compositions of the phases are fixed in the T–X projection by the tie-lines at T_{N1} and T_{N2}. We will use this example to show how the relative arrangement of the invariant points in the P–T–X phase space affects the geometry of the diagram. The systems of Fig. 25 and Fig. 26 are of the same type; the difference between them is that the high-pressure invariant N_2 in Fig. 25 is the lowest (in temperature) among the other invariants, whereas in Fig. 26, it is the highest. Comparison of these two figures shows considerable transformation accompanied by this change. The phase equilibria and crystallization routes are visualized in several sections (Fig. 25,I–IV and Fig. 26,I–IV) and are hoped to be self-explanatory.

An interesting variety of this type of systems is that represented in Fig. 27. Here the γ-form dissolves both of the components to some extent up to high pressures. In the system of this type three four-phase equilibria are observed: $\alpha V\gamma L$, $\gamma VL\beta$, and $\alpha\gamma V\beta$. In the P–T projection they are seen as quadruple invariants N_1, N_2, and N_3 connected by the univariant curves. The corresponding phases change in composition along the v, γ and l lines between T_{N1} and T_{N2}; along γ, v and β between T_{N2} and T_{N3}, and along α, γ, and v lines between T_{N1} and T_{N3}. At pressures within $P_{N3} - P_S$ the γ-form dissolves both components to a limited extent (q.v. isobars Fig. 27,I–III), and it is only at high pressures $P > P_S$ that the single-phase volume of γ reaches the unary plane of A (Fig. 27,IV). In addition, a pressure minimum K_1K_2 is shown in the two-phase area γV in Fig. 27, which corresponds to the congruent sublimation $\gamma = V$ of the γ-form in the temperature interval within $T_{K1} - T_{K2}$.

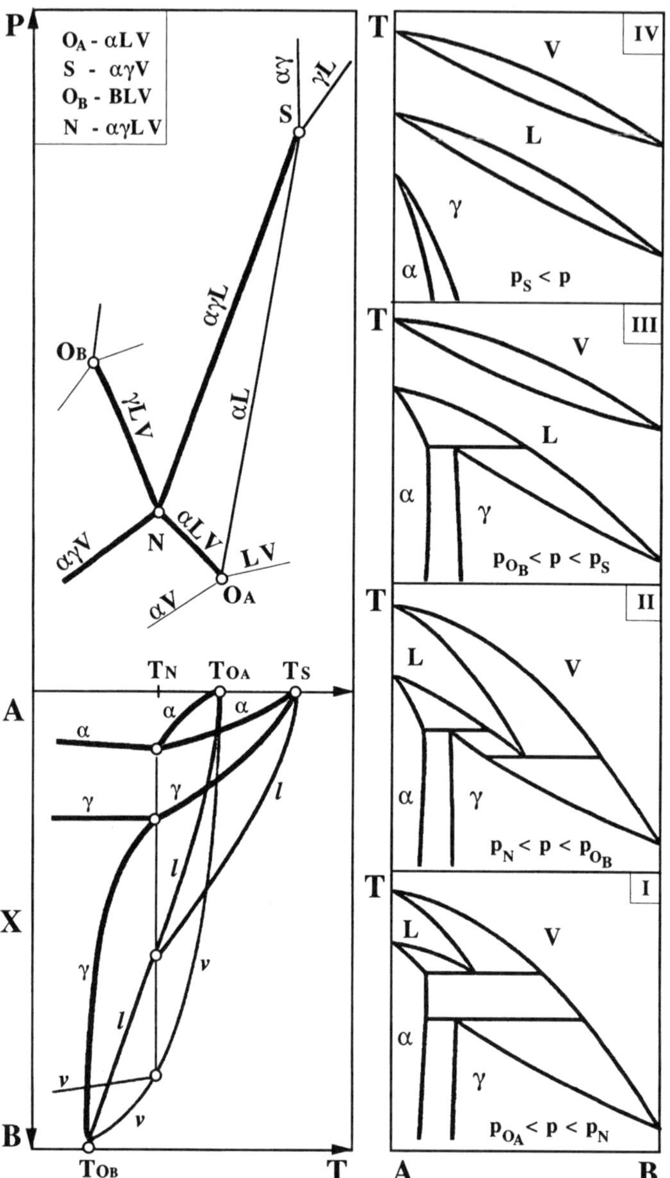

Fig. 24. *P–T–X* diagram of a binary system with Type III non-parallel polymorphism. The solubility of B is complete in γ and restricted in α

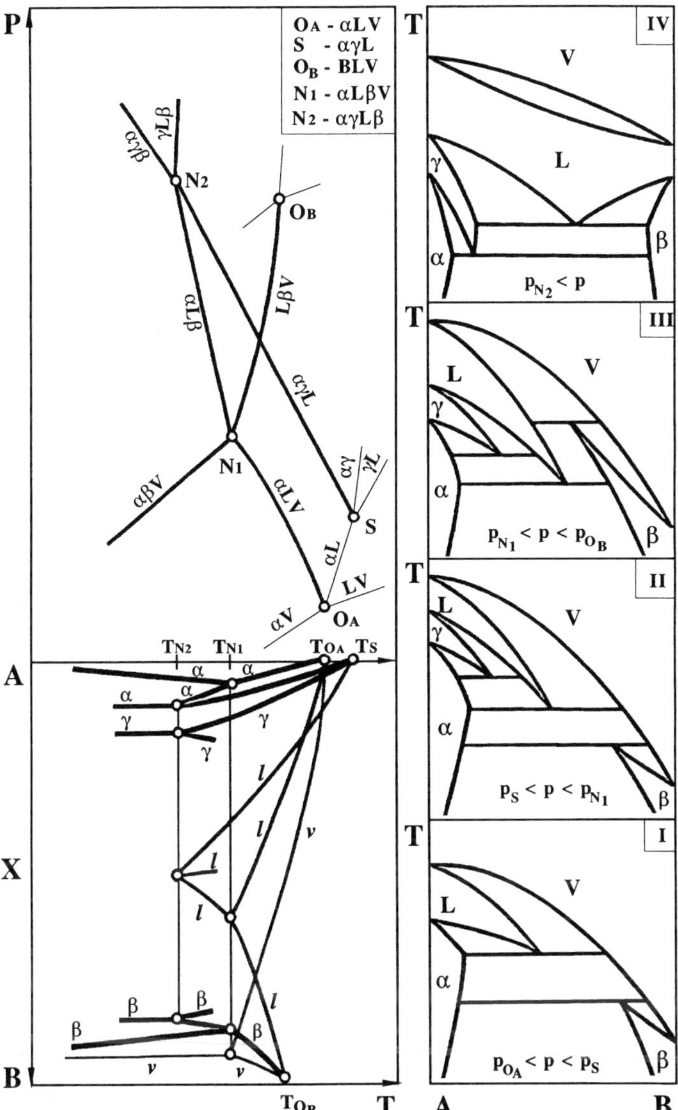

Fig. 25. *P–T–X* diagram of a binary system with Type III non-parallel polymorphism and restricted solid state miscibility. The solubility of B in γ is greater then that in α. T_{N2} is the lowest invariant temperature

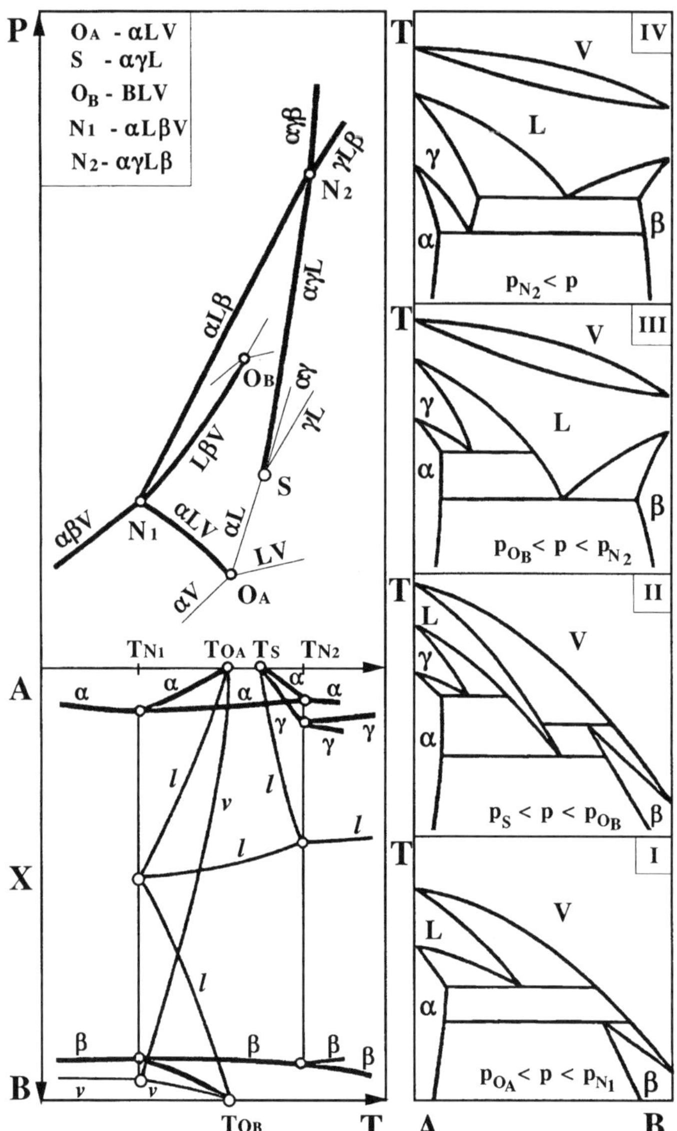

Fig. 26. *P–T–X* diagram of a binary system with Type III non-parallel polymorphism and restricted solid-state miscibility. The solubility of B in γ is greater than that in α. T_{N2} is the highest invariant temperature

The diagrams of the type, Fig. 27, are of substantial applied interest, in particular, in connection with the problem of stabilizing cubic zirconia because some of the $ZrO_2 - M_mO_n$ systems are of this type. The relevant details will be given in a subsequent section.

Thus, we have examined various types of equilibrium $P-T-X$ phase diagrams in a binary system A–B involving two polymorphs of component A. We did not consider miscibility gaps in liquid and solid solutions: such systems have already been discussed in detail [6,21]. Some of the types of systems discussed in this section have been partially presented elsewhere. For example, Ricci [6] discussed "condensed diagrams" for systems with complete miscibility of component B in one of the polymorphs of A and parallel polymorphism, as well as diagrams of the types of Figs. 21 and 22 and $P-T$ projections for Figs. 18–20 and Figs. 24–26. However, the exposition in [6] has its limitations, viz., all of the diagrams are constructed on the assumption of complete immiscibility of the solid phases. As a consequence, the presented diagrams are sometimes difficult to relate to experimental results. Meanwhile, $P-T-X$ diagrams with polymorphism discussed in [24] are essentially confined to the types in Figs. 21 and 22.

1.2.2.7 Metastable states in the *P–T–X* phase space

In this section systems will be examined, in which metastable states appear as a consequence of the polymorphism of the components. Same as for one-component diagrams (Fig. 5), the equilibrium system will be represented by solid curves; metastable states will be given in projections and isobaric sections as dashed lines with the corresponding conjugated phases spelled out and set in brackets. Four-phase metastable states will be shown as quadruple points labeled M in the $P-T$ projection and as dashed tie-lines at the temperature T_M in the $T-X$ projection. The unary metastable point (γLV) will be labeled O'_A. Thus, all of the subsequent diagrams are in fact a superposition of two diagrams: equilibrium (solid lines) and metastable (dashed lines). Therefore, depending on specific experimental conditions, phase transformations in these systems might follow either equilibrium or metastable phase relations. In this section we will explicitly discuss only formation processes for metastable phases, assuming that the reader will follow the equilibrium phase transformations by himself, using the corresponding projections and sections of the diagram.

1.2.2.7.1 Component A with Type IV polymorphism

In this case the γ-polymorph of component A is completely metastable. Three groups of diagrams result from various miscibility patterns.

(1) The miscibility of the α-phase and component B is unlimited. If the metastable (γ)-form is also completely miscible with component B, then the system is that of Fig. 28a. The unary triple point O_B gives rise to the equilibrium univariant curve αLV, and also to the metastable three-phase curve (γLV). The compositions of the phases in the $T-X$ projection follow the corresponding equilibrium α, l, v

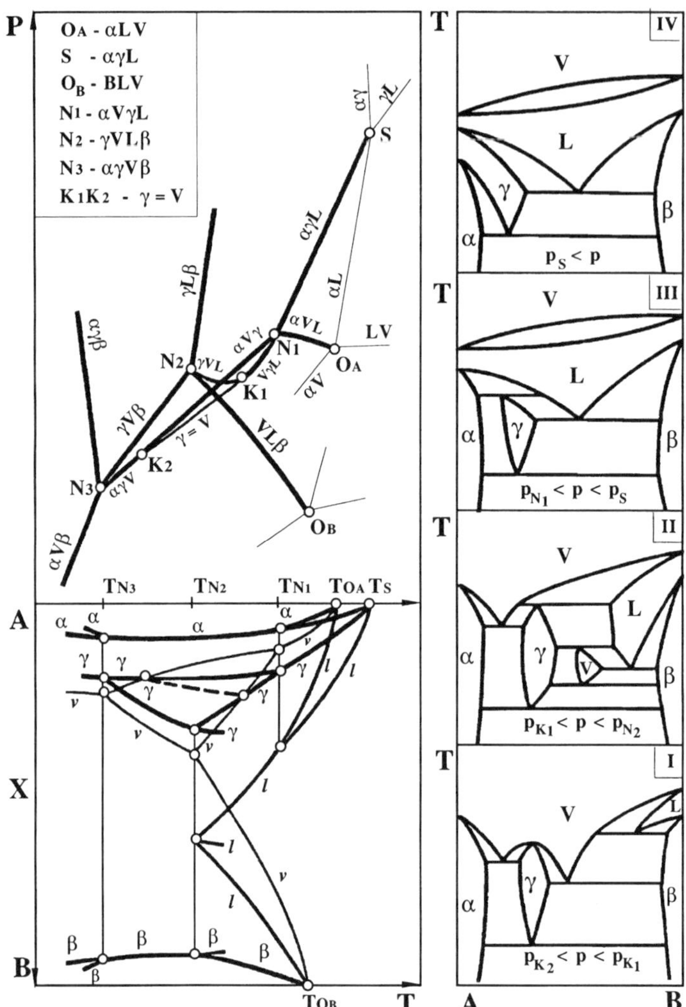

Fig. 27. $P\text{–}T\text{–}X$ diagram of a binary system with Type III non-parallel polymorphism and restricted solid state miscibility. The solubility of B in γ is greater than that in α. Congruent sublimation of γ

and metastable (γ), (l), (v) lines. As can be seen in Fig. 28a, the (γ)-form can be obtained in isobaric conditions at $P < P_{O'A}$ only by condensation from the supercooled unsaturated vapor. The region of the metastable vapor at pressures within $P_{O'A} - P_{OA}$ (Fig. 28a,I) is between the equilibrium and metastable vaporus lines. Depending on the composition, this metastable vapor (V) is isobarically condensed either directly to the (γ)-polymorph (for B-rich compositions) or via prior liquefaction through the metastable (LV) loop. On subsequently decreasing the temperature, all vapor condenses, and in a specific temperature interval the substance becomes a single-phase metastable liquid (L), from which the (γ)-phase finally precipitates. On the other hand, if a seed of the α–form is introduced, any of the metastable states, (γ), (γL), (L), (LV), or (V), may turn into the stable α-polymorph or the αV equilibrium. At higher pressures (Fig. 28a,II) the (γ)-phase may also be isobarically crystallized from the metastable melt (L) or from the metastable loop (LV) by passing through the three-phase metastable state (γLV). The only difference between this system and that with the parallel polymorphism (Fig. 17) is that in Fig. 28a only one triple point appears for component B.

If the metastable polymorph (γ) dissolves only a limited extent of B, then the diagram is that of Fig. 28,b. One more metastable state is formed, solid solution (β). Of course, no changes occur in the equilibrium diagram. The four-phase metastable state (γLβV) is observed in this system together with a set of corresponding metastable lines in projections. The routes of metastable crystallization can be readily followed in isobars of Fig. 28b.

(2) Now consider a case of restricted solubility of component B in the equilibrium polymorph α. Furthermore, if the miscibility of B in (γ) is unlimited, then the diagram will be either that of Fig. 29a, or Fig. 29b, depending on the relative position of the metastable unary triple point O'_A with respect to the equilibrium univariant points. To realized the system of Fig. 29a, it is necessary that the metastable solid solution melts congruently, i.e. there should be an internal tangent (β = L) of two metastable volumes, (β) and (L).

In the P–T projection the (LβV) curve is the metastable extension of the LβV equilibrium. When it touches the (β = L) line at point K', where (β) becomes equal to (L) in composition, the order of the compositions of the three conjugated phases in this metastable state undergoes a change, as seen in P–T and T–X projections. In the T–X projection the l, β, and v composition lines originating from T_{OB}, continue, after reaching the tie-line temperature T_N, as metastable curves (l), (β), and (v), which run through the temperature minimum T_{min} and finally meet at the metastable triple point O'_A. The curves (l) and (β) intersect at point K' where (β) = (L) in composition. The (β = L) curve originates from the point K', which represents the temperature dependence of the congruent melting composition of the (β)-phase. In isobaric conditions falling within the Fig. 29a,I interval, the (β)-phase can be precipitated, depending on its composition, both directly from the vapor in the (βV) loop and from the metastable melt (L). At higher pressures (Fig. 29 a,II) the (β)-phase can also be obtained by supercooling the melt from the LV equilibrium to temperatures below the three-phase equilibrium αLV, i.e., to the metastable state (LβV). If a seed of the α-form is introduced or the system is annealed in suitable conditions, then the metastable states could transform into equilibrium $\alpha\beta$, αV (isobar I) or $\alpha\beta$, αV, αL (isobar II).

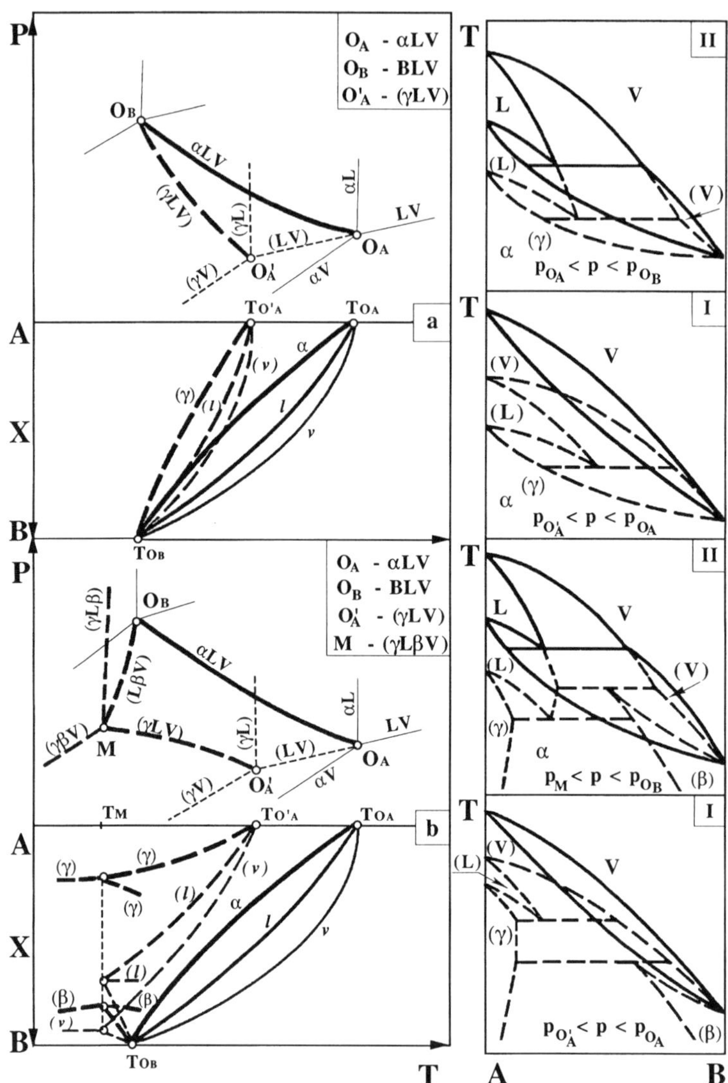

Fig. 28. *P–T–X* diagram of a binary system with Type IV non-parallel polymorphism and complete miscibility of B in the equilibrium polymorph α. (**a**) Metastable polymorph γ is completely miscible with B; (**b**) limited γ–B miscibility

When $T_N > T_{O'_A}$ (Fig. 29b), the metastable extension of the LβV equilibrium comes directly to the point O'_A. In such a system, the metastable solid solution (β) melts incongruently.

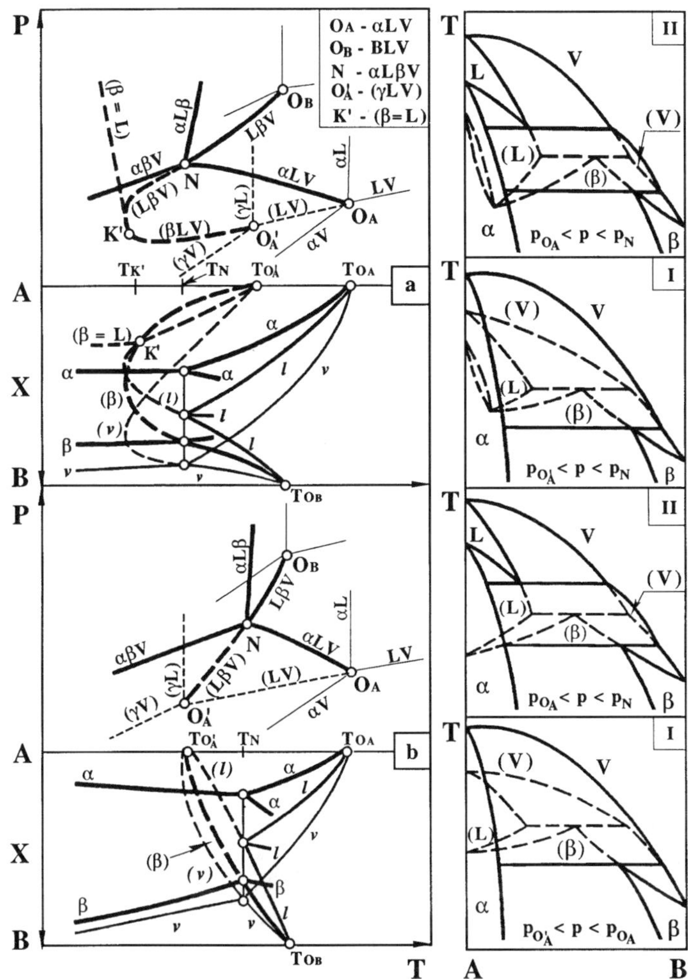

Fig. 29. *P–T–X* diagram of a binary system with Type IV non-parallel polymorphism. Limited solubility of B in α and unlimited in γ. (**a**) Congruent fusion of the metastable solid solution γ; (**b**) Incongruent fusion of γ

(**3**) Figure 30 represents a system with restricted miscibility in all solid phases. Two four-phase points appear in this diagram: N is the quadruple invariant representing αLβV equilibrium, and M is for the metastable state (γLβV). Between them the three-phase surface (LβV) is observed, which is projected onto the *T–X* plane as metastable extensions of the curves *l*, β, and *v* down to the temperature T_M. Another metastable state, (γLV), is projected onto the *T–X* plane as a set of three curves (γ), (*l*) and (*v*), running up to the triple point temperature $T_{O'_A}$. In isobaric conditions

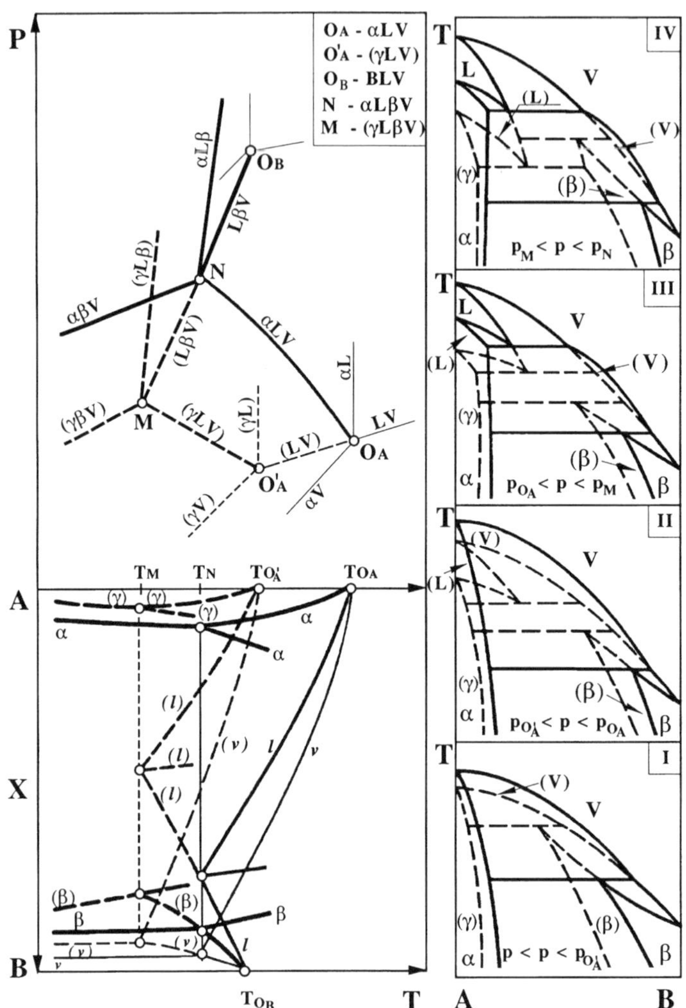

Fig. 30. *P–T–X* diagram of a binary system with Type IV non-parallel polymorphism. Limited solubility of B in α and γ

solid solution (γ) may be crystallized from the vapor of the (γV) loop, Fig. 30,I–III (the metastable vapor area is bounded by the equilibrium and metastable vaporus curves). The (γ)-form can also be obtained from the supercooled melt of the (γL) loop or LV equilibrium, as well as from the β-phase, superheated from the (βV) state up to the ($\gamma\beta$V) tie-line, where the metastable (γ)-phase precipitates instead of the equilibrium α-phase, which should have crystallized on attaining the $\alpha\beta$V equilibrium (sections II and III of Fig. 30). Same as in the above cases, in the presence of the α seed or when annealed, the metastable state may turn into the equilibrium $\alpha\beta$, αV, or αL states.

Thus, we have considered metastable states of a binary system resulting from the metastable state in one of the unary systems (Type IV polymorphism, Fig, 5d). However, the other types of polymorphism of the component (Fig. 5a–c) may also result in forming metastable states in the binary system. We will show it by considering the most complicated case, that of Type I (Fig. 5a).

1.2.2.7.2 Component A with Type I polymorphism

(1) We shall at first refer to the case, in which the α-phase exhibits complete miscibility with component B and that in the γ-form is restricted. In such a system a quadruple invariant $\alpha\gamma LV$ appears (Fig. 31). The equilibrium diagram of Fig. 31 is unlike that of Fig. 18 in the relative arrangement of the invariant points: in Fig. 31 $T_N < T_{S1}$, whereas in Fig. 18 $T_{S1} < T_N$. As a consequence, the order of the compositions of the solid phases in invariant and univariant equilibria involving α- and γ-phases is changed. In Fig. 18 the slope of the univariant curve $\gamma\alpha V$ is $dP/dT > 0$, and it exists at temperatures $T < T_N$, whereas in Fig. 31 the $\alpha\gamma V$ curve appears at $T > T_N$ with $dP/dT < 0$. The difference in the solubility of component B in both polymorphs is clearly seen in sections of Figs. 18 and 31. Metastable states simplify the diagram of Fig. 31 considerably: it turns into a diagram with complete miscibility in all of the aggregate states. Phase transformations of equilibrium and metastable states can be readily followed in the isobars presented in Fig. 31.

(2) If the solubility of component B in both polymorphs is limited, then the system may have either two or three invariant equilibria.

A system with two quadruple points is represented in Fig. 32. The difference between this system and that of Fig. 19 is in the order of compositions of the crystalline phases at the invariant point N_1, which is a consequence of the change in the relevant position of the invariants N_1 and S_1. It can be seen in the isobars of the Fig. 32 that the γ-form dissolves more B than the α-form. Two different metastable diagrams could be formed in such a system. In Fig. 32 they are superimposed. If the metastable state (αLV) is observed between N_1 and O'_A instead of the $\alpha\gamma V$, $\alpha\gamma L$, and γLV equilibria, then the only metastable phase is α, and the system becomes that of the simple eutectic type with an invariant equilibrium $\alpha L\beta V$ at the quadruple point N_2 with the temperature T_{N2}. The corresponding changes in phase relations are seen in the isobars of Fig. 32: above the temperature of the three-phase equilibrium αLV the single-phase volume of the γ-form is no longer observed. It is replaced by the metastable extensions of other fields: (α), (V), and (αV) in section I; (α), (V), (L), (αL), and (αV) in section II; (α), (L), and (αL) in isobars III and IV of Fig. 32.

On the other hand, if the γ-phase becomes metastable, then a metastable diagram of the eutectic type is observed, and a metastable eutectic ($\gamma L\beta V$) is formed at an intersection of four metastable curves: ($\gamma L\beta$), ($\gamma\beta V$), (γLV), and ($L\beta V$). It can be seen in the isobars (Fig. 32) that the single-phase volume of the α-form in such a system is no longer observed and neither are the $\alpha\gamma$, $\alpha\beta$, αL and αV equilibria; they

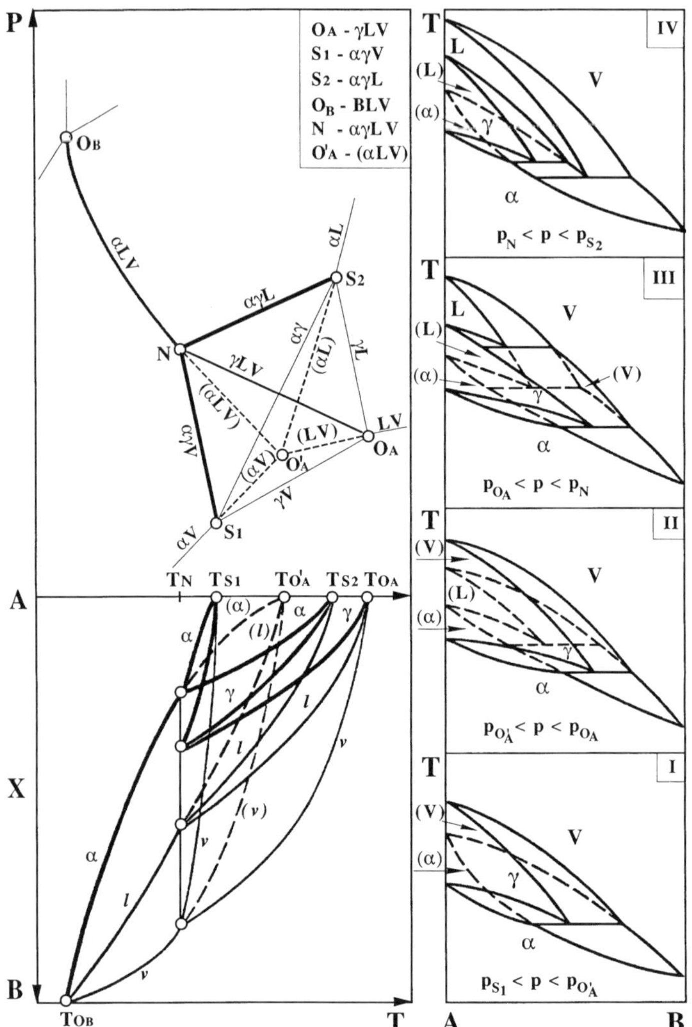

Fig. 31. *P–T–X* diagram of the Fig. 18 type with equilibrium and metastable states

are replaced by metastable areas involving the (γ)-phase. For example, isobars I and II of Fig. 32 exhibit metastable regions (γ), (V), (β), (γV), (βV), and (γβ) instead of the equilibrium fields of α, αV, and αβ. In the pressure interval within the limits of the isobar III (Fig. 32) also missing is the equilibrium melting region αL of the α-polymorph. It is replaced by (γ), (L), and the metastable melting region (γL).

The results presented in this section clearly show serious difficulties to be faced when experimental diagrams of this type are constructed. If an α-seed is introduced, metastable crystallization of γ might be prevented, leaving open, nev-

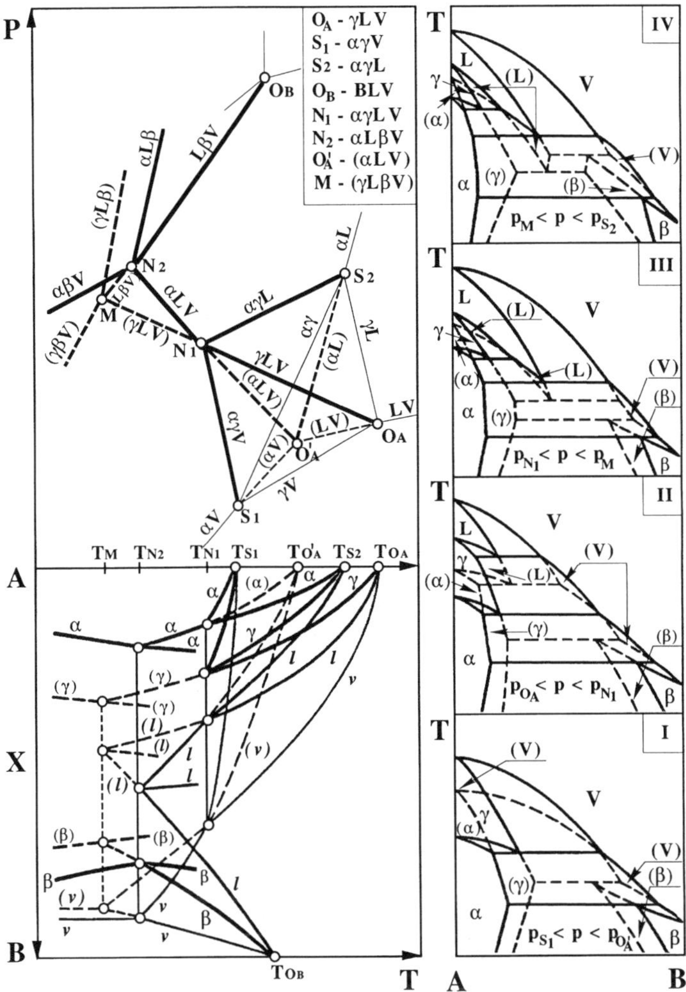

Fig. 32. *P–T–X* diagram of the type of Fig. 19 with equilibrium and metastable states

ertheless, the ways of metastable crystallization of the α-form. The γ-seed, on the contrary, could inhibit only metastable states related to the α-phase.

Figure 33 represents a system with three invariant equilibria. It differs from that of Fig. 20 in the arrangement and slope of the $\alpha\gamma$V curve. The corresponding changes in phase relations are readily seen in the isobaric sections presented in Figs. 20 and 33. The metastable diagram is formed by three-phase metastable curves $(\alpha$LV), $(\alpha\beta$V), $(\alpha$L$\beta)$ and (LβV), and the four-phase quadruple point $(\alpha$LβV). The three-phase curves are metastable extensions of the corresponding three-phase equilibria. Because the γ-phase is not formed in the metastable state, the system is trans-

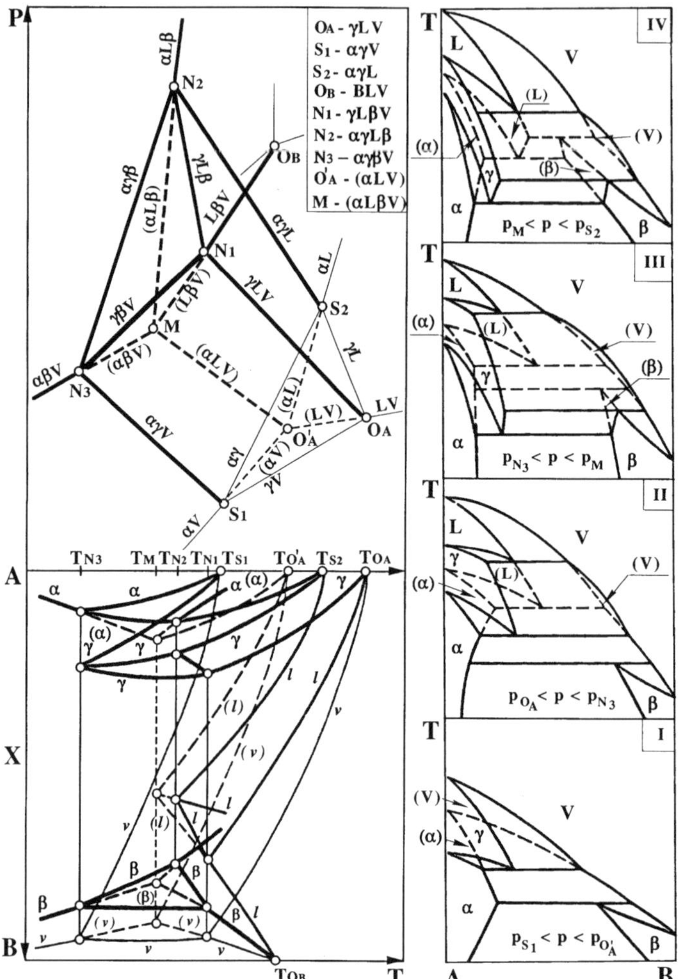

Fig. 33. *P–T–X* diagram of the type of Fig. 20 with equilibrium and metastable states

formed into one of the eutectic type, with limited solid state solubility (Fig. 33). The metastable crystallization routes are readily followed in the isobaric sections.

Thus, we have discussed metastable states of binary systems connected with the polymorphism of a component. However, metastable states may also be observed in the absence of the polymorphism, when a solid state binary compound is formed in the system. Such states will be discussed at a later stage.

1.2.2.8 Formation of a binary compound

The extreme case of negative deviation from the ideal behavior of a solid solution is the appearance of a long-range order at an intermediate composition with the emergence of a new crystal structure and formation of a chemical compound. In the phase space a new single-phase volume appears, which is separated from the others by two-phase spaces. Upon intersection of the respective pairs of two-phase surfaces, three-phase equilibria are observed, in which the crystalline compound C participates. The range of existence of the compound is restricted in T and P. If the high-temperature limit T_Q for the compound is lower than the temperature at which the liquid appears in the system, the compound decomposes to solid components at the vapor pressure of the system. As a consequence, the invariant *peritectoid* equilibrium ACV is observed at $T = T_Q$. If T_Q is the low temperature limit of C, the compound is stable in the interval $T_Q < T < T_m$.

In the two-phase equilibria of the compound C congruent phase transition curves may or may not be involved. These must be tangent to the respective three-phase lines, and the points of tangency are invariant. As shown in section 2.2.2, the phase rule requires that only two phases in a binary system may have identical composition in three-phase equilibria. The third phase is necessarily different in composition; otherwise the number of degrees of freedom for such a state of the system would be negative. Now if we consider the three-phase equilibrium CLV from this point of view, it would be clear that because of the Phase Rule, a binary compound $C = A_m B_n$ cannot have a triple point O_C, at which the compositions of the crystal, liquid and vapor are all identical and equal to $X(O_C) = n/(m+n)$ at temperature $T(O_C)$ and vapor pressure $P(O_C)$.

The compound C may participate in phase equilibria with non-ideal behavior and the resultant extrema in P and T. If solubility in the liquid and vapor is unrestricted and the two-phase spaces AC and CB cross the temperature minimum in CLV, then two invariant eutectic points appear, E_1 (for the ALVC equilibrium) and E_2 (for CLVB). The temperatures of E_1 and E_2 are lower than those of the adjacent invariant points. Such a system is called eutectic. If T_{min} of the equilibrium CLV remains outside the intersection of CLV and the previously mentioned two-phase space, the resulting quadruple point is an intermediate (in temperature) between the adjacent invariants. This is a peritectic system.

1.2.2.8.1 Peritectic systems

The P–T–X space model of a peritectic system is shown in Fig. 34 [6]. As in Fig. 14, B is the low-melting volatile component, and the liquid and vapor are ideal solutions with an ascending liquid–vapor loop. All of the univariant melting curves have positive slopes $dP/dT > 0$. For the sake of simplicity single-phase volumes of the components are reduced to isoplethal planes at $X = 0$ (component A) and $X = 1$ (component B). Conodes, shown in the three-phase equilibria ACL and CLB, define the composition of the conjugated phases in the corresponding equilibria. For example, l is the composition of the liquid at the incongruent melting temperature T_i and l' is the composition of the liquid at the pressure P_i.

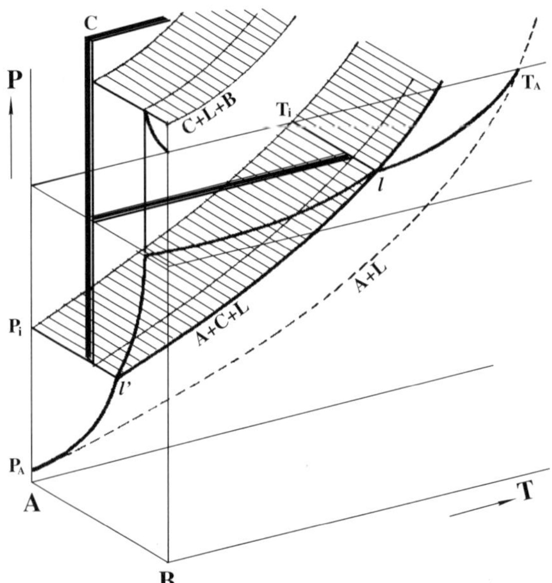

Fig. 34. *P–T–X* diagram of a system with a binary non-stoichiometric compound. Incongruent transition

P–T and *T–X* projections of a peritectic system are presented in Fig. 35. Compound C decomposes (upon heating) at $T = T(P)$ to the solid solution A, liquid, and vapor through a peritectic reaction. This is an invariant four-phase equilibrium ACLV. In the system of Fig. 35, another invariant point CLVB is observed at the eutectic temperature $T(E)$. The pressure dependence of the melting temperatures of the eutectic and peritectic mixtures are projected onto the *P–T* plane as CLV and ACL; the respective compositions of the liquid are given in the *T–X* projection by l' and l''. The compositions of the solids in these equilibria are not marked in the *T-X* projection for simplicity of the figure. Thin lines in the *T-X* projection represent the vaporus in ACV (v' branch), CVB (v''), ALV (v curve at $T(P) < T < T_A$), CLV, and LVB (two v branches at $T(P) > T > T_E$).

The two-phase equilibrium CV of the compound with the saturated vapor appears in the *P–T* projection plane (Fig. 35) as a field within CVB, CLV, and ACV. The *P–T–X* heating route of the sample X_1 (Fig. 35) is seen in the *P–T* projection as the vapor pressure curve I–I and the corresponding *T–X* track of the condensed phase compositions (dot-dash lines in Fig. 35). Four points of discontinuity appear in the vapor pressure curve I–I as a result of the phase transitions in the sample: at T_1 complete evaporation of B is registered, T_2 is the temperature at which the liquid appears, $T(P)$ is the quadruple peritectic point, and at T_3 solid solution A disappears. Thus, if the composition of the vapor is known, a single vapor pressure curve gives the maximum non-stoichiometry of C (solubility of B) at T_1 and T_2, the composition of the liquid in three-phase equilibrium ALV at T_3, as well as the compositions of the

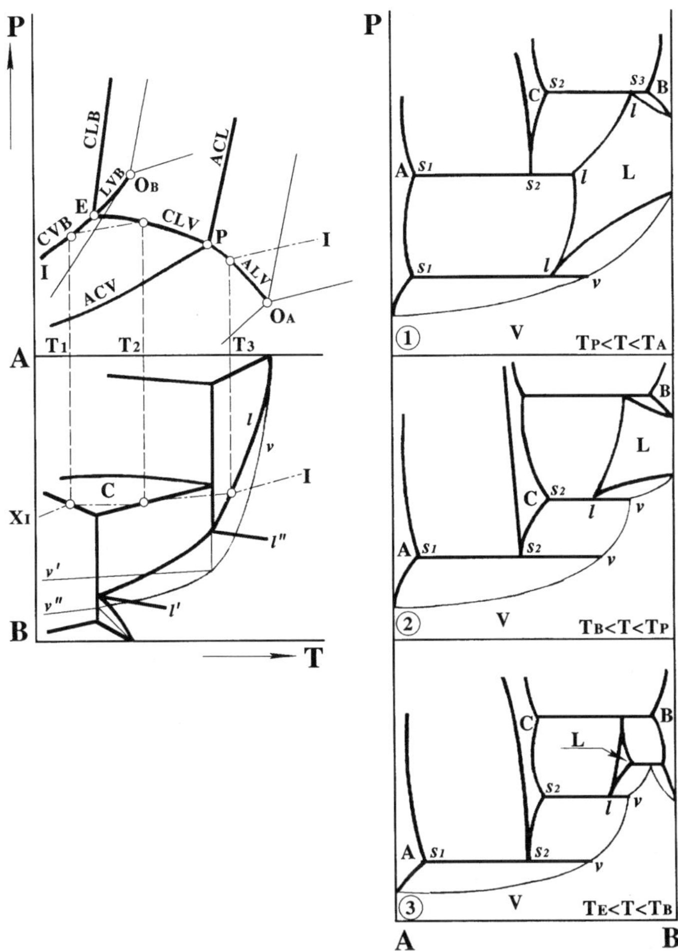

Fig. 35. Complete miscibility in liquid and vapor and compound formation in solid (peritectic)

compound in the CV equilibrium at $T_1 < T < T_2$ and the liquid in LV at $T > T_3$ at the vapor pressures of the experiment. This can be done by substituting the respective experimental data in the corresponding form of Eq. (25).

The crystallization conditions of compound C may be readily followed in the cross-sections of the diagram (the right-hand side of Fig. 35). Above the quadruple point temperature (e.g., T_3 in projections) the compound crystallizes from the liquid (Fig. 35.1) only at elevated pressures. The corresponding two-phase field is restricted in Fig. 35.1 by three-phase horizontals $s_2 l s_3$ and $s_1 s_2 l$. This is a possible (thermodynamically) but inconvenient way of crystal growth. At lower pressures $P(s_1 s_2 l) > P > P(s_1 l v)$ the liquid solidifies to A, and the vapor either liquefies (at

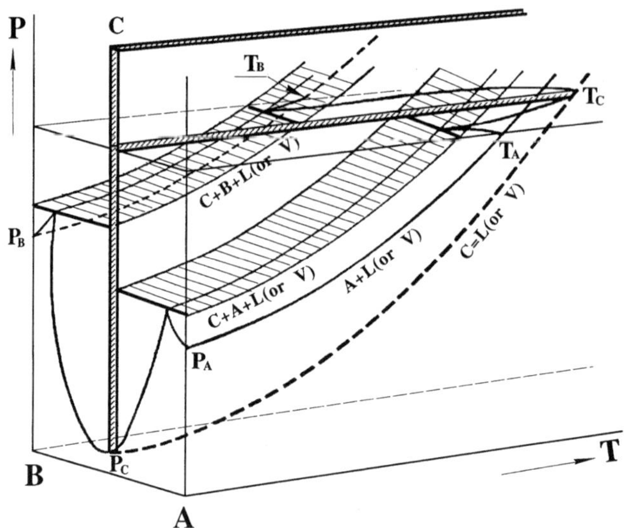

Fig. 36. P–T–X diagram of a system with a binary non-stoichiometric compound. Congruent transition

$X_V > v$) or solidifies (at $X_v < v$). If the vapor composition is $X_V = v$, isothermal compression up to $P = P(s_1 l v)$ at the temperature of Fig. 35.1 (e.g., T_3) results in formation of the solid A (composition s_1) and liquid of the composition $l(T_3)$ calculated from eq. (25) as described earlier.

At $T < T_P$ the compound crystallizes from the liquid and also from the CLV equilibrium (Fig. 35.2, 35.3). For example, at $T = T_2$ compound C with the composition $X_C(T_2)$ given by Eq. (25) crystallizes at pressure $P(s_2 l v)$ (Fig. 35.2). At lower pressures $P(s_2 l v) > P > P(s_1 s_2 v)$ the compound can be grown from the vapor $X_V < v(s_2 l v)$. If the vapor is enriched in B more than $v(s_2 l v)$, it liquefies at the temperature of Fig. 35.2. Below the eutectic temperature T_E the compound crystallizes only from the vapor phase at pressures $P(\text{CVB}) > P > P(\text{ACV})$. It is noteworthy that in all of the cases described the crystallized compound is enriched in the non-volatile component A compared to the crystallization matrix: in the system of Fig. 35, congruent processes C = V and C = L are absent.

1.2.2.8.2 Eutectic systems

The three-dimensional P–T–X phase diagram of a eutectic system (Fig. 36 [6]) is shown for temperatures above the triple points or below them, if the vapor V is substituted for the liquid L. The single-phase field of existence of the vapor (or the liquid) is below (in pressure) and higher (in temperature) than the $T_B P_B \text{CBL} P_C \text{CAL-} P_A T_A$ surface. The two-phase equilibrium volumes are between the C-surface $P_C CT_C P_C$ and the corresponding branches of the liquidus (or vaporus) surfaces. For the LC (or VC) equilibrium the liquidus is below BLC and above C = L (in pres-

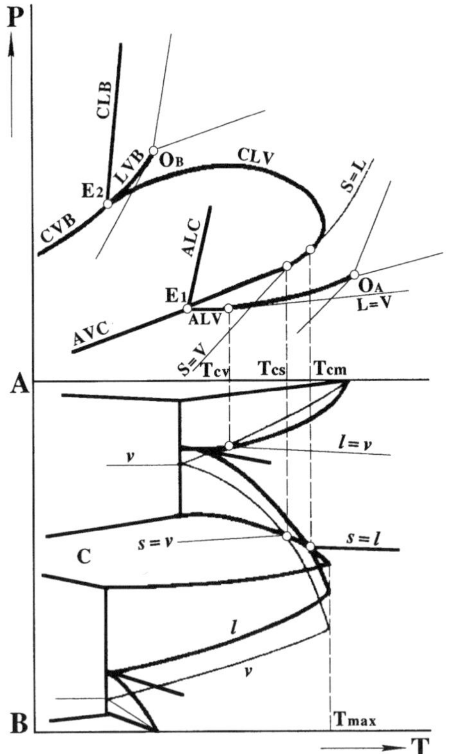

Fig. 37. Complete miscibility in liquid and vapor and compound formation in solid (eutectic)

sure), on the B-side of the C-surface, and for CL it is below CLA and on the A-side of the C-surface. A congruent univariant line C = L (or C = V) appears in the corresponding two-phase equilibrium as a curve of internal tangency between the C-surface and the liquidus (vaporus). It is readily seen in Fig. 36 that the compositions of compound C and the liquid (vapor) coincide along the congruent curve.

Fig. 37 represents the system of Fig. 36 in projections onto the P–T and T–X planes. The maximum melting point T_{max} of C in Fig. 37 is higher than the adjacent invariant temperatures $T(E_1)$ and $T(E_2)$, although it is lower than the triple point O_A. In solid–liquid and solid–vapor equilibria of the compound C in the system shown in Fig. 37, P_{min} and T_{max} are present due to the congruent melting C = L and congruent sublimation C = V of the compound in specific limits of the parameters. Moreover, in the liquid–vapor equilibrium an azeotropic curve L = V appears with the compositions outside the crystallization range of C. Upon meeting CLV and AVL, the curves S = L, S = V, and L = V in Fig. 37 generate three points of congruent transition with the coordinates (P_{cm}, T_{cm}, X_{cm}), (P_{cs}, T_{cs}, X_{cs}), and (P_{cv}, T_{cv}, X_{cv}).

As a result of P_{min} in the two-phase equilibrium of the compound and its saturated vapor, the P–T projection of the CV equilibrium is of dual character. Sublimation of C enriched in B (in respect to S = V compositions) is projected in the P–T

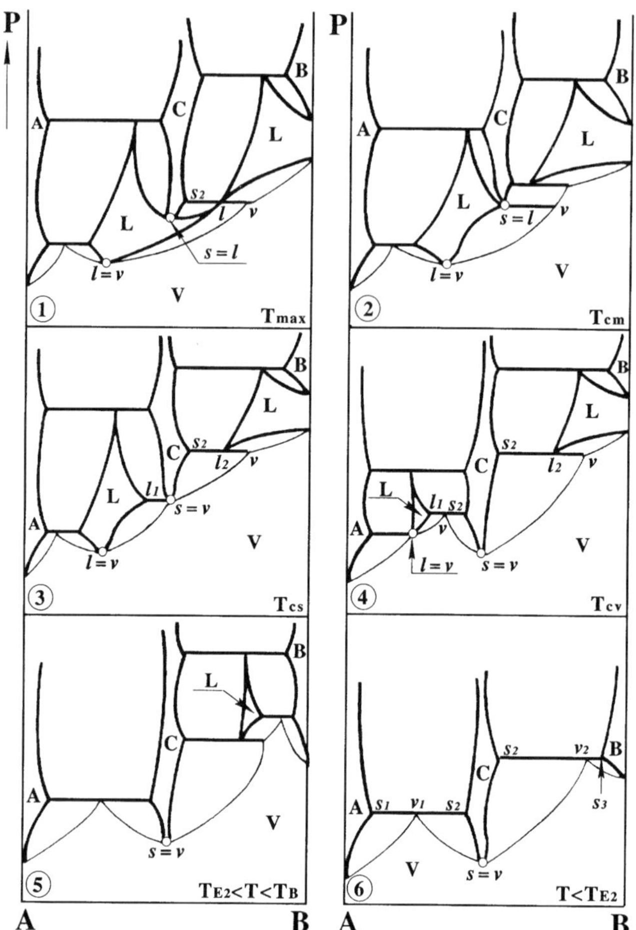

Fig. 38. Isothermal sections of the diagram of Fig. 37

plane within the CVB, CLV, and S = V curves, whereas the field within AVC, LVC, and S = V is the projection of CV with C enriched in A. T_{cs} is the highest temperature of the congruent sublimation of C. At higher T the compound sublimes incongruently. Likewise, T_{cm} is the lowest congruent melting temperature for C, below which no points $X_S = X_L$ are observed, and the compound melts incongruently.

Different possibilities for the crystallization of C are seen in cross-sections of the diagram, presented in Fig. 38 for the whole melting region of C. At the maximum melting point (Fig. 38.1) crystallization of C with composition $X_S(T_{max})$ at the vapor pressure of the system $P(T_{max})$ is observed from the liquid $X_L(T_{max})$ and vapor $X_V(T_{max})$ (q.v. T-X projection , Fig. 37). The vapor pressure $P(T_{max})$ determines the position of the crystallization horizontal s_2/v in Fig. 38.1 (it is interesting to note that two liquid fields coalesce at T_{max} upon a gradual increase of the cross-section temperature up to T_{max}). In the absence of the vapor, C may be crystallized from the liquid at T_{max}, both congruent (point $s = l$) and enriched in A or B (see the respective LC and CL fields in Fig. 38.1). Crystallization of C from the vapor is impossible at T_{max}: the CV space is not crossed by the plane T_{max}.

It is clear from Fig. 38.1 that the composition of C depends on the crystallization procedure: the crystal s_2 grown at T_{max} from the melt at the vapor pressure of the system is different from that grown from the congruent liquid $l = s$. Crystal growth of C from the congruent melt at its own vapor pressure is possible at T_{cm} (Fig. 38.2). The vapor in this $s=lv$ process is enriched in B. At the temperatures of Fig. 38.2 crystals of C can be grown from the melt (in the pressure intervals of LC and CL equilibria) and also from the vapor. The limit in P for this incongruent process is within $s=lv$ and CLV. For congruent crystallization from the vapor the temperature should be reduced at least to T_{cs} (Fig. 38.3). At this temperature the solid $s = v$ is formed along with the liquid l_1. At $T < T_{cs}$ the vapor pressure in S = V is lower than that in the adjacent three-phase equilibria (l_1vs_2 and s_2l_2v, Fig. 38.4). Therefore, C may be grown from the vapor at $X_V = X_S$. Usually crystals, grown by this method, have high structural perfection. It is to be noted that at $T < T_{cs}$ solid C can be obtained with the congruent C = V composition, and also enriched in A or B. For this purpose, crystallization conditions should follow the two-phase VC or CV parameters (Fig. 38.4 – 38.6). If the vapor $v(s_2l_2v)$, Fig. 38.4, is subjected to isothermal compression, solid C of the composition s_2 crystallizes from the melt l_2, s_2 being the maximum non-stoichiometry (solubility of B) at the temperature of Fig. 38.4.

The T_{cv} isotherm (Fig. 38.4) is the lower extremity of the azeotropic equilibrium L = V. Below T_{cv} evaporation is incongruent, and at T_{cv} the azeotrope is in equilibrium with the solid solution A (Fig. 38.4). At $T(E_2)$ the liquid is completely solidified, the L space shrinks to a point, and below this temperature only four single phases emerge in the cross-section: V, A, C and B (Fig. 38.6). Isothermal compression of the vapor v_1 results in solidification into A and C, and at $X_V = v_2$ a mixture of C and B is crystallized. The isobarically invariant equilibria $s_1v_1s_2$ and $s_2v_2s_3$ (Fig. 38.6) are called *vapor eutectics*.

1.2.2.8.3 Polymorphism of a compound

If two different crystal structures of a binary compound, α and β, are stable, then the single-phase volume of C falls into two parts. The α and β single-phase spaces appear in the P–T–X diagram together with a number of two- and three-phase equilibria, which are shown in projections in Fig. 39. Along with two eutectic invariants, E_1 and E_2, two more quadruple points are seen in Fig. 39: Q_1 involves four phases L$\beta\alpha$V, and Q_2 is for $\alpha\beta$VB. In the system in Fig. 39, the triple point O_B is higher in both T and P than O_A, and the high-temperature form β melts congruently at temperatures between O_A and O_B. In all the univariant condensed phase equilibria dP/dT > 0. The vaporus v in the T–X projection, Fig. 39b, is given by thin lines. The v-branch corresponds to the ALV equilibrium, which originates in the melting point of A and extends down to the eutectic temperature $T(E_1)$. Between $T(E_1)$ and T_1 the v-curve is for the LαV equilibrium. At T_1 the vaporus splits up. One branch extends

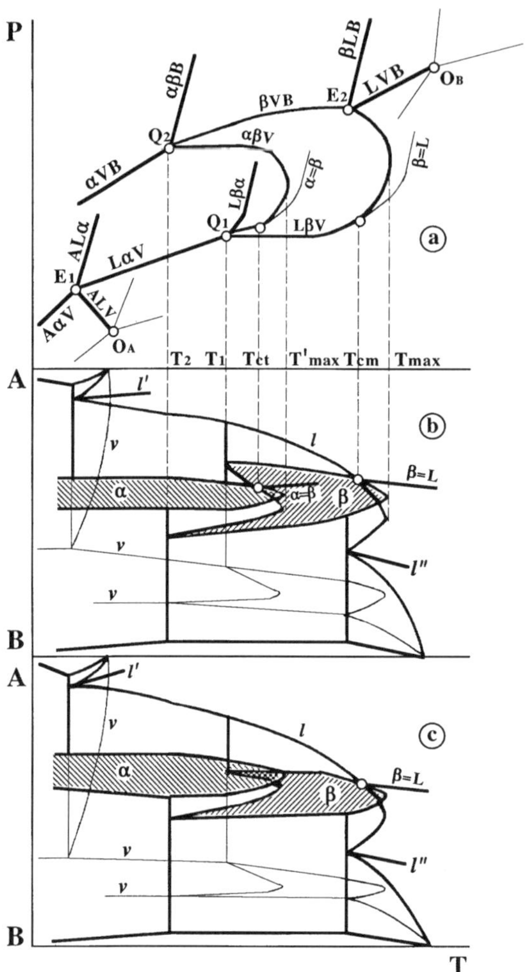

Fig. 39. Projections of a diagram with a binary compound. Congruent melting and phase transition with T_{max}. (**a, b**) Congruent phase transition; (**a, c**) (without α=β) Incongruent phase transition

from T_1 through T_{max} (for LβV) down to $T(E_2)$ (for βLV) and then either to T_2 (vapor in the βVB equilibrium) and $T < T_2$ (for αVB), or up to the melting point of B (the vapor composition in LVB equilibrium). The second branch of v, originating in T_1, is for three-phase equilibrium βαV. It extends from T_1 to T_2 through the temperature maximum T'_{max} of the α–β phase transition region.

Now we shall examine the phase spaces generated by equilibria that comprise the α and β forms of the compound C. Since α→β is the first order phase transition, our analysis will be valid for the melting process as well.

Congruent melting (phase transition) and T_{max} on the liquidus curve – no relationship. It is a general assumption that if a temperature maximum is observed in the melting region, then the compound melts congruently. Vice versa, the absence of T_{max} in the melting region is regarded as proof of an incongruent mode of melting. These types of phase diagrams can be found in any textbook on phase equilibrium (see, for example, [1,4,6]). In this section it will be demonstrated that in fact there is no such relationship.

As has been pointed out, from the geometrical point of view a congruent process involves an internal tangent of the phase surfaces in the corresponding two-phase equilibrium. According to the Gibbs–Konovalov law, this tangent is a pressure and temperature extremum in the two-phase equilibrium. Two phases of identical composition (e.g., $\alpha = \beta$ or S = L) may be involved in equilibrium with the third phase (e.g., vapor). Then the congruent line is tangent to the curved three-phase surface. We recall here that this three-phase surface is orthogonal to the P–T plane. Sometimes it is called a ruled surface because it is generated by an isothermal and isobaric straight line moving so that it continually passes in space through three related curves (see, for example, three-phase surfaces BLC and CLA in Fig. 36). As for the shape of this surface, it is arbitrary (either involving extrema or not), and it does not depend on the existence or absence of the congruent curves in two-phase equilibria.

As an example, a three-phase surface $\beta\alpha V$ involving T_{max} is shown in Fig. 39 whereas in Fig. 42 this surface has no T_{max}. Nevertheless, in both cases systems with similar shapes of this surface may involve either congruent (Fig. 39a,b and Fig. 42a,b,) or incongruent (Fig. 39a,c, and Fig. 42a,c) α–β phase transitions (or melting, if L is substituted for β). Moreover, the three-phase surface may even be corrugated (as will be seen in a subsequent section, when the α–β phase transition in Zn_3As_2 is discussed), i.e., involves multiple extrema, whereas the melting is incongruent.

The arrangement of phases in the P–T–X space is elucidated, if cross-sections of the three-dimensional model are examined. On the left-hand side of Fig. 40 isotherms of Fig. 39a,c, are shown for a system with T_{max} in $\beta\alpha V$ and an incongruent α–β phase transition (the $\alpha=\beta$ line in Fig. 39a is, of course, absent in this case). For all condensed phase equilibria $dP/dT > 0$ is assumed. Due to the pressure dependence of the melting temperature of β, $\beta = L$ is not a vertical, as we have already seen, and as a result, $T_{cm} < T_{max}$, and the composition of β at T_{max} is on the A side of $X_L(T_{max})$. The characteristic feature of the cross-section of Fig. 39a,c at T_{max}, compared to Fig. 38.1, is appearance of the single-phase α-space at high pressure together with corresponding two- and three-phase equilibria. As the temperature is lowered (Fig. 40.1), the βLV equilibrium is observed twice in the cross-section at two different pressures. Therefore, upon crystallization from the melt under the vapor pressure of the system, two different compositions of the β-form can be grown, depending on the chosen conditions: crystals enriched in A are obtained from the A-rich liquid L_1, whereas crystallization from L_2 results in β enriched in B (compared to X_β at T_{max}). Of course, in both cases composition of β is on the B-side of the congruent melting point (q.v. Fig. 40.1).

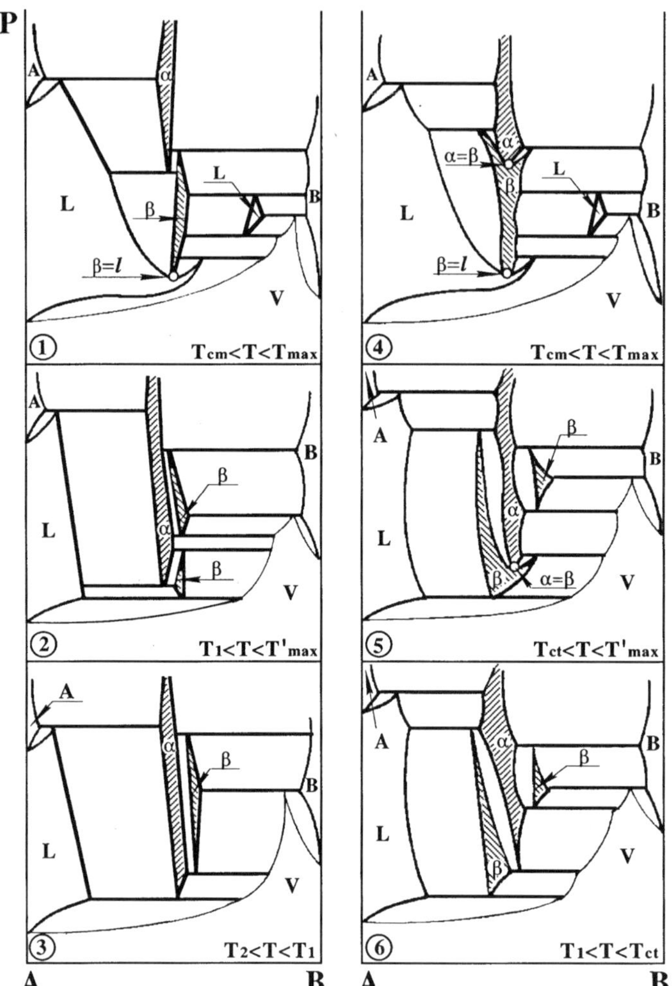

Fig. 40. Isothermal sections of the diagrams of Fig. 39a,c (1–3) and Fig. 39a,b (4–6)

At the temperatures of Fig. 40.2 the isothermal plane cuts the β-space twice: at low pressure within the limits of the LβV and $\alpha\beta$V equilibria and at higher pressures, up to $\alpha\beta$B. In cross-sections between the quadruple points Q_1 and Q_2 (Fig. 40.3) the β-space appears at pressures from $P(\alpha\beta V)$ to $P(\alpha\beta B)$. At higher pressures the β-form is unstable. T_2 is the lower temperature of existence of the β-phase, and therefore at $T < T_2$ the β-space is not cut by the isotherms. The compositional sequence of phases is characteristic for the sections of Fig. 40.1 – 40.3; for all temperatures and pressures $(X_\alpha < X_\beta)_{P,T=const}$, i.e., the α-space is on the A-side of the β-space. Meanwhile, a fixed composition X of the solid may be stable as the α-form in

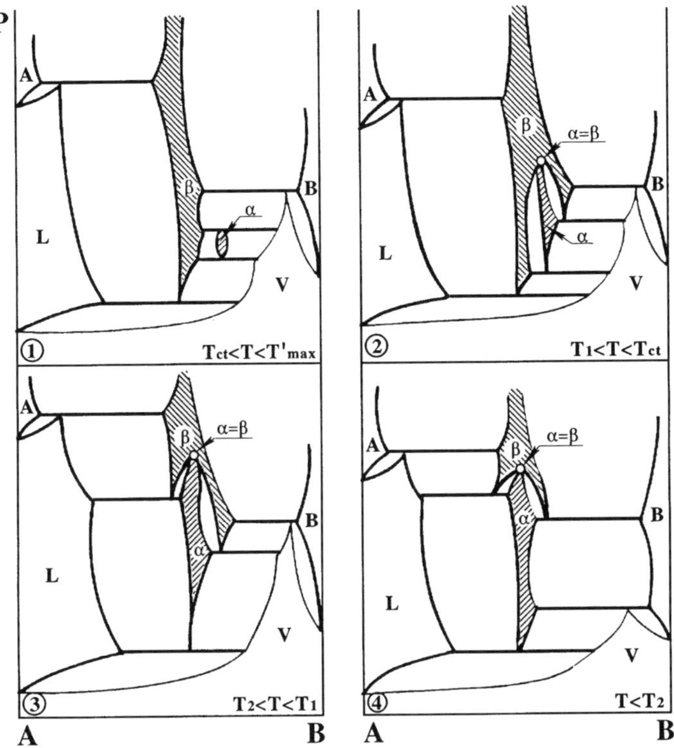

Fig. 41. Isothermal sections of the diagram of Fig. 39a,b with $dP/dT < 0$ for condensed phase equilibria

a specific $P–T$ range, whereas outside of it β is the stable structure. This is the consequence of the partial juxtaposition of the α-and β-fields in the $T–X$ projection of Fig. 39c.

Sections of the system of Fig. 39a,b are shown in the right-hand side of Fig. 40. Figure 40.4 corresponds to the temperatures of Fig. 40.1. The difference between these two systems is that in the former a pressure minimum (and temperature maximum) is observed in the $\alpha\beta$ loop. It is seen in Fig. 40.4 as the point $\alpha=\beta$. In the cross-section by the $T = T_{cm}$ plane point $\beta = l$ appears on the βLV horizontal, which marks the lower limit of the congruent fusion of β. At $T < T_{cm}$ fusion is an incongruent process (the βL field in Fig. 40.5 and 40.6). T_{ct} is the lower extremity for the congruent phase transition. At T_{ct}, point $\alpha=\beta$ impinges on the $\alpha\beta V$ horizontal, whereas at $T < T_{ct}$ the $\alpha=\beta$ curve is not crossed by the isotherms (Fig. 40.6). It is noteworthy that at $T > T_2$ the β-space embraces α in compositional sequence, so that $X_\beta < X_\alpha < X'_\beta$. Below Q_2 the systems of Fig. 39a,b and Fig. 39a,c are identical.

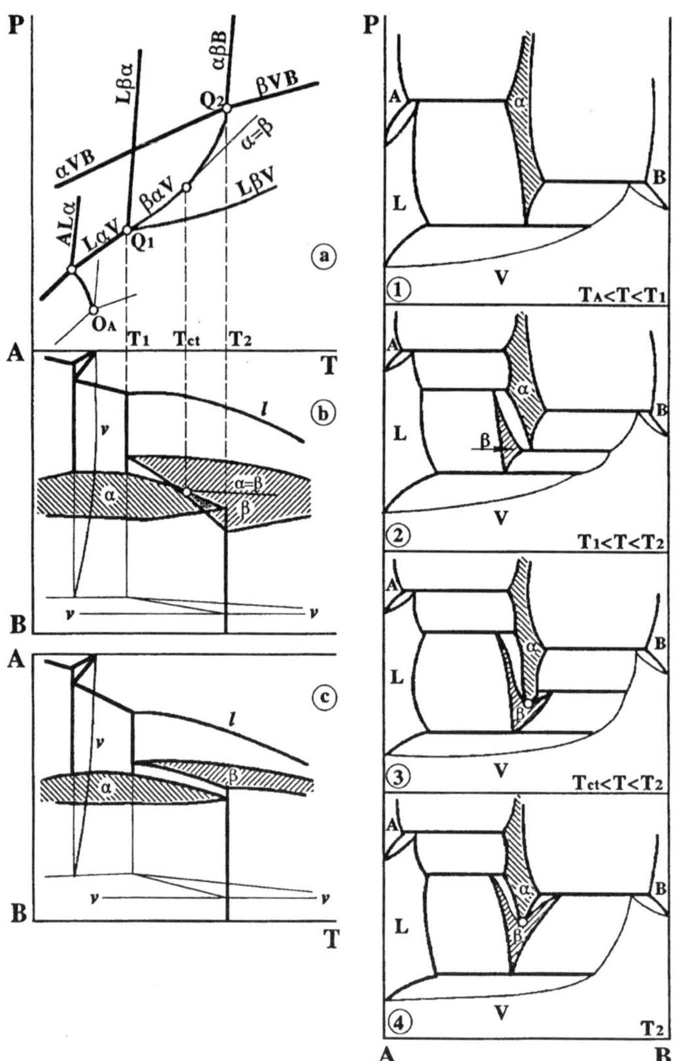

Fig. 42. Phase transition without T_{max}. (**a, b**) Congruent transition; (**a, c**) (without $\alpha=\beta$) Incongruent transition

The arrangement of the univariant equilibria for condensed phases in the P–T–X space may be different from that of Fig. 39a. If they are characterized by a negative P–T slope ($\mathrm{d}P/\mathrm{d}T < 0$), then the congruent phase transition is accompanied by P_{max} in the $\alpha\beta$ loop. Isothermal sections of such a diagram are presented in Fig. 41. In this case T_{ct} is the higher extremity of the $\alpha=\beta$ line, and the congruent transition point $X_\alpha = X_\beta$ is on the B-side of $X(T'_{max})$ whereas in Fig. 39b it is on the

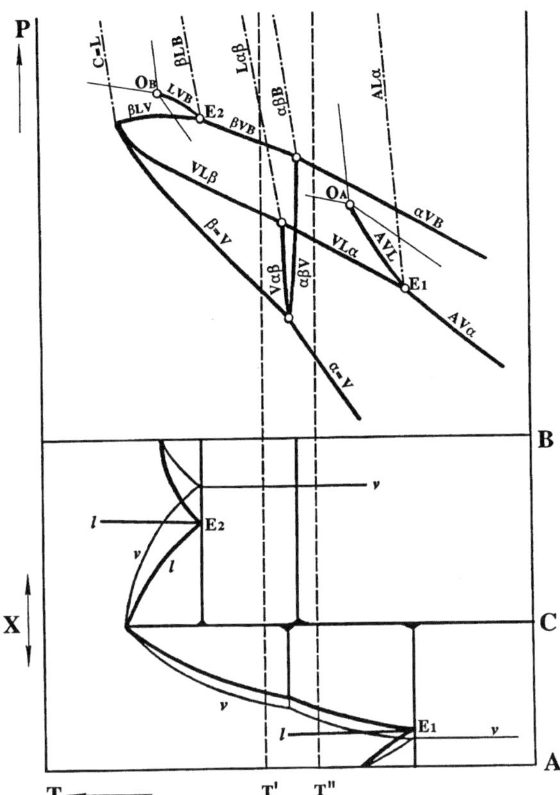

Fig. 43. *P–T* and *T–X* projections of a system with a binary compound

A-side. Furthermore, β becomes the stable form at high pressure instead of α in Fig. 40. In other respects Fig. 41 is self-explanatory.

The phase diagram without T'_{max} in the αβV equilibrium is shown in Fig. 42 for the temperature interval of the phase transition. The shape of the three-phase surface in this case is also independent of the mode of the phase transition (or melting with L instead of β). In the system of Fig. 42a,c the single-phase spaces of α and β do not touch. Accordingly, the α- and β-fields are separated (Fig. 42.2) over the whole range of their coexistence. The β-phase is observed in the cross-sections only at $T > T_1$, and therefore the isotherm, Fig. 42.1, is similar for both systems, Fig. 42a,b and Fig. 42a,c. In the diagram, Fig. 42a,b with the same shape of βαV as in Fig. 42a,c the single-phase volumes of α and β are tangent at $T \geq T_{ct}$. The two-phase αβ equilibrium involves an extreme line (P_{min} and T_{max} for the *P-T* slopes of the condensed phase equilibria shown in Fig. 42). When the isothermal planes cross the α=β curve (or S = L curve with L instead of β), the point α=β emerges in Fig. 42.3. Fig. 42.4 is the cross-section of the diagram at the temperature of the quadruple point Q_2 involving four phases: α, β, V, and B.

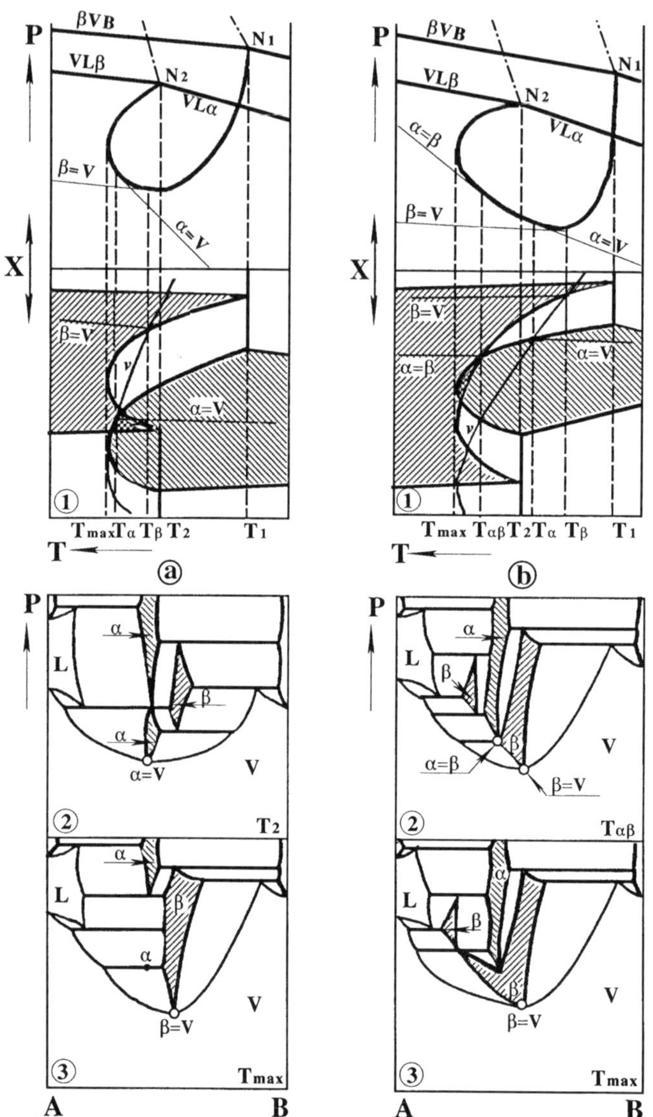

Fig. 44. Projections and isothermal sections of the phase transition region with T_{max}. (**a**) Incongruent transition; (**b**) Congruent transition

A type of system, very important for materials science, is shown in Fig. 43 [35]. The main difference between this and Figs. 39 and 42 is the congruent sublimation of both α and β polymorph forms of compound C. This type of diagram is quite frequent for binary semiconductors (for example, II–V and II–VI compounds). The dot-dash lines in Fig. 43 represent univariant condensed phase equilibria. All of

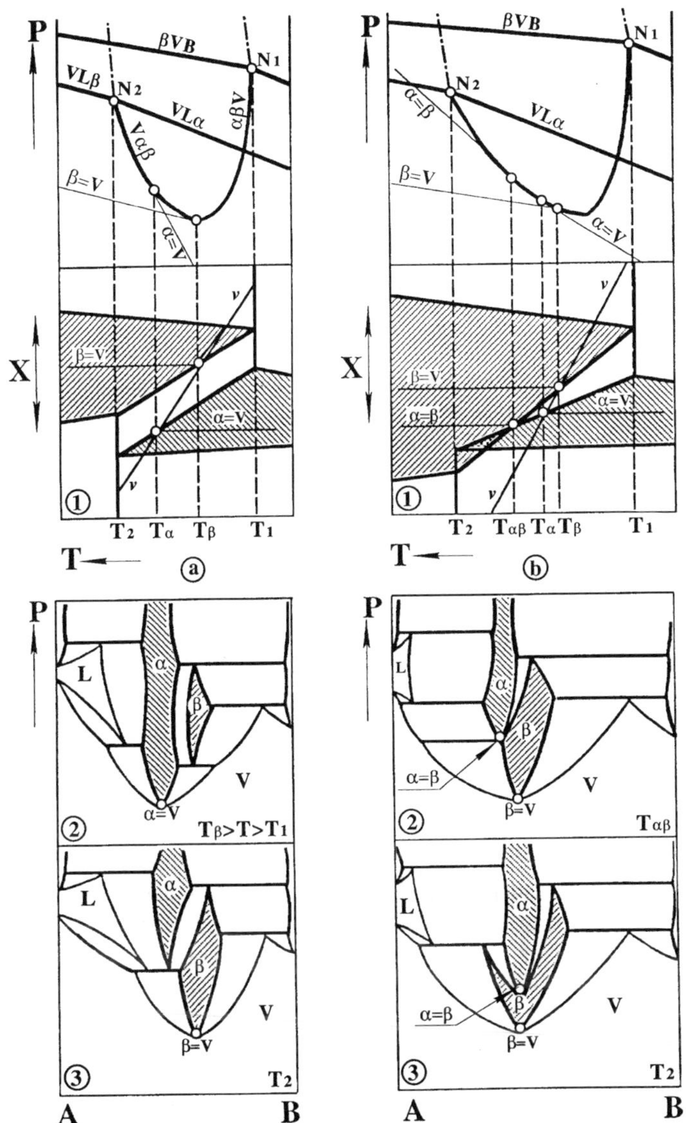

Fig. 45. Projections and isothermal sections of the phase transition region without T_{max}.
(**a**) Incongruent transition; (**b**) Congruent transition

them, as well as those for pure components A and B, have a positive slope $dP/dT > 0$. Details of the phase arrangement in the P–T–X space are given in Fig. 44 and Fig. 45 for the phase transition region $T'T''$ between the dashed lines in Fig. 43.

The three-phase surface $\alpha\beta V$ involving T_{max} is shown in Fig. 44, whereas in Fig. 45 this surface is without T_{max}. Nevertheless, in both cases systems with similar shape of this surface may involve either congruent (Fig. 44b and Fig. 45b), or incongruent (Fig. 44a and Fig. 45a) α–β phase transition. The arrangement of phases in the P–T–X space for this system is elucidated, if cross-sections of the three-dimensional model are examined. On the right-hand side of Fig. 44 isothermal sections of the system of Fig. 43 are shown for temperatures $T' < T < T''$, with T_{max} in the $V\beta\alpha$ equilibrium and the congruent α–β phase transition. Due to the pressure dependence of the phase transition temperature, $\alpha=\beta$ is not a vertical, and as a result, $T_{ct} < T_{max}$, because the composition of β in the $V\beta\alpha$ equilibrium is on the A-side of α at T_{max}. The characteristic feature of the cross-section, Fig. 44b,1, at T_{max} is the congruent sublimation point $\beta = V$ of the β-form. It has to be seen in Fig. 44b,3 that only the β-phase is in two-phase equilibrium with the vapor at T_{max} because the αV space is not cut by this isotherm. Consequently, only the β-form may be crystallized from the vapor, both of the $X_{\beta=V}$ composition, and enriched either in A or B (see the corresponding βV and $V\beta$ spaces in Fig. 44b,3). When the temperature decreases, the $V\alpha$ equilibrium appears in the section. In particular, in Fig. 44b,2 the $V\beta\alpha$ three-phase equilibrium is cut twice by the isothermal plane, and the congruent transition point $\alpha=\beta$ is on the low pressure horizontal.

Figure 44a,1 represents the same P–T shape of the $\alpha\beta V$ curve as in Fig. 44b,1. Nevertheless, no congruent transition is involved in this system. As a result, the two single-phase spaces, α and β, are separated in the P–T–X space, and α is always on the A-side of β. This can be clearly seen in the sections of the diagram (Fig. 44a,2 and Fig. 44a,3). As in Fig. 39 and Fig. 42, a fixed composition of the solid may be stable as an α-form in a specific P–T range, whereas at other P,T β is the stable structure. This is the reason of the partial superposition of the α- and β- fields in the T–X projection, Fig. 44.

The phase diagram without T_{max} in the $\alpha\beta V$ equilibrium is shown in Fig. 45 for the temperature interval of the phase transition. It can be seen that the shape of the three-phase surface in this case is also independent of the phase transition mode. In the system, Fig. 45a,1, the single-phase spaces of α and β do not come into contact. Accordingly, the α- and β-fields are separated (Fig. 45a) over the whole range of their coexistence. In the diagram of Fig. 45b with the same P–T shape of $\alpha\beta V$ as in Fig. 45a, single-phase volumes of α and β are tangent at $T < T_{ct}$. The two-phase equilibrium $\alpha\beta$ involves an extreme line (P_{min} and T_{max} for the P-T slopes of the condensed phase equilibria, shown in Fig. 43). When the isothermal planes cross the $\alpha=\beta$ curve, point $\alpha=\beta$ appears in Fig. 45b,2. Figure 45b,3 is the cross-section of the diagram at the temperature T_2 of the quadruple point that involves four phases: α, β, L, and V.

Thus, we have seen that the congruent solid state phase transition (or fusion) involves an extreme univariant curve in the corresponding two-phase equilibrium, whereas the shape of the three-phase surface, tangent to this curve, is arbitrary, irrespective of the type of the diagram.

1.2.2.8.4 Metastable states

In this section we will consider metastable states, which arise from formation of a binary compound in a system. To facilitate the subsequent exposition, we shall consider systems with a binary non-stoichiometric solid state compound C in which no polymorphism is observed. Compound C will be partially miscible with both components.

Depending on the phase transformations that involve the compound, four types of compounds will be referred to:

- equilibrium sublimation and fusion of the compound;
- equilibrium sublimation and metastable fusion of the compound;
- equilibrium fusion and metastable sublimation of the compound;
- metastable sublimation and metastable fusion, i.e., the compound is completely metastable in the P–T–X phase space.

Accordingly, four types of phase diagrams will be discussed.

Equilibrium sublimation and fusion of the compound. Such a system is presented by the diagram of Fig. 46. The single-phase volume of the compound C is bounded in the P–T–X phase space by four invariant quadruple points, N_1, N_2, N_3, and N_4. If compound C does not precipitate within this P–T–X interval, then the system becomes metastable, of the simple eutectic type, with a metastable four-phase point ($\alpha L \beta V$) (q.v. the dashed lines in the projections and sections of Fig. 46). The metastable states could be turned to equilibrium by annealing or by introduced a seed into the system.

Figure 46 represents a case in which compound C shows no kinds of the congruent behavior. As has already been mentioned, such behavior is manifested by the invariant curves of equal composition, C = V, C = L, or L = V. In the P–T–X phase space the homogeneous volumes of phases that participate in a congruent process touch each other along a tangent, which is in no way restricted in parameters, i.e., it is not expected to be an isobar, isotherm, or isopleth (an X=const curve). However, if the region of existence of one of the phases is localized in the P–T–X space, then the corresponding congruent line should be confined to a certain interval of the temperature, pressure, and composition. And because the single-phase volumes are bounded by corresponding three-phase curves, the congruent curves for the localized phases should span two different three-phase surfaces.

Now we shall examine modifications in phase equilibrium for the system of Fig. 46 that result from congruent sublimation or congruent fusion of compound C. To facilitate visual perception, metastable states in the corresponding Figs. 47 and 48 are shown only in P–T projection. In addition, in the T–X projections three-phase equilibria that meet at the quadruple point N_3 (Fig. 47) or N_4 (Fig. 48) are not shown either.

Consider a system, in which the solid C, along with equilibrium sublimation and fusion, exhibits <u>congruent sublimation.</u> In this case (Fig. 47) the heterogeneous volume CV, which separates homogeneous volumes C and V, is bound in the P–T–X phase space by the surfaces that intersect along the three-phase curves αCV, CVβ, and VCL. As a consequence, the congruent sublimation curve C = V may be located in the P–T–X space in three ways.

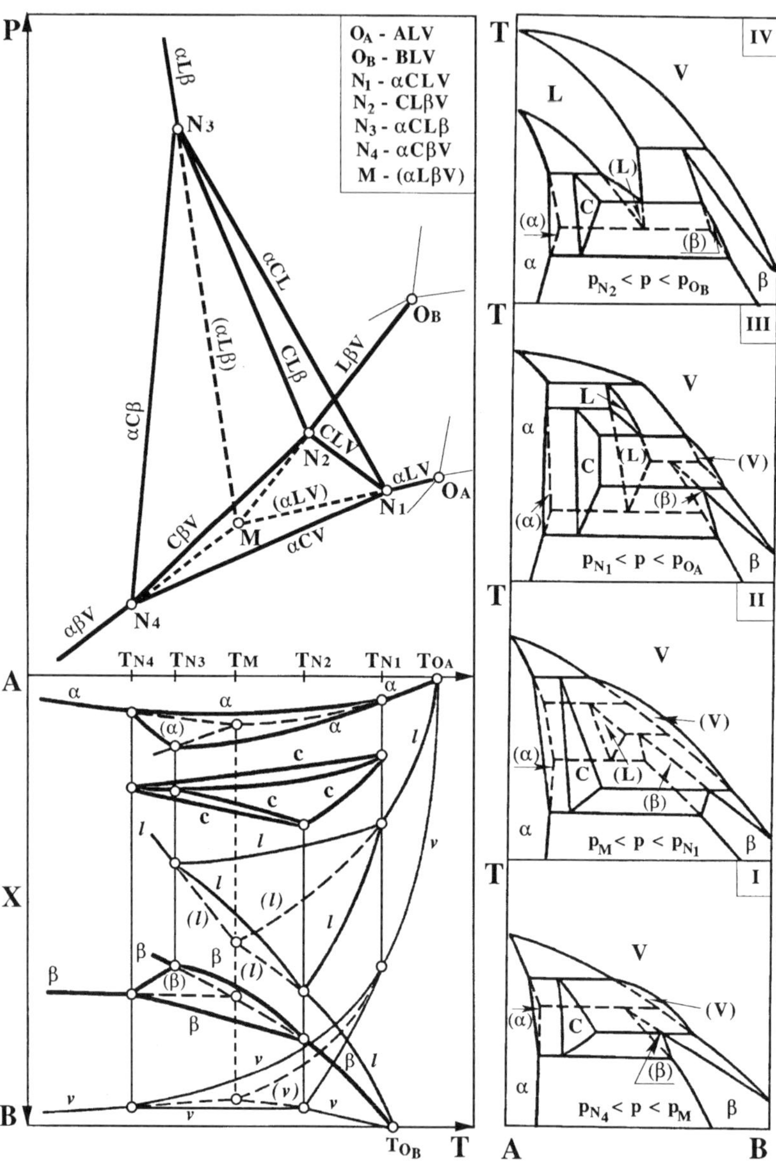

Fig. 46. *P–T–X* diagram of a system with binary compound C. Equilibrium sublimation and melting of C without congruent behavior

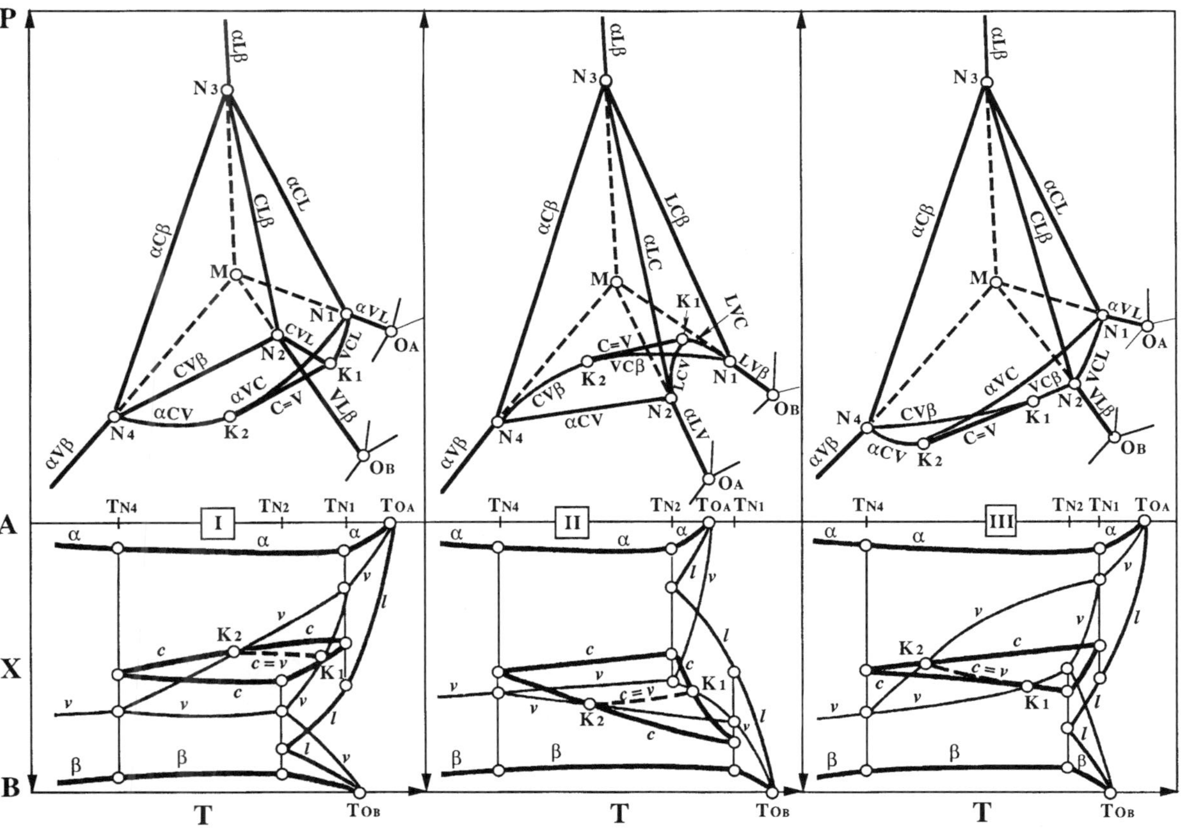

Fig. 47. *P–T–X* diagram of the type in Fig. 46. Congruent sublimation of C

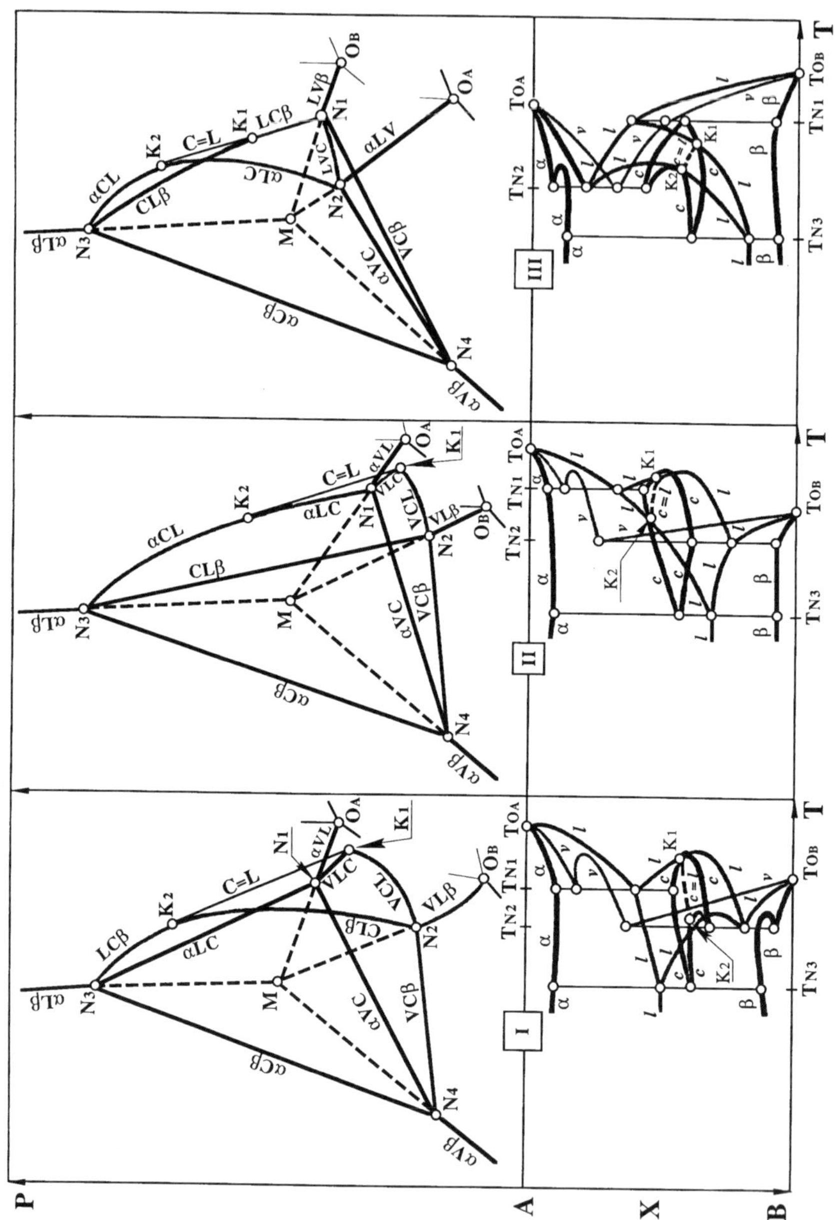

Fig. 48. *P–T–X* diagram of the type Fig. 46. Congruent melting of C

(1) The congruent sublimation line K_1K_2 in Fig. 47,I runs between αCV and CVL surfaces and divides the heterogeneous sublimation volume into two parts, CV and VC. The former is projected onto the P–T plane as the field within $K_1N_2N_4K_2K_1$ and the latter, as $K_1N_1K_2K_1$. The T–X projection of the C = V curve is represented by the dashed line K_2K_1. The lines of the composition of phases C and V in three-phase equilibria cvl and αcv intersect at points K_1 and K_2, which results in a change of the order of compositions of these phases in CVL and αCV equilibria.

(2) In Fig. 47,II the congruent sublimation curve K_1K_2 is between $CV\beta$ and LCV. Here K_1 is the intersection point of the solidus and vaporus lines for the LCV equilibrium, whereas at K_2 the solidus and vaporus of $CV\beta$ intersect. Accordingly, the change in the order of phases in these equilibria is observed.

(3) If the congruent sublimation curve K_1K_2 is confined to the αCV and $CV\beta$ surfaces, then the diagram of Fig. 47,III is realized.

Now suppose that compound C exhibits <u>congruent fusion</u> (the C = L curve in Fig. 48). The heterogeneous melting volume CL separates the homogeneous volumes C and L and is restricted in the P–T–X phase space by three three-phase surfaces, αCL, $CL\beta$, and CLV. Consequently, three different arrangements of the congruent fusion line may be envisaged. In Fig. 48,I the K_1K_2 line is between VCL and $CL\beta$. At point K_1 the solidus and liquidus of VCL intersect, and at K_2, solidus and the $LC\beta$ equilibrium. Corresponding changes in the sequence of phases are noted. Fig. 48,II is an example of the congruent line K_1K_2 running from VCL to αCL, and Fig.48,III, of that from $CL\beta$ to αCL.

It should be stressed that congruent processes (or the absence of those) do not influence the metastable diagram. It is also worth mentioning that points N_4 and K_2 (Figs. 46 and 47) might be outside the real temperature scale, and points N_3 and K_2 of Figs. 46 and 48 might be attainable only at very high pressures.

Equilibrium sublimation and metastable fusion of the compound. At temperatures $T > T_{N2}$ the equilibrium diagram of Fig. 49 shows the eutectic, with limited miscibility in the solid state. However, at $T < T_{N2}$ a new phase, C, appears, which is involved in three univariant equilibria, αCV, $C\beta V$ and $\alpha C\beta$, and in one invariant equilibrium, $\alpha C\beta V$. This substance can be obtained at low pressures (section I) by condensing the vapor over the sublimation loop CV. At pressures $P > P_{N2}$ it can be obtained only via solid state synthesis from the components (the three-phase line $\alpha C\beta$ in the isobars II, III, and IV, Fig. 49). This solid is not involved in equilibrium with the melt, although in the metastable state compound C may coexist with the melt. Two metastable diagrams can be envisaged in such a system. One is formed by metastable extensions of the equilibria that involve the α-phase; these meet at point M_1. This diagram is represented in Fig. 49 by dashed lines (αLV), (αCV), (αCL), (CLV), (LCV), and (C = L). The other diagram results from the metastable extensions of the equilibrium curves that involve the β-phase. It is shown only in the P–T projection of Fig. 49 by dot-and-dash lines ($L\beta V$), ($C\beta V$), ($CL\beta$) and (CLV), which intersect at point M_2 of the four-phase state ($CL\beta V$). An interesting specific feature of the former metastable diagram

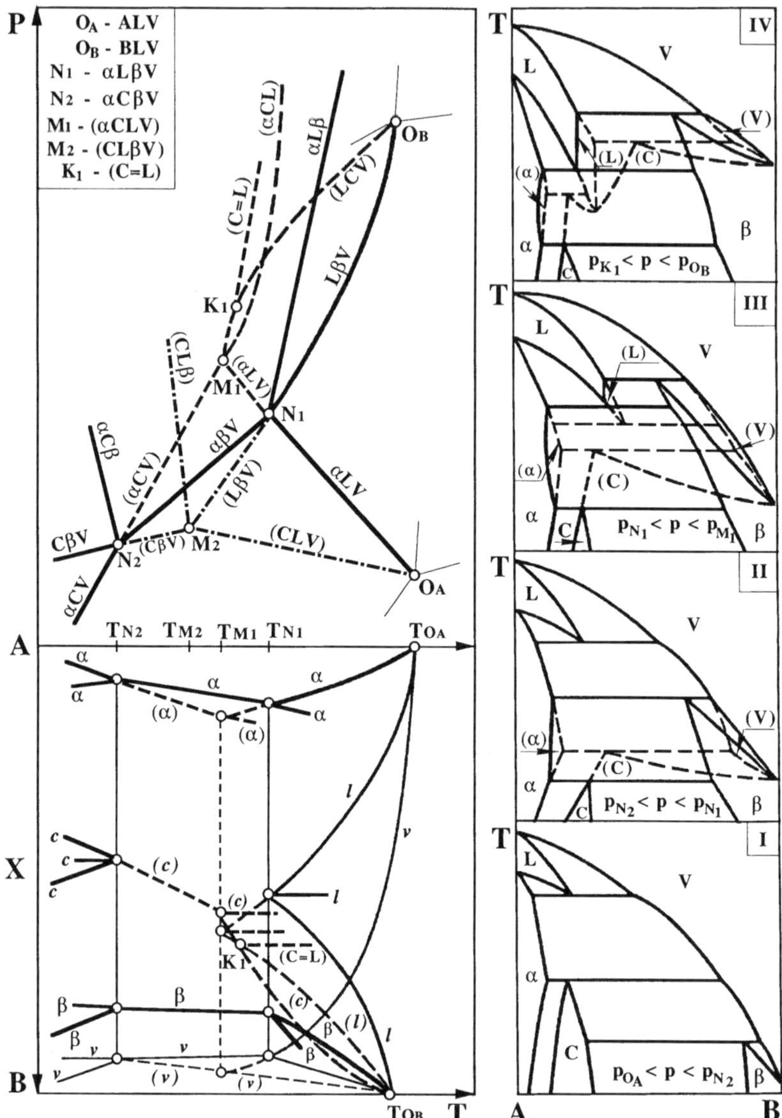

Fig. 49. P–T–X diagram of a system with binary compound C. Equilibrium sublimation and metastable melting of C

that involve the α-phase is the fact that the non-stoichiometric compound C is completely miscible with the component B in the metastable state. At pressures $P > P_M$ such a compound not only coexists with the metastable melt, but even shows congruent behavior (isobar IV, Fig. 49).

Equilibrium fusion and metastable sublimation of the compound. This type of diagram is presented in Fig. 50. It is a simple eutectic system where compound C appears only at high pressures $P > P_{N2}$, in three-phase equilibria not involving the vapor phase: $\alpha C\beta$, $CL\beta$, and αCL. The quadruple point N_2 is a four-phase invariant $\alpha CL\beta$. Although compound C is not involved in equilibrium with the vapor, these two may coexist metastably. Two metastable diagrams can be expected for this system. One is given only in the P–T projection of Fig. 50 by the dot-and-dash lines (αCV), (αLV), (αCL), (CLV), and the four-phase quadruple point M_1. The other is shown in both projections and sections by the dashed lines. Those are the three-phase curves ($L\beta V$), ($CL\beta$), ($C\beta V$), (CLV) and the four-phase quadruple point M_2 ($CL\beta V$). It can be seen in Fig. 50 that both metastable diagrams are of the eutectic type; the main difference between them is that in the former the C-phase is completely miscible with B, whereas in the latter, with A. Phase relations in these diagrams are expected to be self-explanatory from the isobaric sections, Fig. 50.

Metastable sublimation and fusion of the compound. A metastable compound (D) may appear in the system (Fig. 51) that is analogous to that of Fig. 46. Four metastable four-phase points are observed in this system: $M_1(LD\beta V)$, $M_2(\alpha LDV)$, $M_3(\alpha LD\beta)$, and $M_4(\alpha D\beta V)$. As a consequence, a confined metastable single-phase volume (D) is formed in the P–T–X phase space. It is projected onto the T–X plane as a figure bounded by the dashed (d)-lines, and in the P–T projection it is within the lines that run between the four previously mentioned points. The metastable diagram in this system arises, if the metastable compound (D) is formed instead of the equilibrium crystallization of the C-phase. It can be precipitated from different matrices. At pressures within the interval of isobars I–III, Fig. 51, this phase may be obtained from the supercooled melt in (LD) or (LDV) states of sections II–IV, Fig. 51. It can also be prepared via the solid state reaction of the components described by the ($\alpha D\beta$) states in isobars I–IV.

Two general principles that are characteristic of the examined types of diagrams with metastable states should be pointed out.

(1) If polymorphism of the components is observed in a system or a binary solid state compound is formed, then metastable states may be anticipated in such a system. The real physical meaning of this statement, which at a superficial glance might seem a platitude, is revealed only after detailed analysis of the relative arrangement of the single-phase volumes in the P–T–X phase space, especially when the polymorphs or the compounds appear only in metastable states.

(2) Regions of existence of all of the phases in metastable states expand compared to the equilibrium conditions, which can be readily seen in sections of Figs. 28 through 33. This phenomenon should be kept in mind, in particular when indirect experimental methods (metallography, "quenching" of the equilibrium, etc.) are applied to studying the non-stoichiometry in crystals.

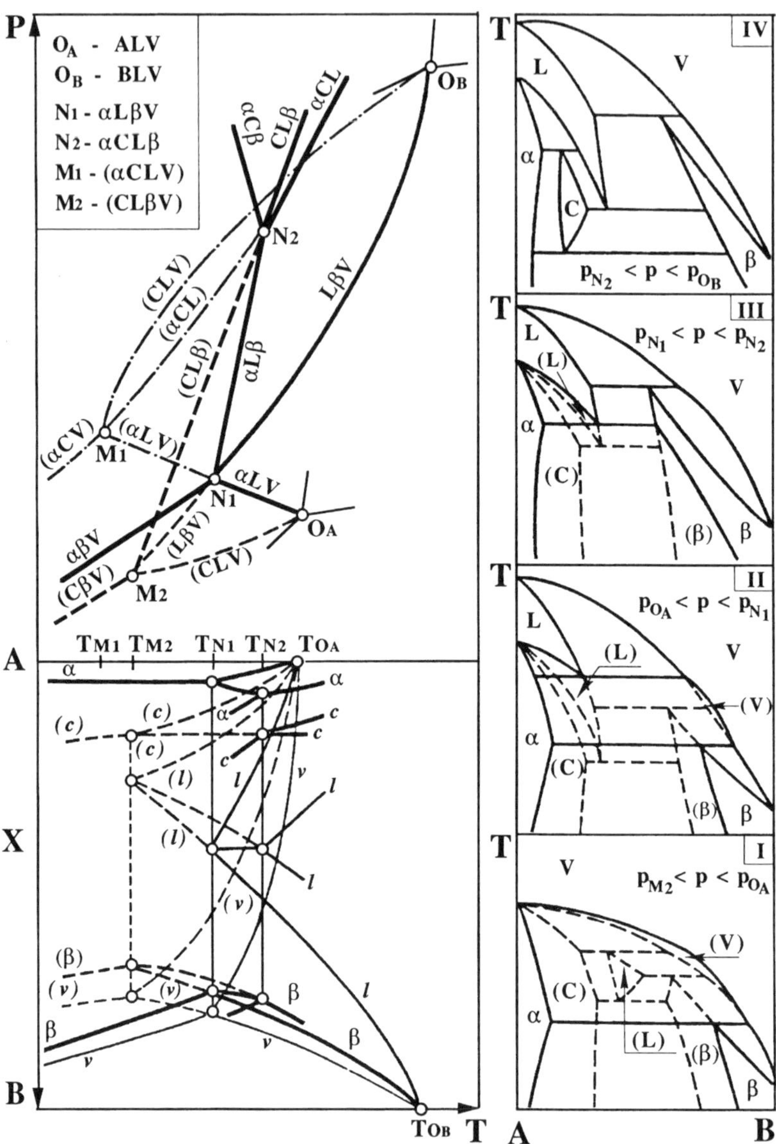

Fig. 50. *P–T–X* diagram of a system with binary compound C. Equilibrium melting and metastable sublimation of C

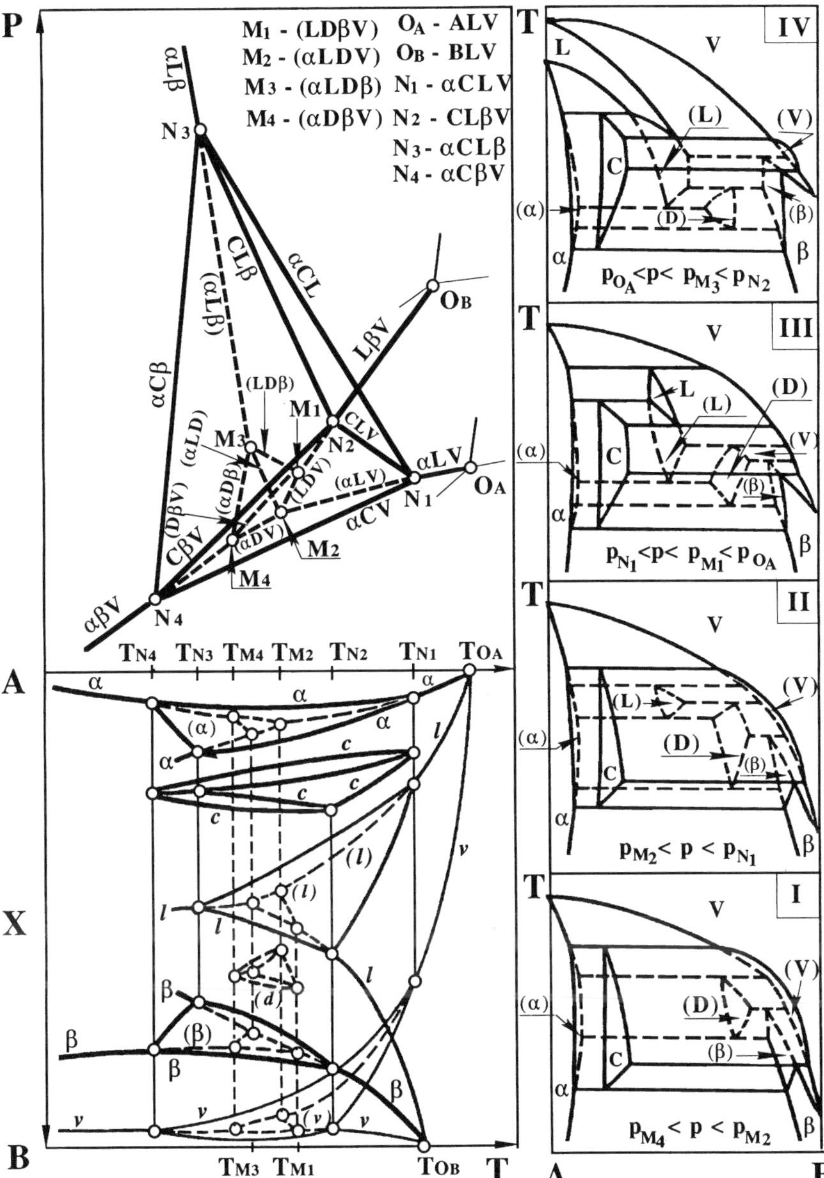

Fig. 51. *P–T–X* diagram of a system with two binary compounds, C and D. Equilibrium sublimation and melting of C, metastable sublimation and melting of D

The main instrument of investigation in this chapter was representation of the phase equilibria as geometrical images in the three-dimensional phase space P–T–X. Such an approach does not require the thermodynamic characteristics of the phases involved (which as a rule are in very short supply) and facilitates direct analysis of the experimental data even for very complicated cases encountered in multiphase heterogeneous systems.

A final remark (or rather a reminder) should be made in conclusion of this discussion of the types of phase diagrams. All of the diagrams were shown on an arbitrary scale, convenient for illustration. It is to be remembered that the laws of thermodynamics, whose geometrical representation is the phase diagram, are independent of the compositional extension of the phases, whether it is 100% or only a small deviation from stoichiometry. Consequently, an appropriate experimental technique is required for each specific case.

2 Experimental Methods of Investigating *P–T–X* Phase Equilibrium

Experimental study of *P–T–X* phase equilibrium consists of determining the relationships between the temperature, pressure, and composition of the phases that are involved in a particular equilibrium. According to the phase rule, two-phase equilibrium in a binary system is fixed by two independent parameters and in three-phase equilibrium only one parameter can be changed independently without disturbing the state of the system. Consequently, a complete study of the *P–T–X* phase diagram involves measuring the functional dependences $P = P(T,X)$, $T = T(X)$, and $P = P(T)$.

Differential thermal analysis (DTA) is the standard method of measuring $T = T(X)$, the temperatures of the first order phase transition in a sample with known composition X. DTA fixes the temperature, which corresponds to the energy consumption (endothermal) or release (exothermal processes). If a differential scanning calorimeter (DSC) is used for this purpose, the energy change is measured quantitatively. In DTA, sample with a fixed composition is heated or cooled in an open or closed system. In the former case DTA is carried out either in air or in an inert gas atmosphere, depending on the chemical nature of the system. From these measurements an isobaric $T–X$ section of the *P–T–X* phase diagram is obtained at a constant pressure, usually 1 atm. When DTA is carried out in an evacuated closed system, the measured phase transition temperature corresponds to the state of the system under its own saturated vapor pressure. Consequently, $T–X$ projection is obtained in this way. Sometimes the moment of melting is fixed visually. In addition, if the sample is held in an ampoule placed in a two-temperature furnace, the colder end of the ampoule may contain the pure volatile component at a chosen "cold zone" temperature. In this case the phase transition temperature is measured along with the partial pressure of the volatile component at this temperature because the temperature dependence of the saturated vapor pressure for the pure component is usually available from standard tabulated data.

A suitable vapor pressure measurement method is to be used to measure the temperature dependence of the pressure in univariant and bivariant equilibria. A comprehensive description of the existing methods for measuring the vapor pressure, along with their merits and deficiencies, can be found in [36–39]. Some details of measuring the vapor pressure of phosphorus in phosphides, sulfur in sulfides, mercury in amalgams are described by Kubaschewski et al. [40]. Certain specifics of vapor pressure measurement for chalcogenides are presented by Novosyolova and Pashinkin [41].

In this section a short description is given for a number of vapor pressure measurement methods most commonly used for different classes of inorganic materials.

Like any other property, vapor pressure can be measured directly or indirectly. Various types of manometers are used for direct measurements. When vapor pressure is determined indirectly, a certain property of the system is measured, and a suitable correlation between this property and the vapor pressure is used to calculate the vapor pressure.

2.1 Indirect methods

Most of the indirect techniques of determining vapor pressure are either static or dynamic. In static conditions, homogeneous (gas phase) or heterogeneous reactions are studied in a closed volume. Dynamic methods are used only for heterogeneous equilibria, when the condensed phase evaporates in vacuum or inert gas or reacts chemically with a flowing gas.

2.1.1 Static methods

The main static methods are the "dew point," transfer, weight loss, and optical absorption techniques. In all of these methods, the property of the system is measured in a closed evacuated vessel. This most important feature of static methods ensures that the system is in equilibrium, irrespective of the kinetic peculiarities of the processes because the time to attain equilibrium is virtually unlimited. Studies can be carried out in a wide interval of temperature and pressure, if an appropriate container is available. Static methods can be used for both homogeneous and heterogeneous equilibria.

The "dew point" method. Vapor pressure in a heterogeneous system with a single-component vapor can be measured in the following way. An evacuated and sealed tube containing the sample is placed in a two-zone furnace. The sample is heated up to a certain temperature T_1 and held in isothermal conditions. In the second zone the temperature T_2 is slowly reduced. At a specific moment, condensation of the volatile component is observed at T_2 (the "dew" is formed). The vapor pressure over the sample at T_1 at this moment corresponds to the saturated vapor pressure of the pure volatile component at T_2 because the system is held in stationary conditions. Cyclic heating-cooling measurements of T_2 for appearance-disappearance of the condensate can produce quite accurate results because high precision data for the temperature dependence of the vapor pressure of the pure component is usually available. The obvious limitation of this method is that it can be used only for systems with a single-component vapor. In this way, the vapor pressure of mercury for amalgams and phosphorus for phosphides was measured [40], as well as zinc for brasses, some chalcogens for chalcogenides [41], etc.

The transfer method. In this method two components are loaded into the opposite ends of an ampoule, which is evacuated and sealed. The ampoule is placed in a two-temperature furnace so that the non-volatile component is held at a higher temperature T_1 and is saturated with the more volatile component, evaporated at a lower temperature T_2. When equilibrium is reached, the total vapor pressure in the system is defined by the saturated vapor pressure of the pure volatile component, fixed by temperature T_2. The composition of the sample at T_1 is determined by direct chemical analysis or from the weight loss of the volatile component and its mass in the vapor phase. This method had only a limited application, for example, for measuring the zinc vapor pressure for brass [42] and the cadmium vapor pressure for its alloys with silver [43].

The weight loss method. This method is based on determining the vapor density, i.e., the mass of the vapor G in a measured volume v. The mass is obtained from the weight loss of the sample ΔG, and the vapor pressure P is calculated from the ideal gas equation

$$P = \Delta GRT/Mv,$$

where M is the molecular mass of the single-component vapor. The weight loss can be measured either continuously or periodically (the "quenched equilibrium" method). The former approach involves direct weighing of the sample, held at a measured temperature, using a torsion balance. An advantage of this procedure is that one sample can be measured in a single experimental run over the whole temperature interval, using step-by-step heating and measuring the weight loss every time the system attains equilibrium. This method was quite frequently used for vapor pressure measurements in halogenide systems [37].

The "quenched equilibrium" method is much more time-consuming. The sample is annealed at a measured temperature in an evacuated and sealed tube, which is quenched abruptly after reaching equilibrium. Then the tube is opened, and the weight loss of the sample or the mass of the condensed vapor is determined. Sometimes chemical or mass spectrometric analysis of the condensate is also done. It is clear that in this way only one (P,T) point can be obtained for each sample. This procedure was used for a number of chemical vapor deposition (CVD) systems: germanium – iodine (bromine) [44], III–V semiconductors with iodine [45] or water vapors [46], beryllium silicates with halogens [47], etc. In publications [44–47], small concentrations of halogens were used so that all of the halogen was in the vapor phase. That is why two additional correlations could be used, along with the ideal gas equation, to calculate the partial pressures: the mass conservation of the halogen and the stoichiometric composition of the solid compound. Consequently, the partial pressures of three vapor phase species could be calculated in this way.

Optical absorption of vapors. Optical methods are used usually for qualitative identification of certain vapor phase species. If the intensities of the incident and transmitted beams are measured, then the concentration of the vapor phase species can be calculated from the Lambert–Beer equation,

$$D_\lambda = K_\lambda CL.$$

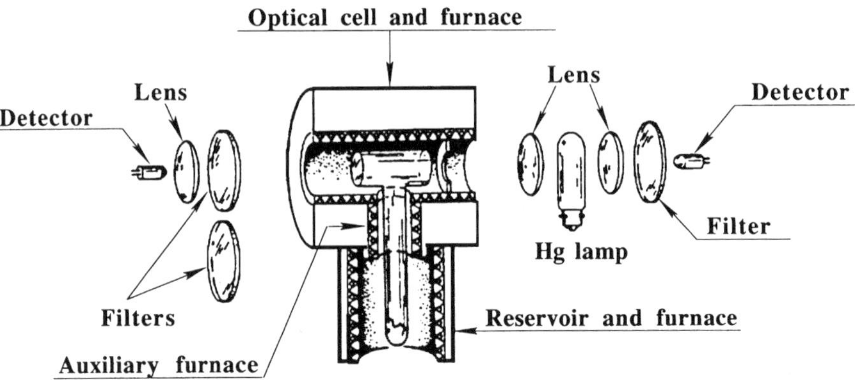

Fig. 52. Optical cell for measuring the optical absorption of saturated vapor

Here D_λ is the optical density of the vapor at the corresponding wavelength λ, C is the concentration, L the length of the optical path through the vapors, and K_λ is the absorption coefficient. Optical density is related to the intensities I_0 and I of the incident and transmitted beams with a wavelength λ:

$$D_\lambda = \log(I_0/I)$$

The concentration of the absorbing species C is readily recalculated to its partial pressure. A serious source of errors in partial pressures determined by the optical method is the temperature dependence of the absorption coefficient K_λ. Brebrick et al. [48–57] eliminated these errors by measuring absorption at constant temperature. A schematic apparatus used by Brebrick for vapor pressure studies of II–VI and IV–VI semiconductors is shown in Fig. 52. The sample of known composition was loaded into the sidearm of the silica cell, which was pumped and sealed. This *T*-shaped cell was placed in a specially designed furnace with two independently operated zones. The optical cell was held at a constant temperature, while the low-temperature reservoir containing the sample was heated stepwise. For each temperature step, the intensities of the incident and transmitted beams were measured in the horizontal cell by the optical system (Fig. 52). The light from the Hg lamp was collimated by a lens and went through the heated silica cell containing the vapors produced in the sample reservoir. The desired wavelength of light was provided by suitable interference filters, and the intensities were measured by detectors and recalculated to partial vapor pressures. This procedure was used to determine the vapor pressure in two-phase and three-phase equilibria and to construct *P–T–X* phase diagrams for a number of binary chalcogenide systems.

A serious methodological achievement is the study the multicomponent system

$$2GaAs(s) + GaI_3(g) = 3GaI(g) + 0.5As_4(g)$$

by optical absorption at 2500–12000 Å [58]. The equilibrium constants of this reaction reported in [58] are in good agreement with those obtained by other techniques.

2.1.2 Dynamic methods

In dynamic processes the condensed phase evaporates in an open system, in a flow of an inert or chemically reacting gas. The main dynamic methods are the "boiling point" and flow techniques. The principal concern with dynamic conditions is to transform the experimental results into equilibrium data. Special measures are taken to ensure that the mass transfer corresponds to the diffusion region because in this case it is possible to extrapolate the results to the equilibrium conditions. Dynamic methods are used only for heterogeneous systems.

The flow method. This method was initially designed to investigate the vaporization of pure metals [59]. In this procedure a static stream of an inert gas flows over the sample heated up to a specific temperature. Then the vapors are transferred to another part of the system where they are condensed or absorbed. In theory, saturation is attained at a zero flow rate. Therefore, the experiment is conducted at several flow rates and the results are extrapolated to the zero rate. This extrapolation may be the source of serious errors. It has been shown [60] that the evaporated amount of the material becomes proportional to the flow rate only above a specific limiting rate. Below this limit the diffusion term of the mass transfer from the hot zone to the condenser becomes ever more significant. On the other hand, saturation of the flowing gas with the sample vapors is reached only at low flow rates. A compromise between these two trends is achieved by different means, depending on the vapor pressure and chemical properties of the system. Special precautions are taken against backward diffusion and thermodiffusion that are appreciable at low flow rates.

Vapor pressure P is calculated using the ideal gas approximation from the equation

$$P = P_{tot}(N/N_0)$$

Here the total vapor pressure $P_{tot} = (P + P_{in})$ is the sum of the partial pressures of the sample and the inert gas pressure P_{in}, $N_0 = (N + N_{in})$ is the sum of the moles of the evaporated substance N and the inert gas N_{in}.

The experimental arrangement for this method is quite simple. The sample (in powdered form, to provide the maximum possible reaction surface) is loaded into a flow reactor. If the evaporation rate is high and the vapor pressure is low, the sample may be held in a boat, which can be weighed before and after the experimental run. Otherwise, the condensate is to be analyzed. To prevent backward diffusion and thermodiffusion, both sides of the reaction zone are narrowed to increase the flow rate locally. To calculate the vapor pressure from the experimental data, the molecular mass of the vapor must be known.

The flow method was used for investigating equilibrium in a number of CVD systems, in particular, semiconductors. In [61–63] the chemical vapor transport of gallium arsenide in the hydrogen flow was reported for different transporting agents: iodine, hydrogen iodide, hydrogen chloride, water vapors, aluminum chloride. CVD systems of spinels with chlorine and hydrogen chloride were studied in [64,65], and in [66–68] refining of aluminum and gallium via chemical vapor-phase reactions with chlorides and iodides was investigated. Mass conservation equations for individual elements were used along with the ideal gas equation to calculate the partial vapor pressures.

The flow method is applicable when vapor pressures is not higher than 100 mmHg. The lower limit depends on the weighing technique. The accuracy of the results is typically not very high.

The boiling point method. The boiling point is the temperature, at which the saturated vapor pressure of a substance is equal to that in the system. The measurement of vapor pressure by this method is based on fixing the moment when the substance starts boiling. This technique was initially used by Greenwood [69,70] and Ruff [71,72] to study vaporization of metals. Two modifications of the boiling point method have been described: isothermal and isobaric. In the former, the temperature is kept constant, and the pressure is gradually changed, whereas in the latter, the constant parameter is the pressure and the variable is the temperature. The boiling point can be determined in different ways. Visual observation of the liquid metals led, not surprisingly, to significant distortions of the results because the samples contained gaseous impurities, which evaporated long before the proper boiling point was reached. More reliable results were obtained when the temperature arrest was recorded at the beginning of boiling [73–75] or sharp movement of a mercury drop in a capillary connected to the reaction volume was observed [76]. The boiling point can also be recorded indirectly, by registering the change in the evaporation rate [71,72]. In this experiment the weight loss of the sample was continuously recorded, and two branches of the curve were extrapolated to the intersection point. It has been proved [77] that the isothermal modification resulted in more reliable data. Novikov et al [78,79] used a differential thermocouple to register the drop in temperature when the boiling point was reached. In [80,81] radioactive isotopes were used to record the evaporation rate. Theoretical analysis of the mass exchange in the boiling point method was reported in [82]. This technique was proved to be useful for investigating the vaporization of metals and also for studying of dissociation processes of a number of solids [83–85].

Reliable results can be obtained by the boiling point method for saturated vapor pressure at high temperatures and pressures. The main source of errors is in fixing the moment when boiling starts.

2.2 Direct vapor pressure measurement

Vapor pressure can be measured directly, if a sample loaded into the reaction vessel, is heated up to a specific temperature. The reaction chamber is connected to a manometer, which records the vapor pressure. Radiation or ionization manometers are usually used for low vapor pressures (10^{-7} to 10^{-1} mmHg). The radiation instrument measures the energy loss of a heated metal wire, which is proportional to the vapor pressure, and the ionization manometer measures the ion current, resulting from the ionization of the vapor by the constant flow of electrons. These methods are applicable only to substances that do not react with metals heated up to high temperatures. Low pressures (up to 5×10^{-5} mmHg) can also be measured with a mercury McLeod manometer [36]. High sensitivity of the instrument results from measurements of the pressure for a vapor species compressed with a mercury column to a fixed volume. Pressures above 1 mmHg can be measured with a standard mercury manometer, which is a U-shaped tube, filled with mercury. One end of the tube is sealed, and the other is connected to the reaction chamber that contains the sample. The difference in mercury levels corresponds to the vapor pressure of the sample.

Mercury manometers can be used for volatile substances, which do not react with mercury, and only at low temperatures. The temperature interval can be extended, if a compensating procedure is used. The U-shaped tube in this case is filled with the liquid to be studied rather than mercury. The liquid evaporates into the sealed end of the tube, and the resulting vapor pressure is compensated for with an inert gas. The pressure of the inert gas is then measured with a mercury manometer. This procedure was used for measuring the vapor pressures of liquid metals [36].

Another compensation procedure used for measuring the vapor pressure at high temperatures involves the *isoteniscope*. This is a U-shaped tube filled with a low-boiling liquid; one end is connected to the reaction chamber and to the other an inert gas can be admitted to compensate for the vapor pressure in the reaction volume. The isoteniscope can be held at a high temperature. Different manometric liquids have been used, depending on the chemical properties of the substance to be studied and the temperature interval. Evaporation of amalgams was studied in [86] at temperatures up to 500°C with the ($NaNO_3$ + KNO_3) eutectic; reactions between sodium, potassium, and aluminum halogenides were studied with an isoteniscope filled with liquid tin [75] or gold [87,88]. In [87,88] the measurements were made up to 1200°C. Borshchevsky et al. [89] measured the vapor pressures of II–IV–V_2 semiconductors with an isoteniscope filled with liquid B_2O_3 and placed in a furnace together with the samples.

The use of the isoteniscope is limited because specific liquids are required for this purpose. The temperature interval is limited from below by the melting point of the manometric liquid and from above by its volatility. The accuracy of the method is not very high.

2.2.1 Membrane manometers

If the reaction vessel is separated from the environment by a thin membrane, then the vapor pressure inside the reactor deforms this membrane. The reactor is placed in a furnace with a flat temperature zone, and the shift of the membrane from the initial position is used as a measure of the vapor pressure. When the membrane is made of quartz, the measurements can be conducted with high accuracy over a wide temperature range. As a rule, vapor pressure up to 1 atm is measured by the membrane manometer, although Ugai et al. [90] reported an experimental setup working at pressures up to 15 atm. The temperature limit is typically about 1000°C. Different membrane shapes have been reported, as well as different ways to register the deflection. A highly sensitive instrument with an accuracy of about 0.01 mmHg was described in [91]. A mirror was attached to a flat membrane, which moved together with the membrane and reflected a light beam. Deflection of the beam was a measure of the vapor pressure in the system. The Bodenstein manometer [92–94], in which a spiral membrane is used, is also very sensitive. Both types can be used for absolute reading of vapor pressure or as zero instruments. In the former case, the manometer is calibrated against standard vapor pressures, whereas in the latter case, the membrane is returned to its initial position by an equal external pressure, which is measured with a standard gas manometer. Both these manometers have very low mechanical stability and, consequently, had rather limited application.

The Bourdon gauge is a widely used modification of the membrane manometer. Vapor pressure in a number of semiconductor systems was measured in [95,96] with a quartz Bourdon gauge. The membrane had a sickle shape, and the manometer was used as a zero instrument. The apparatus is shown in Fig. 53 together with the temperature profile of the furnace. The sample 1 is loaded into the reaction volume through the tube 8, the system is pumped from both sides 6 of the membrane 3, and the sample can be heated to an appropriate temperature to desorb volatile impurities effectively. After this treatment the reaction volume is sealed at point 9, and the gauge is put into the isothermal zone of the furnace. The deflections of the quartz fiber 4, attached to the top of the spoon, are caused by the vapor pressure in the reaction vessel, and are balanced by an equal pressure of argon admitted to the upper section of the gauge through the tube 7. Blank experiments showed that no diffusion of argon through the membrane was detected up to about 1100°C at vapor pressures below 2.5 atm. The argon pressure was measured simultaneously by two U-shaped manometers. One was a standard mercury instrument, MBP type, with an accuracy of 0.1 mmHg, and the other was filled with a high boiling point liquid (equilibrium vapor pressure at room temperature about 10^{-7} mmHg, density ρ (298 K) = 0.98 g/cm^3). The sensitivity of the second manometer was about 0.07 mmHg, approximately the same as that of the Bourdon spoon, and it was used for pressures up to 30 mmHg.

During the measurements the Bourdon gauge was placed in a vertical furnace so that the reaction chamber and the spoon were maintained at the same temperature (Fig. 53). The temperature was measured at different points of the reaction vessel with Pt-Pt/Rh thermocouples placed in special wells 2. The apparatus was standardized by comparing the measured vapor pressures of several chemical elements with recommended tabulated values.

Membrane manometers are used for measuring total saturated and unsaturated vapor pressure. This method has been used in the majority of studies of the *P–T–X* phase diagrams. Novosyolova et al. reported results for a number of binary and ternary semiconductor systems: IV–VI materials were studied in [97–99]; the IV Group elements were germanium, tin, or lead, and the VI Group elements were sulfur, selenium or tellurium. V–VI–VII systems with antimony or bismuth as the V Group; sulfur, selenium, or tellurium as the VI Group; and iodine as the VII Group element

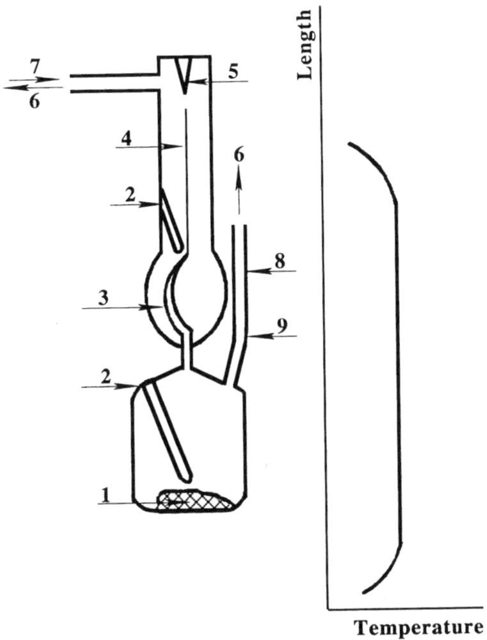

Fig. 53. Bourdon gauge and the temperature profile of the furnace. *1* – sample, *2* – thermo-couple wells, *3* – quartz spoon, *4* – fiber, *5* – zero pointer, *6* – to vacuum, *7* – argon inlet, *8* – tube for introducing the sample, and *9* – sealing place

were studied in [100–106]. Membrane manometers are applied for investigating phase equilibrium and chemical equilibrium as well. For this purpose, partial pressures of all species, comprising the system, are to be calculated from the measured total vapor pressure. Some mathematical aspects of this problem were discussed in [37,38,95,107–109]. Results for a great variety of halogenide systems were compiled by Suvorov [37]. A number of CVD semiconductor systems were also reported, for example, germanium–chlorine [110], germanium–bromine [111], germanium–iodine [112], silicon–chlorine [113], zinc selenide–iodine [114], cadmium sulfide–iodine [115], some III–V compounds with halogens [116,117].

Measurement of the vapor pressure with a membrane manometer has two significant advantages: it is a direct and a static method. Because of that even very slow processes can be studied by this method. It is also the main experimental technique for constructing the P–T–X phase diagrams.

The main experimental methods of vapor pressure measurement are listed in Table 1 together with the corresponding measurement limits.

Table 1. Pressure limits of experimental methods of vapor pressure measurement [36,37]

Methods	Pressure interval (mmHg)
Static methods	
Mercury manometer	$10^{-3} - 10^{3}$
Isoteniscope	$1 - 10^{3}$
Spiral manometer	$10^{-5} - 1$
Membrane manometer	$10^{-1} - 10^{3}$
Optical absorption	$10^{-3} - 10^{3}$
Weight loss	>10
Dynamic methods	
Flow	$10^{-4} - 10^{2}$
Boiling point	$10^{-1} - 10^{3}$

Depending on the expected pressure interval for the system to be studied, as a rule, it is possible to choose a suitable method for vapor pressure measurement. Usually direct static methods are preferable. In this way one can be sure that the data correspond to the equilibrium conditions because the measurement can be made in heating-cooling cycles with practically unrestricted time for isothermal exposure.

Gauges of various types for different vapor pressure intervals are commercially available (see, for example, Edwards Vacuum Products Catalogue 1998–1999).

2.3 Vapor pressure scanning

As we have already seen, crystalline non-stoichiometric compound $AB_{1\pm\delta}$ in equilibrium with vapor may have a congruent composition $s = v$ and may be enriched in constituent components A or B (according to the relative position of the solidus and vaporus volumes), if an incongruent sublimation is considered. The composition of the solid X_S at a fixed temperature and pressure is given by Eq. (25), which we will rewrite here for convenience:

$$X_S = (N_B - n_B)/\left[(N_A + N_B) - (n_A + n_B)\right] \tag{25}$$

The composition X_S can be determined from the vapor pressure experiment. The procedure consists of calculating the numbers of gram-atoms n_A and n_B in the vapors from the total vapor pressure P at temperature T. If upon heating the sample in a closed volume, only one component is vaporized (e.g., B), then $n_A = 0$, and n_B in Eq. (25) is readily calculated from the ideal gas equation and the equilibrium constants for polymerization of B in the vapors, if these reactions are actually observed. Such systems are by far not exceptional in inorganic materials science. Suffice it to mention here III–V and III–VI semiconductors, V–VI–VII ferroelectrics, high-T_c superconductors, etc. This mode of evaporation was considered, in particular, in [95,105] in connection with the non-stoichiometry of SbSI. In subsequent sections the non-stoichiometry of In_2Se_3 and Cr_2Se_3 will be discussed in detail, which also involves this type of sublimation.

Of course, it is to be understood that a single-component vapor in a binary system is only an abstraction, although a useful one, if the amount of the second component in the vapors does not appreciably affect the crystal composition. Routine experimental errors in vapor pressure measurements ($\pm$ 1 mmHg, $\pm$ 1 K) introduce uncertainties into the estimated mass of the vapor, which are smaller than those of conventional weighing techniques. Therefore, the accuracy of the crystal composition calculated from Eq. (25) is expected to be limited by the errors in weighing the initial sample, and the single component vapor model is applicable, if the associated errors do not exceed those resulting from weighing.

In a general case, when a binary crystal sublimes incongruently, both components are evaporated to form different homo- and heteroatomic vapor species. Calculation of partial pressures in multireaction systems from the total vapor pressure has been discussed in detail elsewhere [95,107–109]. In this section it will be shown, how to use vapor pressure data on the incongruent sublimation of a binary compound $AB_{1\pm\delta}$ to calculate the analytical composition of the vapor X_V and eventually the crystal composition [96,118].

The composition of the vapor X_V in equilibrium with the solid X_S in a closed volume v at the temperature T is equal to

$$X_V = n_B/(n_A + n_B). \tag{26}$$

It is determined by three equations: the total vapor pressure P, which is the sum of all the partial pressures P_j

$$P = \sum_j P_j, \tag{27}$$

and two equations (i = A,B) that represent the mass conservation law:

$$n_i = (v/RT) \sum_j \mu_{ij} P_j, \quad i = A, B. \tag{28, 29}$$

Here μ_{ij} is the number of atoms of component i in the vapor phase species j. All of the partial pressures are related by the equilibrium constants of the vapor phase reactions

$$K_P = \prod_j P_j^{\nu_{ij}}. \tag{30}$$

Here ν_{ij} is the stoichiometric index of the vapor species j in the reaction i.

Two crucial points should be stressed in connection with the procedure described:

1. In this way the compositions of the conjugated phases, crystal X_S and vapor X_V, are obtained *directly at a high temperature.*

2. The confidence intervals for the compositions can be rigorously calculated by applying the error accumulation law because all of the associated experimental errors are known. Relevant calculations show that with typical vapor pressure measurement equipment, the crystal composition can be determined with an accuracy as high as 10^{-3} to 10^{-4} at.%, and sometimes even better.

It is clear that the calculated composition X_S, Eq. (25), corresponds to an individual crystalline phase rather than to a mixture of condensed phases only within the two-phase equilibrium field SV. Figure 54 is a close-up of the near-solidus in *T–X* and *P–T* projections. In *T–X* (Fig. 54b) the range of existence for the solid compound is shaded, and in *P–T* (Fig. 54a) the sublimation of this solid is within the three-phase loop. If the composition of the initial two-phase sample is S_1, S_2, or S_3, the evaporation route is the following. In the *T–X* projection, predominant evaporation of A leads to a shift in the condensed phase composition toward B. At a certain point (a′, b′, or c′), the entire component A evaporates, and the only remaining condensed phase is compound $AB_{1\pm\delta}$. Compositions X_1 at these points correspond to the maximum non-stoichiometry of the compound at temperatures $T(X_1)$. In *P–T* projection, as long as the sample contains two condensed phases, the vapor pressure changes along the three-phase curve. At the transition point (a, b or c), where a break in the vapor pressure curve is observed, the system goes into the two-phase equilibrium SV, and the calculated X_S corresponds to an individual crystalline phase.

Direct calculation of the non-stoichiometry of X_S from Eqs. (25–29) runs into a serious technical difficulty. This set of five equations contains six unknowns: two compositions (X_S and X_V), two numbers of gram-atoms in the vapors (n_A and n_B) and two independent vapor pressures because the equilibrium constants Kp, Eq. (30), reduce the total number of Pj's to only two independent partial pressures. Consequently, the system of Eqs. (25–29) is underdetermined. To resolve this problem, the phase rule had to be applied. The two-phase equilibrium SV in a binary system is bivariant according to the Gibbs phase rule. Consequently, only two parameters can be chosen arbitrarily. If these two are P and T (a point on the vapor pressure curve), then all of the other (X_S and X_V, in particular) become fixed. Now, suppose that this point $\{P(X_2),T(X_2)\}$ belongs simultaneously to two different curves (1 and 3 in the *P–T* projection, Fig. 54) which, of course, is an intersection point. Then X_S and X_V are fixed for both experiments (with two different N_A, N_B, and v), and because five new Eqs. (25–29) can be written for the second run, whereas only four new independent unknowns are added, the resulting system of ten equations with ten unknowns can be solved. It means that X_S, X_V, and all P_j's are found directly at the intersection point for the measured T and P. Transformation of Eqs. (25–29) leads to a very important result:

$$X_S = (N_{B1}v_2 - N_{B2}v_1)/\left[(N_{A1} + N_{B1})v_2 - (N_{A2} + N_{B2})v_1\right]. \tag{31}$$

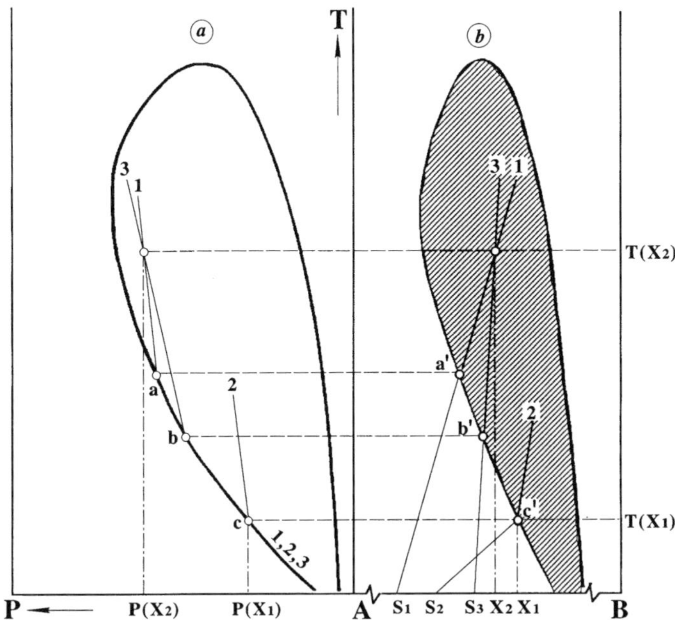

Fig. 54. *P–T* and *T–X* projections of the non-stoichiometry range for the compound $AB_{1\pm\delta}$. 1–3–projections of the evaporation routes of the samples S_1, S_2, and S_3

This equation relates X_S to the initial experimental conditions (N_A, N_B, and v) of the two intersecting curves, 1 and 2, and involves no information about the vapor. It means that Eq. (31) is applicable to an arbitrary composition of the vapor, unknown, in particular. If all of the region of existence of the solid is covered by a net of vapor pressure curves, then the result is a data file of intersection points that scan the whole of the solidus and from which the solidus surface is reconstructed in the *P–T–X* phase space. That is why this method was called **vapor pressure scanning** of the solidus. The data obtained in this way in a specific temperature range comprise the composition of the solid and the partial vapor pressures of all of the species as a function of temperature. Because the corresponding partial pressures in the saturated vapor for pure components are available from Standard Thermodynamic Tables, the activities of the components are readily calculated, i.e., complete thermodynamic characterization of the $AB_{1\pm\delta}$ phase can be given. The earlier quoted accuracy of the method (10^{-3}–10^{-4} at.%) implies that by vapor pressure scanning crystalline phases with narrow (sub-0.1 at.%) range of existence can be studied; this is of specific interest in semiconductor materials science.

This procedure can also be used as a high-precision analytical tool. If the solidus of a certain crystal has already been scanned, then all one has to do to analyze a sample of this compound is to measure the vapor pressure for this sample, find a suitable intersection point, and calculate the composition of this sample from Eq. (31).

3 Experimental Data on *P–T–X* Phase Diagrams and Non-stoichiometry

3.1 Semiconductor systems

3.1.1 II–VI compounds

II–VI semiconductor compounds are used for infrared, X-ray and γ-ray detection, in thin film solar cells, photo-refractive and blue/UV emission devices [118]. CdTe and Cd–Zn telluride, used as detectors for direct transformation of high-energy radiation to electrical signals, create a new generation of efficient detectors for medical applications, such as tomography. CdZnTe is used as a buffer layer in heteroepitaxy of mercury–cadmium telluride and as a substrate for epitaxial technology of (Hg,Cd)Te, the most important semiconductor material for infrared detector applications [119]. The next generation of infrared detection devices demands further improvement in material technology. These applications require high-quality multilayer heteroepitaxial structures with buffer layers serving to overcome the adverse effects of lattice mismatch, in particular, between Si or GaAs substrates and the active II–VI layer [120]. The lattice constant of (Cd,Zn)Te can be adjusted by changing the content of ZnTe to match the lattice of the (Hg,Cd)Te epilayer [121]. Some of the applications of II–VI bulk single crystals, according to P.Rudolph [122], are presented in Table 2.

Table 2. Selected applications of II–VI binary and ternary single crystals [122]

Material	Application
ZnO	Fitted substrates for GaN epitaxy
ZnS	Electro-optical modulators and switches
ZnSe	Substrate for homoepitaxial blue LDs
CdS	Electro-optical modulators and switches
CdSe	Optical parametric oscillators
ZnTe	Substrates for green LEDs and LDs
CdTe	Substrates for (Hg,Cd)Te epitaxy, X-ray and γ-ray detectors, windows, photorefractive devices, solar cells
(Cd,Zn)Te	Fitted substrates for IR detectors, X-ray detectors for computer tomography
Zn(S,Se)	Fitted substrates for quantum confined blue laser multilayer systems

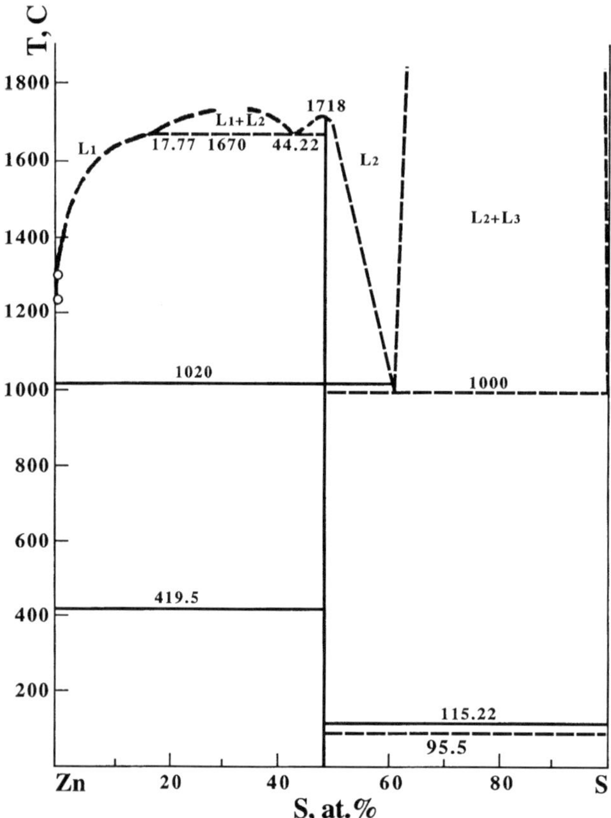

Fig. 55. *T–X* phase diagram of Zn–S system

3.1.1.1 Zinc chalcogenides

In the zinc–chalcogen systems, a single compound is formed with a congruent melting temperature that is considerably higher than the melting points of the components.

The *T–X* diagram of the **zinc–sulfur** system is presented in Fig. 55 [123]. The melting temperature of ZnS is 1718°C, and the eutectic temperatures are 419.5°C (for zinc) and 115.2°C (for the sulfur eutectic). Both eutectics are degenerated. At 1020°C the solid-state phase-transition wurtzite → zinc blend is observed. Miscibility gap in the liquid is seen in Fig. 55 at two temperatures: at 1670°C two liquids, 17.77 at.% S and 44.22 at.% S in composition, are in invariant equilibrium with the high temperature polymorph of ZnS, whereas the low temperature modification is in equilibrium with two liquids at about 1000°C. Detailed study of this system is hampered by high temperature and high vapor pressure of the constituent components.

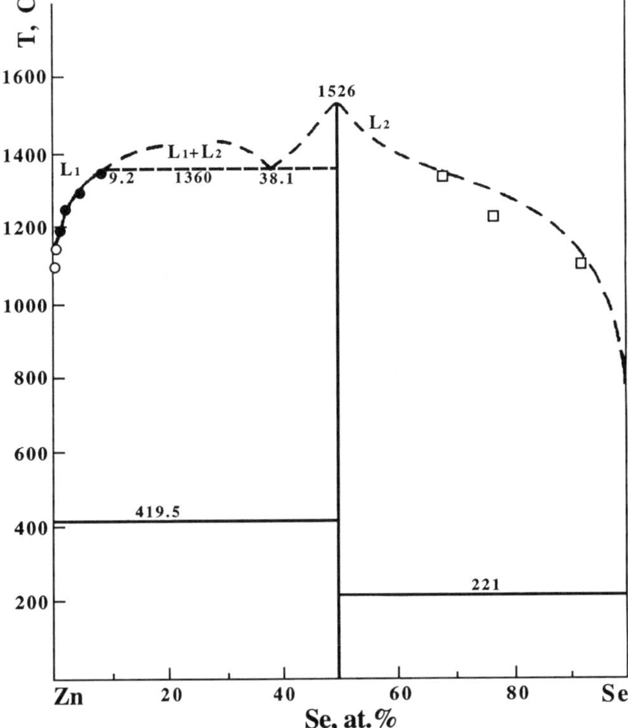

Fig. 56. *T–X* projection of the Zn–Se phase diagram

The **zinc–selenium** system (Fig. 56), quoted by Rudolph et al. [124], is similar in shape to Zn–S. The congruent melting temperature of ZnSe is 1526°C, and two eutectic melting points are 419.5°C (zinc) and 221°C (selenium). In this system, too, both eutectics are degenerated. A miscibility gap in the liquid phase appears in this system at 1360°C, and the compositions of the liquids are 9.2 at.% Se and 38.1 at.% Se in invariant equilibrium with ZnSe. A more recent DTA study of the Zn–Se system in the near-50 at.% range [125] showed that the congruent melting temperature of ZnSe is 1522 ± 2°C, and it corresponds to the composition of 50.1 at.% Se. The melt at the near-melting temperatures was shown to be a regular associated solution. The solid-state phase-transition 2H (wurtzite) → 3C (zinc blende) temperature depends on the composition; at the congruent phase-transition point the temperature is 1411 ± 2°C and the composition is stoichiometric.

The solidus surfaces for ZnS and ZnSe were not studied experimentally. Only an indirect conclusion on non-stoichiometry in these compounds can be made from electrical measurements. It has been shown that in II–VI compounds, an excess of the metallic component results in *n*-type conductivity, whereas the chalcogen non-stoichiometry leads to a *p*-type material. ZnS crystals are usually of the *n*-type, whereas ZnSe is both of *n*- and *p*-type. Consequently, it may be assumed that

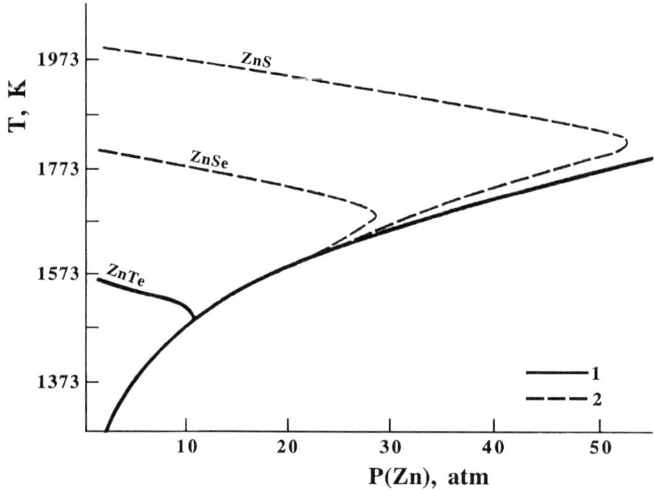

Fig. 57. Experimental (1) and calculated (2) $P(Zn)$–T diagrams of the Zn–S, Zn–Se, and Zn–Te systems

the range of existence for ZnSe includes the stoichiometric composition, whereas for ZnS the single-phase volume is on the Zn-side of the stoichiometry.

No experimental data are known for P–T projection of the Zn–S diagram. Only calculated zinc vapor pressures have been quoted in [22] (Fig. 57, dashed lines). Analysis of the phase equilibria in the Zn–Se system was presented by Brebrick [48]. In both systems the Zn partial pressures in three-phase equilibrium with ZnS(Se) and Zn-rich melt coincide with the saturated vapor pressure of pure zinc (the solid line in Fig. 57) up to about 1300°C. At the maximum melting temperature, the zinc vapor pressure was estimated as about 30 atm for ZnSe and ~50 atm for ZnS.

The T–X diagram for the **zinc–tellurium** system (Fig. 58 [123]) has been studied in more detail. The maximum melting temperature of ZnTe is 1300°C, and the eutectics melt at 419.5°C (almost pure zinc) and 449°C (99.73 at.% Te). The miscibility gap in the liquid has a critical point at $T = 1340$°C and $X_L = 17.8$ at. % Te. The invariant equilibrium $L_1L_2S(ZnTe)$ is at $T = 1215$°C, $X_{L_1} = 5$ at.% Te, $X_{L_2} = 34.6$ at.% Te, which is in agreement with the regular associated solution model of the melt [126]. DTA results of Steininger et al. [127] are also shown in Fig. 58; they quoted a somewhat lower melting temperature for ZnTe, 1290±2°C from [128] and 1295±20°C [129], both measured by DTA.

The solidus surface of ZnTe, calculated by Jordan [126], is shown in Fig. 59 together with the monotectic horizontal and the liquidus (semi-bold, almost flat curve). The single-phase volume of ZnTe is entirely on the Te-side of the stoichiometric plane, which is consistent with the p-type behavior of this material and recent vapor pressure scanning results [130]. The composition at the maximum melting temperature is ~50.0013 at.% Te, and the maximum congruent sublima-

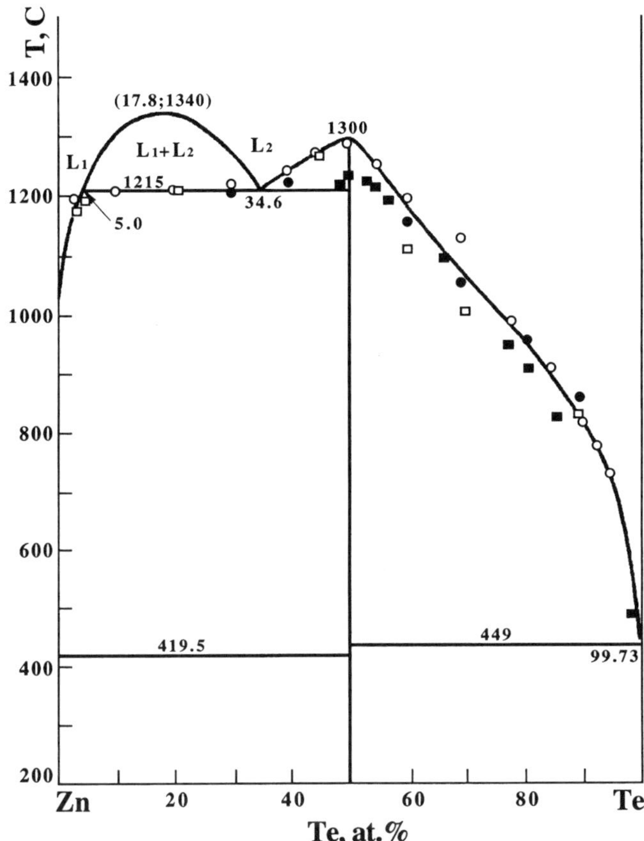

Fig. 58. *T–X* projection of the Zn–Te phase diagram

tion point is significantly lower than T_{max}, with the composition $X_{cs} = \sim 50.002$ at.% Te. The solidus on the Te-side is retrograde, and the maximum Te non-stoichiometry of approximately 4.6×10^{-3} at.% corresponds to 1200°C [126].

The partial pressures of Zn and Te_2 were measured by an optical absorption method [56] for the three-phase equilibria of Zn-saturated and Te-saturated ZnTe with the corresponding melts (Fig. 60). The Zn vapor pressure is indistinguishable from the saturated vapor pressure of pure zinc (solid line in Fig. 60) up to 909°C, the upper limit of the measurements, which is in good agreement with direct vapor pressure measurement [130]. The Te_2 partial pressures in Fig. 60 run below the saturated vapor pressure of pure tellurium in the whole temperature interval, whereas in [130] the vapor pressure in the three-phase equilibrium S(ZnTe)LV coincides with that for pure tellurium up to ~800°C.

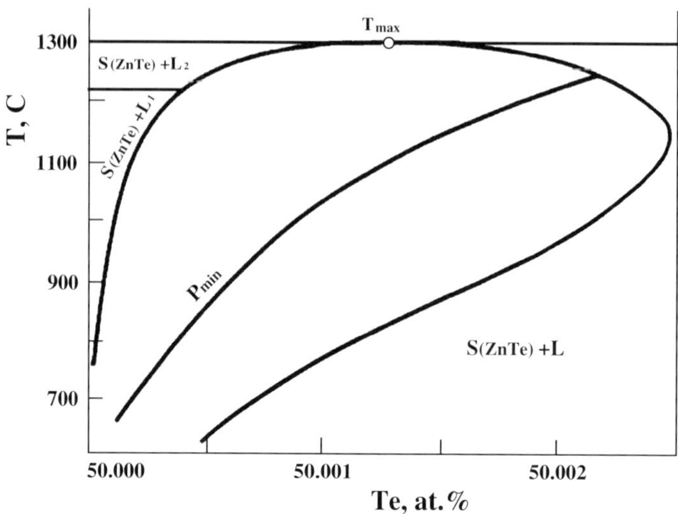

Fig. 59. Calculated non-stoichiometry range of ZnTe

3.1.1.2 Cadmium chalcogenides

In the cadmium–sulfur and cadmium–selenium systems a miscibility gap in the liquid phase was reported on the chalcogen side of the compound CdX. In the *T–X* diagram Cd–Se (Fig. 61 [131]), the monotectic reaction appears at 991°C, and the eutectics are at 317°C (cadmium) and 213°C (selenium), which is only slightly lower than the melting temperatures of the pure components. The maximum melting points are 1405°C for CdS and 1240°C for CdSe [122]. The estimated melting temperatures quoted in [22] are 1410°C (CdS) and 1268°C (CdSe) (Fig. 62), and the corresponding compositions at T_{max} are on the Cd-side of the stoichiometric plane. The maximum non-stoichiometry of CdS was estimated from annealing experiments at fixed temperatures and vapor pressures of the components [22]. The solubility of cadmium in CdSe was assessed from the total vapor pressure results [132] and high temperature measurements of electrical conductivity and the Hall effect at fixed vapor pressures of the components [133]. The solidus for both compounds is asymmetric, and the solubility of cadmium is retrograde (Fig. 62).

Cadmium–tellurium system. The *P–T–X* phase diagram for this system, quoted in [20,22,134,135], was constructed essentially from only two experiments: optical density measurements of the vapors [56] and visual registration of the melting points of Cd–Te alloys at fixed vapor pressures of either tellurium or cadmium [136], along with DTA results [127,128,136]. Non-stoichiometry in CdTe was studied experimentally by Hall effect measurements on samples with various compositions, or calculated on the basis of the quasi-chemical theory of defects [20,22,135]. In [137–140] di-

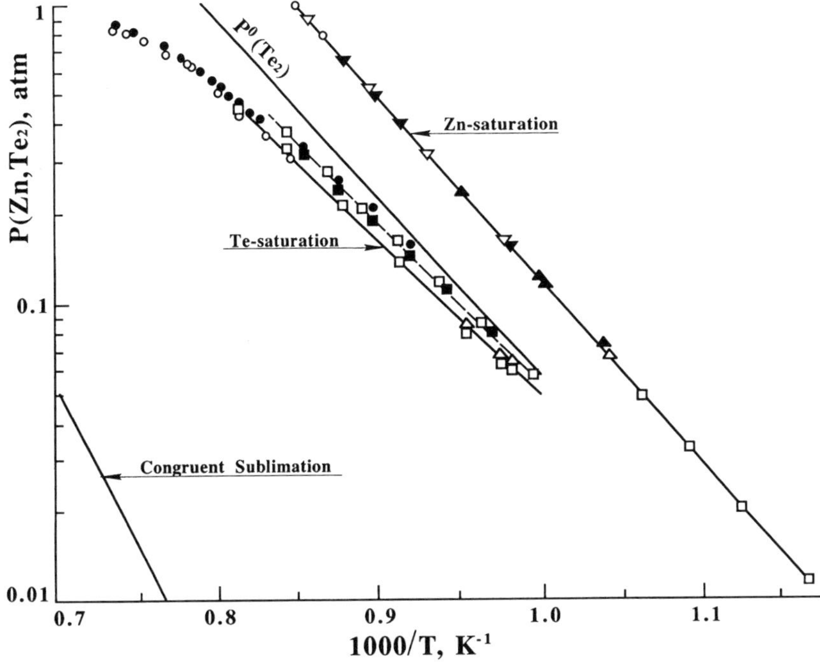

Fig. 60. $P(\text{Zn})$–T and $P(\text{Te}_2)$–T projections of the Zn–Te phase diagram

rect experimental measurements of the total vapor pressure in the cadmium–tellerium system were reported, the P–T–X phase diagram was constructed and experimental results were presented on vapor pressure scanning of the CdTe solidus surface.

***P–T–X* phase equilibrium.** Some typical experimental results of the vapor pressure measurement for Cd–Te are shown in Fig. 63. The general outline of the P-T projection of the phase diagram is in agreement with that constructed from indirect data [20,22,134,135]. Meanwhile, the specifics of the three-phase curves in the P–T–X phase space are best described by direct measurements.

A characteristic feature of the P-T projection is a vapor pressure maximum in the VLS (vapor–liquid–solid CdTe) equilibrium and two vapor pressure extrema in SLV (solid CdTe–liquid–vapor): $P_{\max}$ = 161 mmHg at $T(P_{\max})$ = 1231 K, and $P_{\min}$ = 135 mmHg at $T(P_{\min})$ = 1305 K (throughout the subsequent exposition the order of phases in phase equilibria will follow the increase in Te content in the phases). It should be mentioned that indirect data, presented by Kroger [20], led to quite different corresponding coordinates: $P_{\max}$ = 960 mmHg; $T(P_{\max})$ = 1330 K and $P_{\min}$ = 240 mmHg; $T(P_{\min})$ = 1365 K. The low-temperature portions (T < 900 K) of the VLS and SLV curves are indistinguishable from the respective

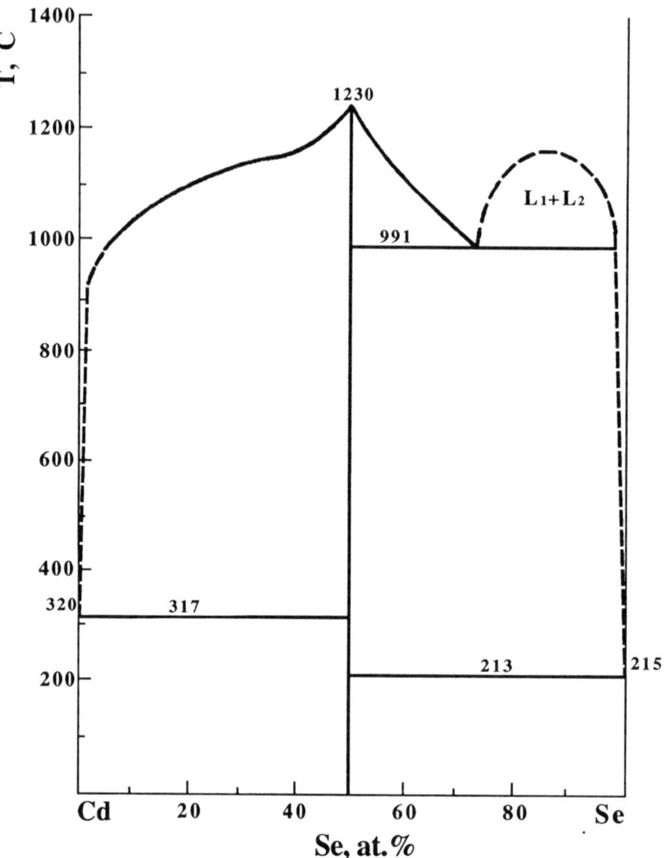

Fig. 61. *T–X* projection of the Cd–Se phase diagram

liquid–vapor equilibria for pure cadmium [141] or tellurium [142], which implies that the mutual solubility of these elements in the liquid phase is very low.

It is noteworthy to compare direct vapor pressure measurements (Fig. 63) with those calculated from indirect data [56,136].

Equilibrium SLV. The total vapor pressure of Te-saturated CdTe in equilibrium with liquid and Te-rich vapor can be described with a maximum 5% uncertainty by polynomials $\log P = \sum a_i (T \times 10^{-3})^i$, $i = -1,0,1,...,n$. The best fit coefficients and the fitting intervals are presented in Table 3. Up to 1200 K the total vapor pressures correspond to the partial pressures $P(Te_2)$ reported by Brebrick [56] within the ±10% uncertainty adopted in [56]. Results obtained by Lorenz [136] for SLV equilibrium in a very limited interval (only 13 K) are too high (10 to 80 times greater than those in Fig. 63), and their intrinsic uncertainty is as high as 50%. At

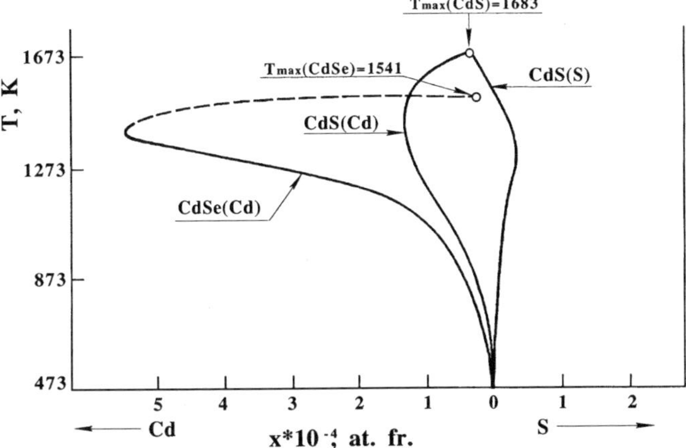

Fig. 62. Non-stoichiometry ranges of CdS and CdSe

elevated temperatures, the sum of the partial pressures $P(Cd)$ and $P(Te_2)$, reported by Brebrick [56], is also within a $\pm 10\%$ limit of the experimental vapor pressure data. At the maximum melting point of CdTe, according to [56], the equilibrium vapor is almost pure Cd ($X_V = \sim 2$ at.% Te).

Table 3. Best fits $\log P(\text{mmHg}) = \sum a_i T^i$ for VLS and SLV equilibria

T (K)	a_0	a_{-1}	a_1	a_2	a_3
		VLS Equilibrium			
885–1000	2.6843	−2.8862	2.8814	—	—
1010–1035	153.2017	−79.6068	−70.9508	—	—
1045–1365	476.0611	−128.8887	−668.6840	426.8594	−102.7022
		SLV Equilibrium			
880–1030	11.4092	−7.4951	−2.3694	—	—
1035–1085	373.4608	−141.1831	−328.3377	97.5417	—
1090–1305	−16.0869	−0.7741	31.2609	−12.9086	—
1315–1365	22433.4289	−9754.5787	−17195.5652	4394.3456	—

Equilibrium VLS. The experimental vapor pressures for Cd-saturated CdTe in equilibrium with liquid and Cd-rich vapor were best fitted within a 3% limit by similar polynomials. The corresponding coefficients are also given in Table 3. Results reported by Lorenz [136] for temperatures $T < 1316$ K were calculated assuming an ideal liquid solution model, and disagree with the experiment (Fig. 63). The high-temperature data [136] (at $T = 1316–1365$ K) were extremely

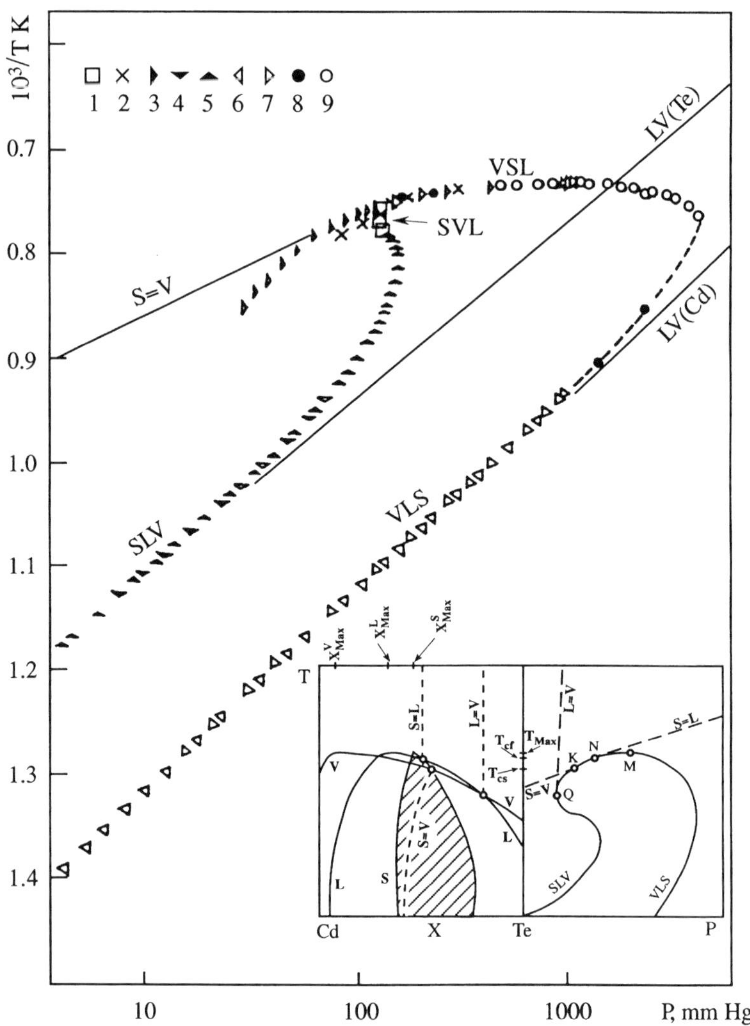

Fig. 63. *P–T* projection of the Cd–Te diagram. Initial compositions (at.% Te): 50.031(1), 50.03(2), 50.02(3), 50.04(4), 52.84(5), 49.7(6), 41.9(7), 8–[56], 9–[136]

poorly reproducible: the differences in vapor pressures for the same temperature were as high as 300 mmHg.

Thus, Lorenz [136] gave only semiquantitative outlines of the *P–T* projection, and Brebrick [56] studied only a portion of it, i.e., the SLV equilibrium.

Direct vapor pressure measurements (Fig. 63) in the entire region of CdTe stability make it possible to carry out a geometrical analysis of the phase equilibria in the Cd–Te system. The compositional sequence of phases at the invariant points (Cd and Te eutectics) requires three congruent processes in the system: S=L, congruent fusion; S=V, congruent sublimation; and L=V, congruent vaporization. All of them are shown in the inset of Fig. 63, which is an enlargement of the $P–T$ and $T–X$ projections of the melting region of CdTe. From the relative spatial positions of these curves, as well as those of the univariant three-phase equilibria and the maximum melting point of CdTe, it follows that

(1) The point of tangency K (Fig. 63) between the S=V curve and the univariant equilibrium belongs to the SLV portion of the three-phase curve; consequently, at the maximum congruent sublimation point, T_{cs}, crystalline CdTe is Te-saturated, and the corresponding temperature is lower than T_{max}(CdTe), the maximum melting point M of CdTe.

(2) The congruent fusion curve S=L also touches at the point N (Fig. 63) the SLV portion of the univariant curve because the melting temperature of CdTe increases with increasing pressure [20]; consequently, at the minimum congruent melting point, T_{cf}, CdTe is also Te-saturated.

(3) The L=V curve is the azeotropic line with a pressure minimum because heteroatomic interaction is predominant in liquid Cd–Te solutions [143]; it means that the point of tangency Q (Fig. 63) between L=V and the three-phase curve also belongs to SLV; and the congruent vaporization composition is on the Te-side of the solidus: $X_{cv} > X_S$ (in at.% Te).

From these considerations, a necessary condition follows for the maximum melting temperature of CdTe (Fig. 63, inset): at T_{max}, the sequence of compositions of the equilibrium phases is $X_V < X_L < X_S$, and for congruent points, $X_{cf} < X_{cs} < X_{cv}$.

The crystallization of non-stoichiometric CdTe under various $P–T$ conditions from different matrices (liquid, vapor, or both), as well as annealing the prepared material, is convenient to follow in sections of the $P–T–X$ diagram. Figure 64 presents a succession of schematic isobaric sections of the CdTe melting region starting with $P > P_{max}$(VLS) down to $P < P_{min}$(SLV). The corresponding $P–T$ coordinates of the univariant equilibria for each Fig. 64 isobar are given in Table 4. At a pressure that corresponds to the maximum melting point (Fig. 64,3), CdTe is usually crystallized by cooling the melt of the composition X_L, $X_V(T_3) < X_L < X_L(T_3)$, from the VL region down to T_3. Material obtained in this way is in equilibrium with the liquid and vapor, both enriched in Cd. The vapor in equilibrium with congruently melting CdTe (Fig. 64,4, tie-line at T_3) is also enriched in Cd. Material saturated with Te to various concentrations can be prepared by cooling the melt under conditions, shown in Fig. 64,6–9, from the corresponding regions LV down to temperatures determined by the T_6 tie-line.

Table 4. Coordinates P(mmHg) – T(K) for univariant equilibria in isobaric sections, Fig. 64

Sec #	P	P (mmHg)	T_1(LV$_{Cd}$)	T_2(VLS)[a]
1	$> P_{max}$(VLS)	> 2985	> 1175	—
2	P_{max}(VLS)	2985[a]	1175	1240
3	P_{max}(VLS)–P_{cf}	1034–2985	1065–1175	1072–1240
4	P_{cf}	1034	1065	1072
5	P_{max}(SLV)	160.5	920	922
6	P_{max}(SLV)–P_{cs}	150.8–160.5	915–920	917–922
7	P_{cs}	150.8	915	917
8	P_{cs}–P_{cv}	129.8–150.8	905–915	907–917
9	P_{cv}	~129.8	905	907
10	P_{min}(SLV)	129.8	905	907
11	$< P_{min}$(SLV)	< 129.8	< 905	< 907

Sec #	T_3	T_4(LV$_{Te}$)	T_5(SLV)[a]	T_6(SLV)[a]
1	—	> 1445	—	—
2	—	1445	—	—
3	1240–1365	1300–1445	—	—
4	1365(VL=S)	1300	—	—
5	1340(VSL)	1105	1230	—
6	1325–1340	1100–1105	1185–1230	1230–1275
7	1325(V=SL)	1100	1185	1275
8	1310–1325	1085–1100	1150–1185	1275–1305
9	1310	1085	1150	—
10	1310(V=S)[b]	1085	1150	1305
11	< 1310	< 1085	< 1150	

[a]Calculated from data in Table 3

[b]Calculated from data [144]

The sections in Fig. 64 also show the Cd and Te saturation limits for CdTe annealed in vapors of the corresponding composition (VS and SV regions in Fig. 64,3–11). Thus, the maximum Cd non-stoichiometry of CdTe on annealing in vapors is defined by the boundary of the solidus S between the tie-lines T_2 and T_3 in Fig. 64,3–6, or between the tie-line T_2 and the temperature T(S=V) in Fig. 64,7–11. The Te non-stoichiometry is determined by the solidus surface in equilibrium SV between the tie-lines T_5 and T_6 (Fig. 64,6–9), or the temperature T(S=V) and the tie-line T_5 (Fig. 64,10–11). The shape of the P–T projection (Fig. 63), as noted by Kroger [20], also provides a rather unusual way of CdTe crystallization, viz., by heating the liquid in isobaric conditions. Such a process corresponds to the transition from the liquid-vapor regions adjoining the pure components (either Cd or Te) up to the three-phase equilibrium VLS (Fig. 64,2–11) or SLV (Fig. 64,5–11). The composition of the liquid in this case should be in the interval X_L–X_V

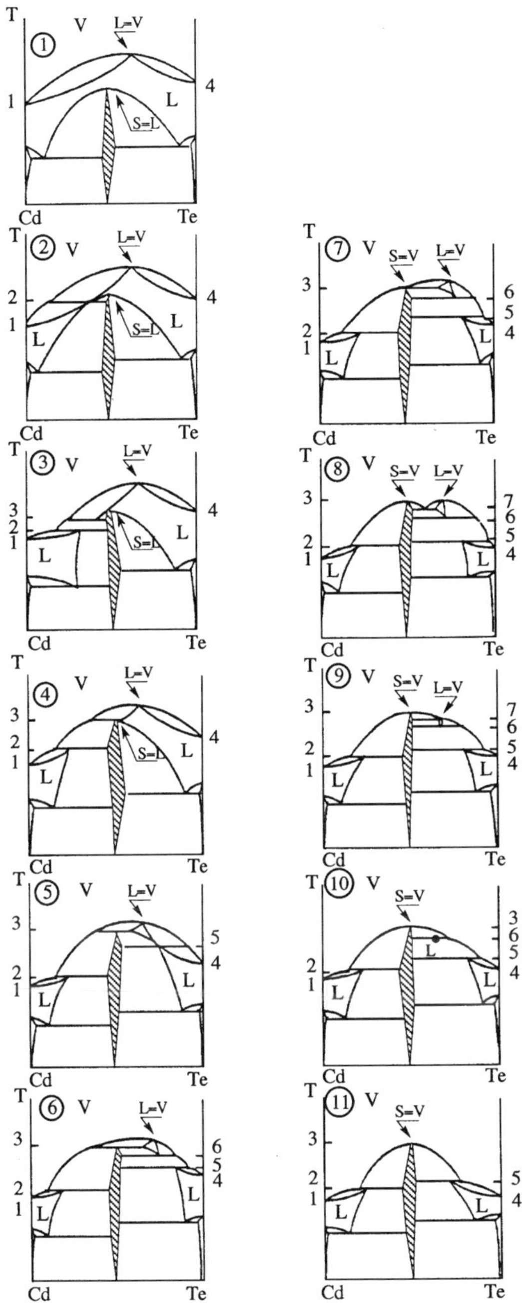

Fig. 64. Isobaric sections of the *P–T–X* phase diagram Cd–Te. Pressure ranges for individual sections and co-ordinates of the univariant equilibria are given in Table 4

for the corresponding three-phase equilibrium. For example, at pressure $P_{max}(SLV)$ (Fig. 64,5), Cd-saturated CdTe can be prepared by heating the liquid with composition X_1 in the interval $X_V(T_2) < X_L < X_L(T_2)$ from the VL region up to $T_2 = 992$ K. If the liquid with composition X_L, corresponding to $X_V(T_5) < X_L < X_L(T_5)$, is heated from the LV region up to T_5, then Te-saturated CdTe crystallizes from it at 1230 K

Thus, the composition of the crystalline material can be controlled by choosing a matrix and *P–T–X* conditions suitable for either crystallization or annealing in the vapors of the corresponding composition.

Non-stoichiometry in CdTe. It has been shown [20] that there was a minimum total vapor pressure for CdTe in the two-phase solid-vapor equilibrium, which corresponded to the congruent sublimation S(CdTe)=V. This congruent sublimation curve divides the sublimation region into two parts: VS, where the vapor V is enriched in Cd compared to the Cd-saturated solid S; and SV, where the crystal is Te-saturated and the vapor is Te-rich. The composition of the crystal X_S in equilibrium with the vapor can be determined from Eq. (25) at every experimental point (P,T) by subtracting the number of gram-atoms of cadmium $n(Cd)$ and tellurium $n(Te)$, evaporated at this temperature from the number of gram-atoms $N(Cd)$, $N(Te)$ in the initial sample:

$$X_S \text{ (at.\% Te)} = \left[N(Te) - n(Te) \right] / \left\{ \left[N(Cd) + N(Te) \right] - \left[n(Cd) + n(Te) \right] \right\} \times 100\% \quad (32)$$

For this purpose, one should know the partial pressures of all of the vapor species. No heteroatomic gaseous molecules were observed in the mass spectra of CdTe [144]; the only Cd species is Cd(g), whereas tellurium forms seven gaseous polymers Te_k, $k = 1$ to 7 [145]. Hence,

$$n(Cd) = P(Cd)v/RT, \quad (33)$$

$$n(Te) = (v/RT)\sum kP(Te_k), \quad k = 1 \text{ to } 7, \quad (34)$$

and the total vapor pressure P, measured at every temperature T, is a sum of eight partial pressures:

$$P = P(Cd) + \sum P(Te_k), \quad k = 1 \text{ to } 7. \quad (35)$$

Eight independent equations are necessary to calculate these eight partial pressures. One of them is Eq. (35); six more are provided by the equilibrium constants K_j of the polymerization reactions,

$$Te_k = k/2 \ Te_2$$

$$K_j = P^{k/2}(Te_2)/P(Te_k), \quad k = 1 \text{ to } 7. \quad (36\text{–}41)$$

The K_j values can be calculated by a standard procedure from $R \ln K_j = \Delta\Phi_j - \Delta H_0/T$ at any temperature because the individual free energy functions Φ_T and standard enthalpies of formation ΔH_{0f} are tabulated for all of the Te_k polymers [145]. The eighth equation is the relation between $P(Cd)$ and $P(Te_2)$ determined by the Gibbs free energy of formation of CdTe, ΔG_T:

$$\Delta G_T = RT \ln\left[a(Cd)\, a^{\alpha}(Te)\right], \tag{42}$$

where $\alpha = X_s/(1-X_s)$. It was assumed that at $T = $ const, ΔG_T remains constant within the homogeneity range of CdTe. The partial pressures are readily calculated from the activities $a(i)$ because the saturated vapor pressures for pure Cd(l) and Te(l) are known [141,145]. Thus, the vapor pressure problem of determining the composition of the crystalline CdTe in the solid–vapor equilibrium at a fixed temperature consists of solving a system of eight equations (35–42) with eight unknown partial pressures. Subsequently, X_s is calculated from Eq. (32), and the composition of the vapor X_v is calculated from

$$X_V \text{ (at.\% Te)} = n(Te)/\left[n(Cd) + n(Te)\right] \times 100\%. \tag{43}$$

In these calculations, the values of $\Phi_T(CdTe,s)$ and $\Phi_T(Cd,g)$ were taken from the IVTANTERMO database [145].

As a result of this treatment, each experimental (P,T) point produces a pair of scanning points, (P,T,X_S) and (P,T,X_V), on the solidus S and vaporus V surfaces in the solid–vapor equilibrium. Correspondingly, the entire experimental data file $\{P,T\}$, treated in this way, results in two sets of scanning points, $\{P,T,X_S\}$ and $\{P,T,X_V\}$, with one-to-one correspondence, which outline the position of the solidus and vaporus conjugated surfaces in the P–T–X phase space.

When the boundary of the homogeneity range of CdTe is reached during the vapor pressure experiment, a break in the P–T curve is observed which corresponds to the change in the phase state of the system, i.e., phase-transition SLV $\rightarrow$ SV for Te-saturated CdTe, or VLS $\rightarrow$ VS for Cd-saturated CdTe.

To determine the P–T–X coordinates of this point (which is actually the maximum non-stoichiometry), P–T and T–X projections of the experimental curves were obtained in an analytical form $P = f(T)$; $X_S = \psi(T)$. The temperature and vapor pressure at the boundary of the homogeneity range can be calculated from a system of equations,

$$\begin{aligned} P &= f(T) \\ P &= \phi(T) \end{aligned} \tag{44}$$

where $P = \phi(T)$ is the temperature dependence of the vapor pressure in the three-phase equilibrium SLV or VLS (Table 3). Then the corresponding composition X_S can be calculated from the individual $X_S = \psi(T)$ polynomials at the phase-transition temperatures calculated from Eqs. (44). The results are given in Table 5 and the inset of Fig. 65, which shows the solidus on an enlarged scale. It can be seen that the solidus is strongly asymmetrical; the maximum Cd non-stoichiometry is almost an order of magnitude less than that of Te; the stoichiometric plane $X = 50$ at.% is within the single-phase volume. The solidus line in Fig. 65 is a best fit of the experimental data $X_S = \Sigma a_i T^i \pm t\sigma(T)$, where t is the Student criterion. The corresponding coefficients are given in Table 6 for Cd- and Te-saturated CdTe along with the confidence intervals for X_S as a function of the temperature. Also in Table 6 the liquidus compositions X_L are presented, which were calculated at the phase-transition SLV $\to$ LV or VLS $\to$ VL temperatures in a way, similar to that described earlier for X_S. From these results, the composition X_{cv} of the azeotropic point $X_L = X_V$ was estimated as 55.1 $> X_{cv} >$ 53.5 at.% Te for a temperature between T_{min} in SLV (1305 K) and T_{max} of the congruent sublimation (1324 K).

Partial thermodynamic functions. To calculate the partial vapor pressures of Cd and Te$_2$ for fixed X_S's, the point solutions of Eqs. (32–42) for each vapor pressure curve were represented by best fit polynomials,

$$\log P(i) = f_i(T), \quad i = \text{Cd or Te}_2 , \tag{45}$$

$$T = T(X_S). \tag{46}$$

For an assumed X_S, the corresponding temperature can be calculated from Eq. (46), and then $P(\text{Cd})$ and $P(\text{Te}_2)$ are obtained from Eq. (45) for every vapor pressure curve. The resulting $\{P(i),T\}$ files were best fitted in the usual form, $\log P(i) = A(i) - B(i)/T$, $i = \text{Cd or Te}_2$. The temperature dependencies of the partial pressures are given in Table 7 separately for solid CdTe in equilibrium with Cd-rich and Te-rich vapors (VS and SV equilibria, respectively). It can be seen that the Te-side surface of the solidus is in the $X_S > 50.003$ at.% Te region, and the Cd side spans $X_S = 49.999$–50.001 at.% Te and crosses the stoichiometric plane $X=50$ at.%. Another characteristic feature of the sublimation region of CdTe reflected in Table 7 is that, in spite of a small X_S step ($2 \cdot 10^{-4}$ at.%), no X_S was found for which $B(\text{Cd}) = B(\text{Te}_2)$. It means that no constant congruent sublimation composition of CdTe exists. These vapor pressure coefficients approach each other at $T > 1150$ K for $X_S = 50.001$–50.002 at.% Te, although $B(\text{Cd})$ in this region is slightly greater than $B(\text{Te}_2)$, implying that the congruently subliming composition, X_{cs}, gradually shifts from Cd toward Te with increasing temperature. The last row of Table 7 was calculated from mass spectrometric data [144] for congruent sublimation of CdTe at low temperatures.

Standard thermodynamic procedure was used to calculate the partial molar enthalpies and entropies from partial vapor pressures. Tabulated saturated vapor pressures for pure liquid Cd and Te (reference states) were taken from [141,142]. The partial molar functions were virtually independent of the composition of the

Table 5. Homogeneity limits of CdTe, liquidus, and congruent sublimation

T (K)	P (mmHg)	X_S (at.% Te)	X_L (at.% Te)	X_V (at.% Te)
		Cd-saturated CdTe		
697.2	3.6	50.0002 ± 0.0002		0
819.6	33.5	50.0001 ± 0.0001		0
864.1	68.2	49.9990 ± 0.0001		0
871.3	76.3	49.9983 ± 0.0006		0
900.8	119.1	49.9990 ± 0.0001		0
939.7	209.0	49.9994 ± 0.0002		0
969.4	316.3	49.9991 ± 0.0002		0
992.0	429.5	49.9967 ± 0.0003		0
1073	1100	49.9966		
1123	1800	49.9946		
1173	2400	49.9957		
1223	2850	49.9967		
		Te-saturated CdTe		
945.5	17.5	50.0006 ± 0.0001		99.9
1016.6	42.4	50.0036 ± 0.0001		99.9
1046.9	59.1	50.0055 ± 0.0004		99.9
1073	74	50.0071		
1076.0	77.6	50.0073 ± 0.0001		99.8
1094.1	90.3	50.0092 ± 0.0003		99.9
1103.3	97.4	50.0098 ± 0.0001		99.7
1121.0	110.8	50.0138 ± 0.0007		99.6
1123	111	50.0138		
1173	141	50.0135		
1223	148	50.0082		
1243.1	159.7	50.0087 ± 0.0008		96.5
1282.2	147.9	50.0050 ± 0.0010		96.0
1285.5	146.3	50.0047 ± 0.0015		90.8
1301.0	137.9	50.0044 ± 0.0030		85.6
1312.0	131.0	50.0038 ± 0.0040		77.7
1359.4	549.0	50.0034 ± 0.0006		2.0
1361.8	654.2	50.0013 ± 0.0004		0.8
		Liquidus		
1034.8	51.4		82.81	100
1124	116.1		73.77	99
1176.3	146.5		66.86	99
1234.9	160.4		61.52	96
1281.3	148.3		56.80	91
1336.3	188.7		52.62	27
1357.3	476.3		50.86	3.5
1364.7	1011		50.0003	0.4
		Congruent sublimation		
1073	2.6	50.00079		
1123	6.8	50.00083		
1173	16.1	50.00109		
1223	35.7	50.00150		
1324	149	50.00218		

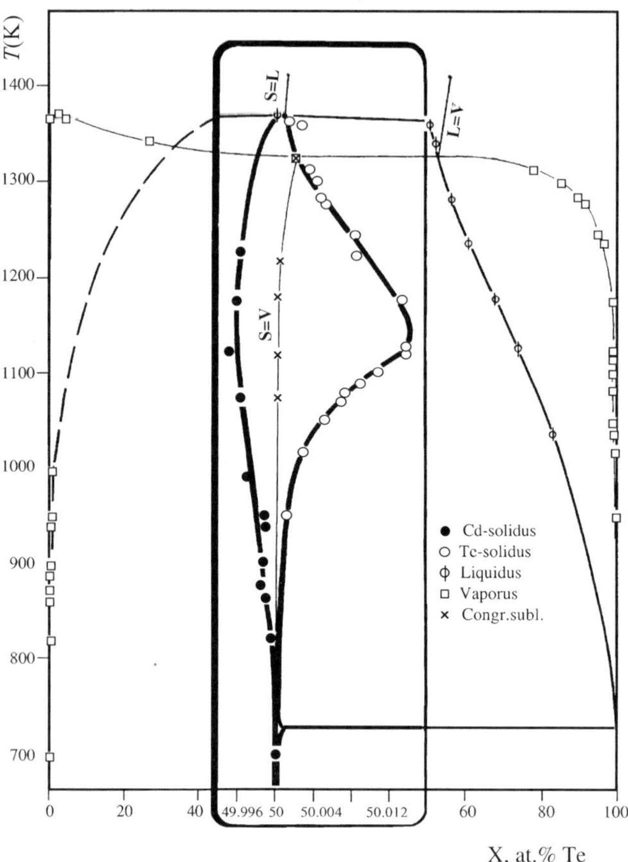

Fig. 65. T–X projection of the Cd–Te diagram. The near-solidus region is given on an enlarged scale

solid for X_S = 50.004–50.01 at.% Te (equilibrium SV) and 50.0008–50.001 at.% Te in VS within the uncertainties typical for the vapor pressure method, ±(4–8) kJ/mol. A rigorous estimate of the uncertainties associated with the values in Table 7 proved to be rather difficult because a repeated best-fit procedure was used in these calculations.

Isotherms of partial pressures. A similar procedure was used to obtain the isotherms of $P(\mathrm{Cd})$ and $P(\mathrm{Te_2})$ within the homogeneity range of CdTe. Only instead of Eq. (46), individual $X_S = \psi(T)$ fits were considered. For an assumed T the corresponding $P(\mathrm{i})$ is readily calculated from the suitable Eq. (45), and the composition of the conjugated solid X_S is calculated from the $X_S = \psi(T)$ fits.

Table 6. Best fits $X(\text{at.\% Te}) = \sum a_i T^i \pm t\sigma(T)$ for Cd- and Te-saturated CdTe, liquidus, and congruent sublimation

	a_0	a_1	a_2	a_3	a_4	$T\,(\text{K})$
$X_S(\text{Cd})$	49.9170	$2.8853\cdot10^{-4}$	$-3.2215\cdot10^{-7}$	$1.1381\cdot10^{-10}$	—	
$\sigma^2(T)$	$6.547\cdot10^{-7}$	$-8.2894\cdot10^{-11}$	—	—	—	700–1365
$X_S(\text{Te})$	65.0398	$-5.3398\cdot10^{-2}$	$7.0437\cdot10^{-5}$	$-4.0903\cdot10^{-8}$	$8.8255\cdot10^{-12}$	
$\sigma^2(T)$	$-5.580\cdot10^{-7}$	$1.6652\cdot10^{-9}$	—	—	—	945–1365
X_L	266.321	-0.23581	$5.6681\cdot10^{-5}$	—	—	
$\sigma^2(T)$	1.507	$-1.074\cdot10^{-3}$	—	—	—	1035–1365
X_{cs}	50.0124	$-2.4376\cdot10^{-5}$	$1.2559\cdot10^{-8}$	—	—	
$\sigma^2(T)$	$9.949\cdot10^{-9}$	$-4.2307\cdot10^{-12}$	—	—	—	1073–1324

The resulting $P(i) = f_i(X_S)$ plots are shown in Fig. 66 for four temperatures. From their best fits, it is also possible to estimate the non-stoichiometry limits and congruent sublimation compositions X_{cs} for these temperatures. It is clear that the former is an intersection point with either the SLV or VLS curve, and the latter should correspond roughly to the $P(\text{Cd}) = 2P(\text{Te}_2)$ condition. The relevant results can be seen in Table 5.

Comments on CdTe non-stoichiometry. The total vapor pressure measurements at high temperatures resulted in two sets of scanning points, $\{P,T,X_S\}$ and $\{P,T,X_V\}$. From them the conjugated solidus and vaporus surfaces were reconstructed in the P–T–X phase space. This procedure, known as vapor pressure scanning of the solidus [96], proved sensitive enough to investigate deviations from stoichiometry in CdTe as small as 10^{-4} at.% directly at high temperatures.

Because the compositional region of stability for CdTe is quite narrow, special attention should be given to the uncertainties, δX_S, associated with the composition of the solid, X_S, obtained by this method. Three main sources contribute to these uncertainties: (1) experimental errors in measuring the vapor pressure, temperature, reaction volume, and initial masses; (2) uncertainties in the thermodynamic functions of tellurium species; (3) the assumption that $\Delta G_T = \text{const}$ at $T = \text{const}$ within the single-phase region of CdTe. All of these factors influence δX_S in different ways, depending on the experimental conditions.

1. The uncertainties δX_S resulting from experimental errors are readily calculated at every experimental point by applying the error accumulation law because all of the experimental errors are known. It was shown that the main source of these uncertainties was the precision of the balance (5×10^{-5} g in the experiments described), which led to a typical δX_S value within $(1–5)\times10^{-4}$ at.%.

2. It is difficult to estimate rigorously the uncertainties of the thermodynamic functions of all of the tellurium polymers Te_k [145]. To avoid the need to do so, an

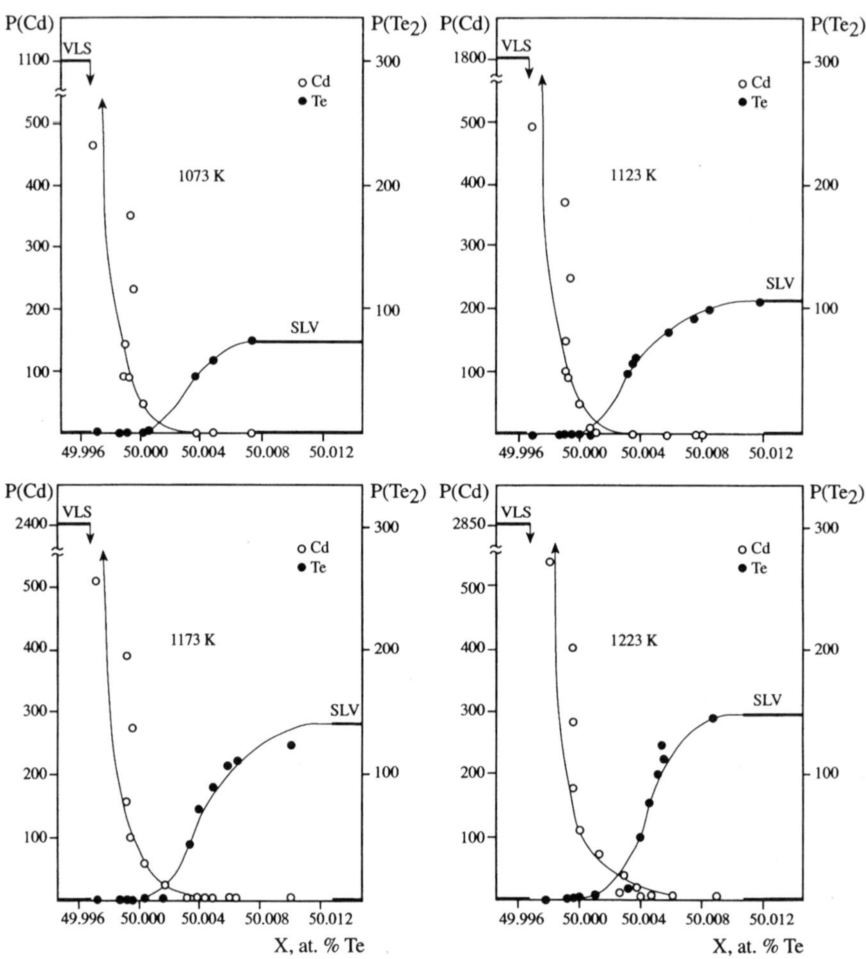

Fig. 66. Isotherms of the partial pressures $P(Cd)$ and $P(Te_2)$. Horizontal lines correspond to three-phase equilibria

additional set of X_S calculations was done with Te_2 as the only tellurium species (the thermodynamics of this molecule is the most reliable). The result was that both Te_k and Te_2 models gave the same X_S values within the uncertainties caused by experimental errors.

3. The influence of the ΔG_T = const assumption on the calculated X_S was checked by fluctuating the ΔG_T values within $\pm(2–4)$ kJ/mole, the maximum expected variation of ΔG_T within the homogeneity range of CdTe [56]. It was shown that in the VS equilibrium, where the vapor was made up of almost pure Cd ($X_V \cong$

Table 7. Temperature dependences of Cd and Te_2 partial pressures $\log P(i)$ (mmHg) $= A(i) - B(i)/T$

X_S (at.%Te	A(Cd)	B(Cd)	A(Te$_2$)	B(Te$_2$)	T(K)
		Equilibrium SV			
50.003	11.74	13400	4.05	2500	1009 – 1324
50.004	11.85	13500	3.83	2200	1027 – 1304
50.005	11.78	13500	3.96	2300	1042 – 1285
50.006	11.74	13500	4.04	2400	1056 – 1268
50.007	11.70	13400	4.12	2400	1070 – 1254
50.008	11.66	13400	4.20	2500	1082 – 1240
50.009	11.65	13400	4.23	2500	1095 – 1228
50.010	11.64	13400	4.24	2500	1108 – 1216
		Equilibrium VS			
49.9990	5.74	3300	16.05	22600	880 – 1334
49.9992	6.76	4900	14.00	19500	973 – 1340
49.9994	8.06	6700	11.41	15900	973 – 1344
49.9996	8.74	7700	10.05	13900	1073 – 1349
49.9998	8.54	7500	10.44	14300	1073 – 1353
50.0000	8.36	7300	10.81	14600	1073 – 1358
50.0002	8.45	7500	10.63	14200	1073 – 1362
50.0004	9.37	8800	8.79	11700	1073 – 1365
50.0006	10.31	10200	6.90	8900	1073 – 1364
50.0008	9.51	9400	8.49	10500	1073 – 1353
50.0010	9.13	9000	9.27	12500	1173 – 1346
[144]	9.27	9760	8.97	9760	—

10^{-4} at.% Te), this influence was undetectable. For the SV equilibrium (Te non-stoichiometry) at $T < 1250$ K the variations in X_S were negligible compared to the δX_S associated with experimental errors. At higher temperatures (1250–1350 K), the $\Delta G_T =$ const assumption influenced the calculated X_S quite noticeably, up to 3×10^{-3} at.% Te in some experiments. All of the X_S values in Table 5, obtained directly at experimental phase-transition points, are listed with the calculated uncertainties, which incorporate all of the contributions mentioned. It is difficult to make a rigorous estimate of the uncertainties in X_S, calculated from the isotherms of the partial pressures, because of consecutive approximations used for this purpose. Nevertheless, a certain measure of confidence in the results can be drawn from comparing X_S, calculated at 1073 K and 1123 K, with those obtained from direct experiments at 1076 K and 1121 K. From Table 5, it is obvious that the agreement is quite good.

As an independent check of the results, X_S was also determined at a number of (P,T) points using the "intersection method", Eq. (31) [96]. In this approach, the composition of solid CdTe is calculated from Eq. (47):

$$X_S = \left[N_1(\text{Te})v_2 - N_2(\text{Te})\,v_1 \right] \big/ \left\{ \left[N_1(\text{Cd}) + N_1(\text{Te}) \right] v_2 - \left[N_2(\text{Cd}) + N_2(\text{Te}) \right] v_1 \right\}, \quad (47)$$

at the intersection point of two vapor pressure curves, 1 and 2, and it is independent of the composition of the vapor. In Eq. (47) N_1(Cd), N_1(Te), v_1 and N_2(Cd), N_2(Te), v_2 are the initial masses and reaction volumes for two experiments, from which the two intersecting vapor pressure curves originate. It was proved that the X_S values, obtained by this method, were the same as those calculated from the ΔG_T = const model to within $\leq 4 \cdot 10^{-4}$ at.%.

Estimating of the uncertainties of the vapor composition X_V is reasonable only for the SV equilibrium, because in VS the vapor is almost pure Cd (see Table 5). Because of uncertainties in the thermodynamics of Te_k polymers [145], δX_V's were estimated by comparing two models of the vapors, Te_2 only and Te_k, $k = 1$ to 7. The resulting δX_V values were within 0.5 at.% at $T < 1280$ K, whereas at higher temperatures, the uncertainties δX_V rose to $\pm(1–3)$ at.%.

Special attention should be given to the last two rows of the Te solubility section of Table 5. The two corresponding vapor pressure curves (for initial samples 49.94 and 49.95 at.% Te) went through the VS region and intersected the steeply ascending portion of the three-phase curve at 1359.4 K, 549 mmHg and 1361.8 K, 654.2 mmHg, respectively. According to Fig. 63, this is the VSL branch of the three-phase curve, where the liquid is enriched in Te compared to the solid. This means that the intersection points correspond to the VS $\rightarrow$ VSL phase-transition, i.e., the compositional sequence of the vapor and solid phases in these equilibria remained unchanged. This proves that these two curves did not meet the congruent sublimation curve S=V. Therefore, the maximum congruent sublimation point of CdTe is lower than 1359 K and corresponds to the composition $X_{cs} > 50.0034$ at.% Te. The actual maximum congruent sublimation temperature was found as a point of tangency of the SLV and S=V curves. The former was taken from Table 3, and the latter was calculated from the thermodynamic functions of CdTe(s) and all of the vapor species. The congruent sublimation point, calculated in this way, is T_{cs}(max) = 1324 K, which is as much as 41 K lower than the maximum melting point of CdTe [20,135]. The T–X projection of the congruent sublimation curve is shown in Fig. 65.

It should be stressed that in the VLS equilibrium near the maximum melting point of CdTe (Table 5 and Fig. 63), the crystal with super-stoichiometric Te is in equilibrium with a Te-rich melt and the vapor, which is virtually pure Cd (0.8 to 2.0 at.% Te). This also implies that at the maximum melting point, CdTe is in equilibrium with the vapor $X_V < 0.8$ at.% Te, which compares quite well with the $X_V \cong 2$ at.% Te at T_{max} reported by Brebrick [56].

The homogeneity range of CdTe, listed in Table 5, differs considerably from that obtained from the Hall effect measurements [135,146], especially on the Te side, where the maximum solubility is one to two orders of magnitude greater than that calculated from the Hall data [135,146]. This can be attributed either to non-equilibrium Hall data or electrical inactivity of the native defects in Te-saturated CdTe. Berding [147] made ab initio calculations of the equilibrium native defect densities in CdTe and showed that the Te-saturated material is p-type and highly compensated. The dominant acceptor is cadmium vacancy and the compensated donor is the tellurium antisite. In Cd-saturated CdTe, the predominant defects are cadmium interstitials, resulting in n-type conductivity. Quantitative results were found in [147] in close agreement with the data presented in Table 5 and Fig. 65.

Crystal Growth of CdTe. It has been argued that the crystallizing matrix in high temperature crystal growth processes is in near-equilibrium conditions with the growing crystal. In particular, for CdTe, fundamental studies of both melt and vapor phase growth [148-151] showed that quasi-equilibrium approach was a very good approximation of the experimental data. Consequently, it might be anticipated that an appropriate adjustment of the parameters in the growth region according to equilibrium data could lead to strict control over the composition of the crystal which is a crucial factor in numerous applications of CdTe, especially in detectors of high energy radiation.

In this section, we show how distinctive features of the phase equilibrium in Cd-Te influence the crystallization of CdTe from the melt and vapors. The crystal growth process will be followed in isothermal P–X sections of the P–T–X phase diagram (Figs. 63 and 65) at characteristic crystallization temperatures, which is an explicit way to determine and, ultimately, control the crystal composition.

Crystal growth from the vapor phase. An advantage of vapor–phase growth is the relatively low temperature of the process (compared to the melting point of CdTe) and, as a consequence, higher crystallographic perfection of the material because a number of defects inherent in melt growth are naturally eliminated. Here we will consider only crystallization from saturated vapors in the solid–vapor SV state or physical vapor deposition (PVD) of CdTe; chemical vapor deposition (CVD) has much lower growth rates and potentially leads to contamination of the material with the transporting agent.

Two crucial factors are to be addressed in PVD of CdTe: the quality of the crystal and the growth rate. Both of them are ultimately linked to the composition of the vapors and the crystallization temperature. In the quasi-equilibrium approach, the – driving force of diffusion – controlled mass transport is the partial pressure gradients of the vapor species in the evaporation and growth regions. The maximum growth rates should be expected to correspond to the stoichiometric vapor composition,

$$P(\mathrm{Cd}) = \sum k P(\mathrm{Te}_k), \quad k = 1, 2, ..., 7, \tag{48}$$

where k is the number of Te atoms in the Te vapor species. If the Cd/Te ratio deviates from that determined by Eq. (48), the rate of mass transport is controlled by minor species. This was experimentally observed by Palosz et al. [148] and Laasch et al. [149]. From the viewpoint of crystal quality the optimum is reached at the congruent sublimation composition X_{cs} (Fig. 63), where $X_{\mathrm{S}} = X_{\mathrm{V}}$ and

$$X_{\mathrm{V}} \text{ (at.\% Te)} = \left\{ \sum k P(\mathrm{Te}_k) / \left[\sum k P(\mathrm{Te}_k) + P(\mathrm{Cd}) \right] \right\} \times 100\%. \tag{49}$$

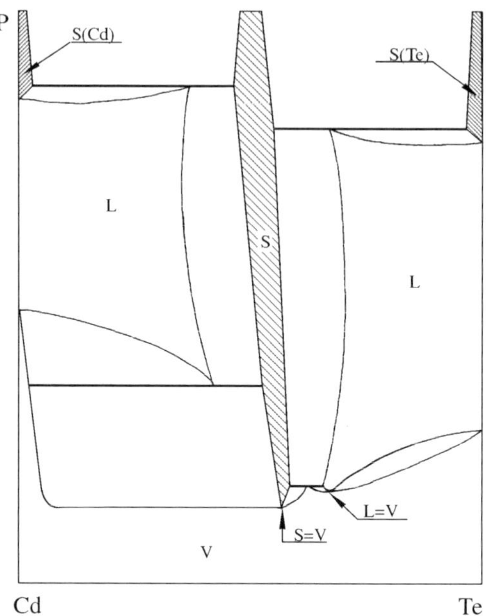

Fig. 67. *P–X* section of the Cd–Te diagram at temperatures between congruent vaporization T_{cv} and congruent sublimation T_{cs}

When $X_S \neq X_V$, interface instabilities are observed that result in a gradual change in crystal composition. As can be seen in Figs. 63 and 65, for CdTe, the congruent sublimation composition $X_{cs} \neq 50$ at.% Te and, moreover, it depends on temperature, especially at high temperatures, meaning that certain deviations from Eq. (48) are to be expected for optimal crystal growth.

Although the diffusion mass transport rate increases with temperature, the PVD of CdTe should be conducted well below the maximum congruent sublimation temperature T_{cs}=1324 K (Figs. 63 and 65) for two reasons: (1) X_{cs} becomes less temperature dependent at $T < 1173$ K [140], and it is easier to maintain steady-state growth conditions, X_S(feed) $\cong X_S$ (cryst), either with [149] or without a "cold tail" [148]. (2) At high temperatures, even a very small deviation in X_V from X_{cs} (Fig. 67) would lead to condensation of the liquid in the SVL three-phase equilibrium. The isothermal section (Fig. 67) of the *P–T–X* diagram (Figs. 63 and 65) corresponds to the temperature interval from 1313 K (congruent vaporization temperature T_{cv} where X_L=X_V) to 1324 K (congruent sublimation temperature T_{cs}, Figs. 63 and 65). The difference in vapor pressure in Fig. 67 between the congruent sublimation (S = V) and the SVL equilibrium is only 2 to 3 mmHg, meaning that even in large volume reactors, it would take just traces of excess Te over X_{cs} to go to the vapor–liquid–solid mechanism of crystallization and all of the detrimental consequences. On the other hand, for $T < T_{cv}$ (Fig. 68), the pressure difference between P(S=V) and SLV (for excess Te) or, still more, VLS (for excess Cd) is much higher. For

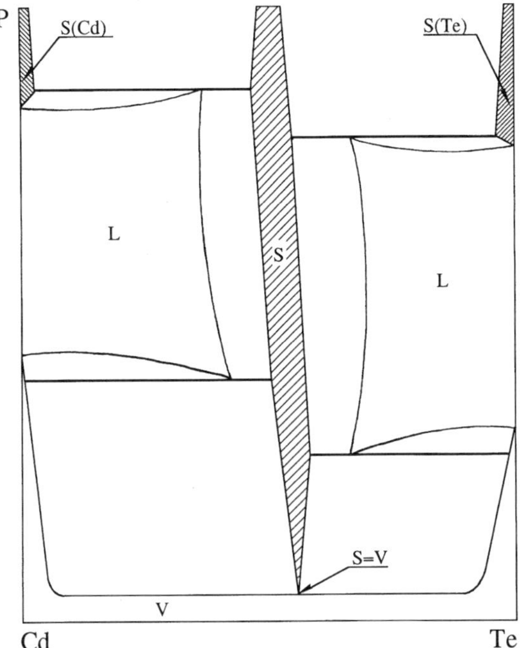

Fig. 68. *P–X* section of the Cd–Te diagram at $T < T_{cv}$

example, from the data [137] at 1153 K, the $[P(SLV) - P(S{=}V)]$ difference is about 120 mmHg, whereas $[P(VLS) - P(S{=}V)]$ is as much as 2130 mmHg. It is clear that a considerable excess of Te is needed to condense the liquid, and the Cd-rich liquid formation is outside the pressure limits of the conventional PVD process for CdTe. These results were observed in the experiments of CdTe growth by PVD at 1153 K [148].

Control over the composition of the crystal grown from vapors is determined by the temperature dependence of the congruent sublimation composition X_{cs} [140]. In the conventional temperature range of CdTe PVD (1123–1223 K [148]), X_{cs} changes from 50.00083 at.% Te to 50.0015 at.% Te [140].

Crystallization from the melt. The best results in CdTe crystal growth from the melt were achieved with Bridgman technology, either vertical or horizontal. The Czochralski method, most commonly used for other semiconductors, failed so far to produce satisfactory results because of specific physical properties of CdTe [150,151].

Vertical Bridgman method. In this technology, crystal growth occurs at the near-maximum melting temperature T_{max} from the (liquid + solid) state of the system. Figure 69 is an isothermal section of the phase diagram (Figs. 63 and 65)

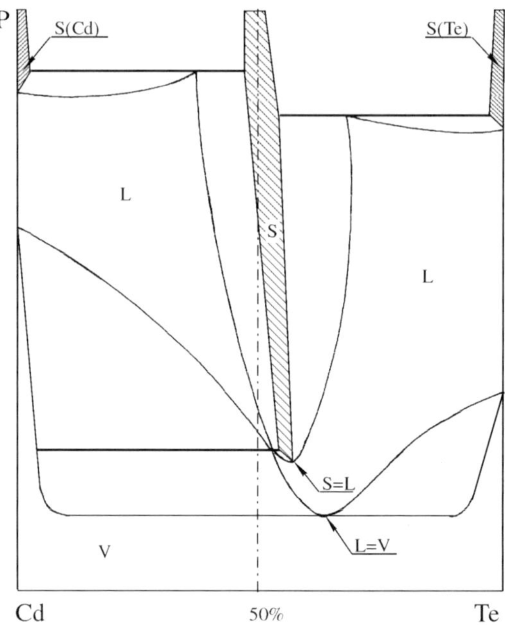

Fig. 69. *P–X* section of the Cd–Te diagram at the maximum melting point T_{max}

at T_{max}. According to vapor pressure scanning data [137–140], the composition of the liquid in the three-phase equilibrium VLS in Fig. 69 is X_L=50.0003 at.% Te, whereas the solid X_S is still richer in Te, both in VLS and, moreover, at the congruent melting point S=L. Consequently, to reduce the Te excess in the crystal, the pressure should be increased above P(VLS) (Fig. 69) which, according to [137], is $P(T_{max})$=1.36 atm. Two methods have been used to this end: a cadmium reservoir with an independent temperature control [150,151] and an inert gas over-pressure [152]. In the former method, the liquid was enriched in cadmium compared to X_L in VLS, whereas in the latter, the initial composition was near-stoichiometric. As a result, high-quality crystals were grown in [152] at a pressure of more than 100 atm (high pressure Bridgman HPB technology), and Rudolph et al. [150,151] managed to decrease the crystallization temperature somewhat due to Cd enrichment of the melt. The corresponding section of the phase diagram (Figs. 63 and 65) at a temperature lower than T_{cf} (congruent melting point) is presented in Fig. 70. The near-stoichiometric composition of the crystal was attained in [150,151] at P=2 atm in the (L+S) state, above the VLS equilibrium. This compares very well with vapor pressure scanning data [137–140], according to which X_S=50 at.% Te at T=1358 K and P=1.56 atm in the VLS equilibrium.

It should be pointed out that, because no direct experimental data have yet been reported for the CdTe solidus in the (L+S) equilibrium, all efforts to control the

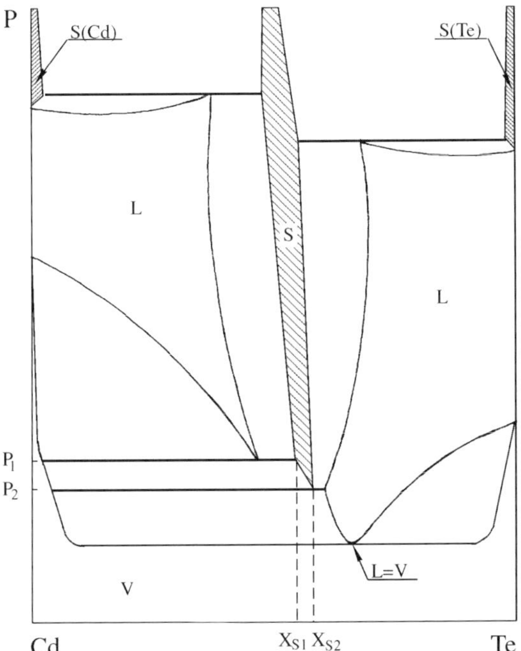

Fig. 70. *P–T* section of the Cd–Te diagram at temperatures between congruent melting T_{cm} and congruent sublimation T_{cs}

crystal composition in the vertical Bridgman technology rely heavily on an empirical approach.

Horizontal Bridgman method. In this method, the crystal is grown from the VLS state (Figs. 69 and 70) at near-melting temperatures. The presence of vapors allows flexible control over the crystal composition. Because the three-phase equilibrium VLS in Cd–Te is univariant, the crystallization temperature as well as the composition of the crystal X_S and melt X_L become fixed, if the vapor pressure in VLS is appropriately chosen. Above the congruent sublimation temperature T_{cs}=1324 K, the vapor in Cd–Te is essentially pure Cd (Fig. 65). Consequently, control over the stoichiometry might be arranged by suitably adjusting the temperature in a separate cadmium "cold tail" with an independent temperature control, which determines the vapor pressure in the growth chamber. Such crystal composition control is expected to be very efficient because high precision direct experimental data on CdTe non-stoichiometry are available [137–140]. In this way, CdTe crystals can be grown both from a Cd-rich melt at a vapor pressure $P_1>P(T_{max})$ (Fig. 70) and from a Te-rich melt at $P_2<P(T_{max})$ (Fig. 70). The corresponding crystal compositions are X_{S1} and X_{S2}.

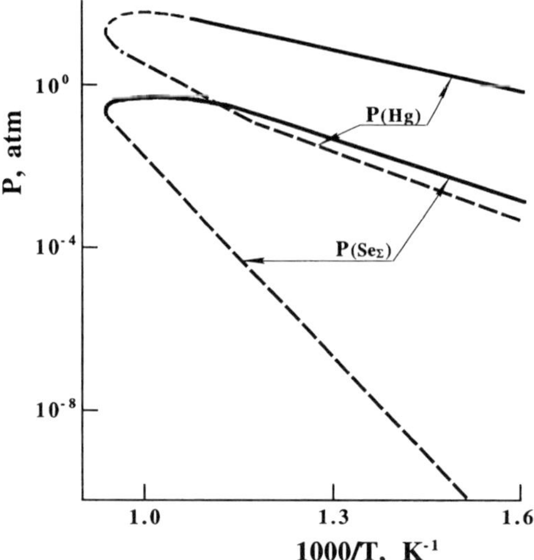

Fig. 71. P(Hg)–T and P(Se$_\Sigma$)–T projections of the Hg–Se system

Post-growth treatment. Cooling down the as-grown CdTe crystal constitutes an additional problem. Because of incongruent sublimation, the solid loses the more volatile component resulting in a progressive shift in composition. This shift can, however, be counterbalanced, if a special cooling process is developed. From the $P–T–X$ point of view, the crystal would not change in composition X_S on cooling, if the conjugated vapor follows in composition X_V the vaporus surface in the iso-plethal (X=const) section of the $P–T–X$ phase diagram (Figs. 63 and 65). The relevant data are presented in Table 7, where the temperature dependences of Cd and Te$_2$ partial pressures are given for a number of solid compositions X_S. The solid–vapor equilibrium is given in Table 7 separately for the Te-rich vapors (SV equilibrium) and Cd-rich vapors (VS equilibrium). These results can also be used for postgrowth annealing of material to create specific non-stoichiometry.

3.1.1.3 Mercury chalcogenides

According to the analysis of non-stoichiometry in mercury chalcogenides reviewed in [22], the single-phase range of existence for mercury sulfide and mercury selenide is on the Hg-side of the stoichiometric plane, whereas for mercury telluride, it is almost completely on the tellurium side. Only in the temperature interval 573 to 653 K is the stoichiometric plane inside the homogeneity region of HgTe. The maximum mercury non-stoichiometry in HgTe was reported to be 1.7×10^{-5} at. %. For HgSe, the maximum melting temperature is on the mercury-side of the 50 at.% plane [22].

The $P–T$ projection of the $P–T–X$ phase diagram for mercury selenide is shown in Fig. 71 in partial pressures of the components measured by the optical absorption

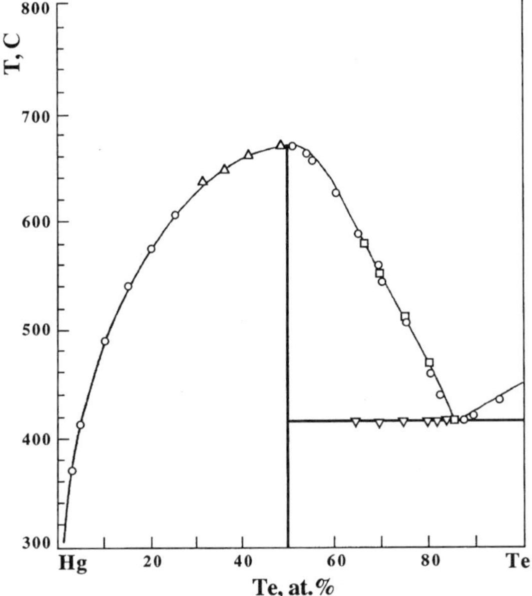

Fig. 72. *T–X* projection of the Hg–Te system

method [52]. The mercury partial pressure at the maximum melting point can be as high as 100 atm, and the selenium vapor pressure is about 1 atm. Partial juxtaposition of the mercury and selenium partial pressures within the sublimation range of HgSe suggests that a vapor pressure minimum is to be expected within the sublimation loop of HgSe, which would correspond to the congruent sublimation S(HgSe) = V.

The *T–X* diagram for the mercury–tellurium system is presented in Fig. 72 according to Brebrick's DTA and optical absorption studies [28,53]. For DTA measurements, the samples of different compositions, shown in Fig. 72, were sealed in evacuated silica tubes, and phase-transition temperatures were recorded under the saturated vapor pressure. Hence, Fig. 72 is the *T–X* projection of the *P–T–X* phase diagram. Note that the tellurium eutectic is 85 at.% Te in composition, meaning quite a noticeable solubility of mercury in the melt. The region of non-stoichiometry of HgTe cannot be resolved on the scale of Fig. 72; as a consequence, it is seen as a vertical line. The maximum melting temperature of HgTe is 670°C [28]. The *P–T* projection of the phase diagram (Fig. 73) was also reported by Brebrick [28,53] in partial pressures of the components from optical measurements. The saturated vapor pressure of pure mercury is the straight line P^0(Hg), which at low temperatures approaches the three-phase equilibrium curve for solid mercury-saturated HgTe, melt, and vapor. The lower portion of the three-phase loop corresponds to Te-saturated HgTe in three-phase equilibrium with the Te-rich melt and vapor. Within this three-phase loop at temperatures below the maximum melting point is the two-phase sub-

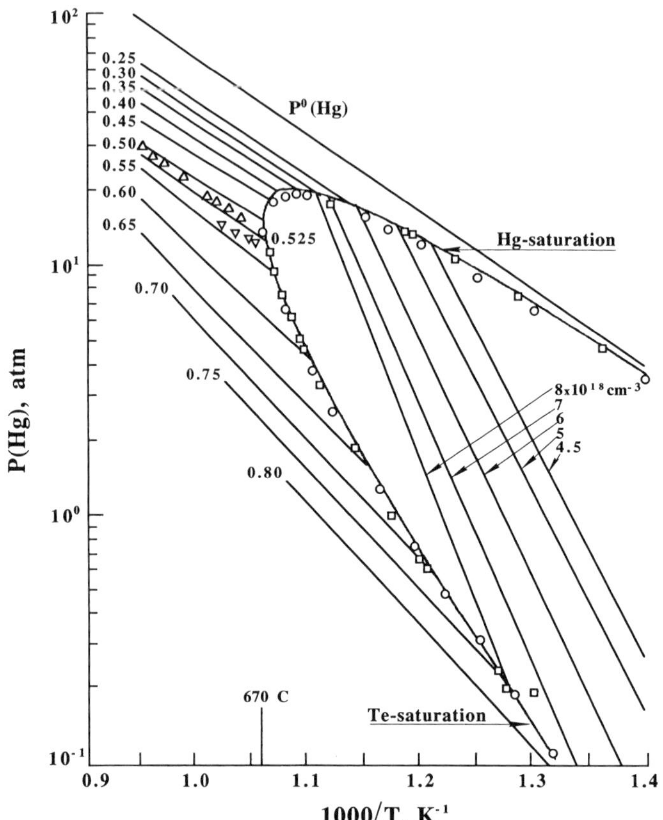

Fig. 73. *P*(Hg)–*T* projection of the Hg–Te system. Liquidus lines (outside of the three-phase equilibrium curve) are labeled according to the atomic fraction of Te. The composition of the solid is given inside the three-phase curve as the difference (*p–n*)

limation range of the non-stoichiometric HgTe in equilibrium with vapor, whereas outside of it is the equilibrium liquid–vapor for the compositions, cited on the individual curves. The points on the three-phase curve correspond to those on the *T–X* projection (Fig. 72) and describe the maximum non-stoichiometry of HgTe. The compositions in the liquid–vapor equilibrium are given in atomic fractions of tellurium, and for the sublimation region, the labels on the curves quote the composition in terms of the excess of the valence band holes over the conduction band electrons [28]. It is seen in Fig. 73 that the crystal composition at the maximum melting temperature is 52.5 at.% Te. The partial pressures of Te_2, also measured by Brebrick [53] but not shown in Fig. 73, are below the Hg partial pressures in the whole sublimation range of HgTe. Because they do not overlap, there is no solid HgTe composition, for which the vapor is 50 at.%. Hence, no congruent sublimation composition was observed for HgTe.

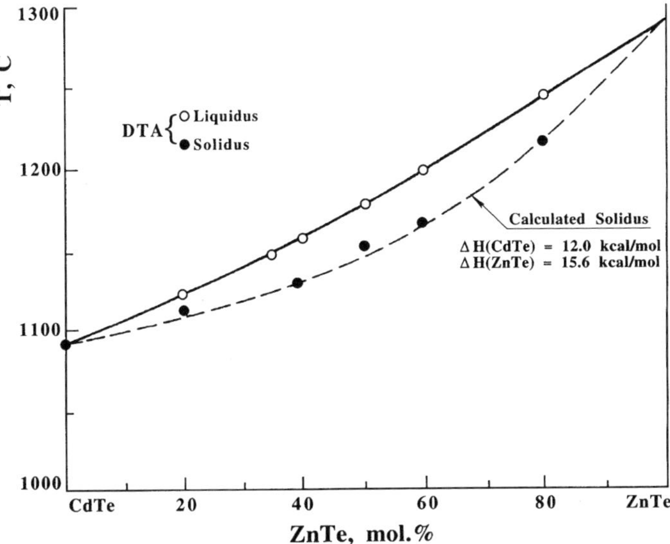

Fig. 74. *T–X* projection of the CdTe–ZnTe system

3.1.1.4 Ternary II–VI systems

The ternary zinc–cadmium–tellurium system was studied by Steininger et al. [127]. Experimental DTA measurements for the ternary Zn–Cd–Te liquidus and quasi-binary CdTe–ZnTe solidus were compared with calculations based on different solution models. The quasi-binary CdTe-ZnTe system [127] is shown in Fig. 74. DTA results for liquidus and solidus are given by the corresponding hollow and solid circles, and the calculations for the solidus (the dashed line) showed that the gap between the liquidus and solidus is very small (less than 0.16 mole fraction). This solidus was calculated in [127] from experimental DTA data for the liquidus and enthalpies of fusion (also quoted in Fig. 74), assuming ideal behavior in both solutions. The deviation of the calculated solidus temperatures from DTA was 3.8°C and resulted from small positive deviation from ideality in both solid and liquid solutions.

In a later publication [153], thermodynamic analysis of the ternary Zn–Cd–Te system was done on the basis of a subregular associated liquid model and the Onda–Ito model [154] in the pseudoregular approximation for quasi-binary solid solutions. A miscibility gap in ZnTe–CdTe was predicted below 428°C and experimentally observed by transmission electron microscopy (TEM) in MOCVD $Zn_xCd_{1-x}Te$ layers grown at 365°C. Subsequently [155], phase separation was also reported from TEM data for $Cd_{0.96}Zn_{0.04}Te$ bulk crystals. Apparently, more detailed studies of this system are called for, especially in connection with the ever growing applied importance of these materials.

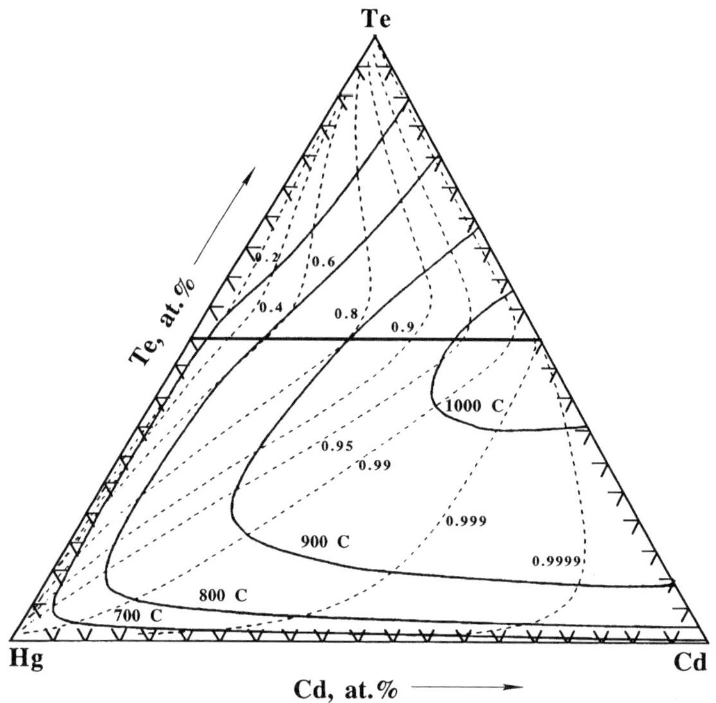

Fig. 75. Cd–Hg–Te system. Solid lines are the isotherms and dashed curves are the isoconcentration lines. Labels correspond to *x* values in $Cd_xHg_{1-x}Te$

The Hg–Cd–Te system (Fig. 75) was reported by Brebrick [28] on the basis of thermodynamic calculations using the associated solution model for the liquid phase. The solid lines in Fig. 75 are projections of the liquidus isotherms on the Gibbs triangle for temperatures listed on the curves, and the dashed lines correspond to solid solution isoconcentration lines. The compositions are given as *x* values in the formula $Cd_xHg_{1-x}Te$. CdTe–HgTe is assumed to be a quasi-binary system with no miscibility gap in the solid phase. Partial pressures of mercury along the three-phase curves for a number of $Cd_xHg_{1-x}Te$ solid solutions are presented in Fig. 76 according to Brebrick [156]. The compositions are given on the corresponding curves. Calto Brebrick [156]. The compositions are given on the corresponding curves. Calculations, based on the ideal solution of the Cd, Hg, Te, CdTe, and HgTe species in the liquid phase, are compared in Fig. 76 with the relevant available experimental data. Figures 75 and 76 are believed to be self-explanatory.

Some aspects of phase equilibrium in quaternary II–VI systems were discussed by Yu and Brebrick for Hg–Cd–Zn–Te [51] and Cd–Zn–Te–Se [53].

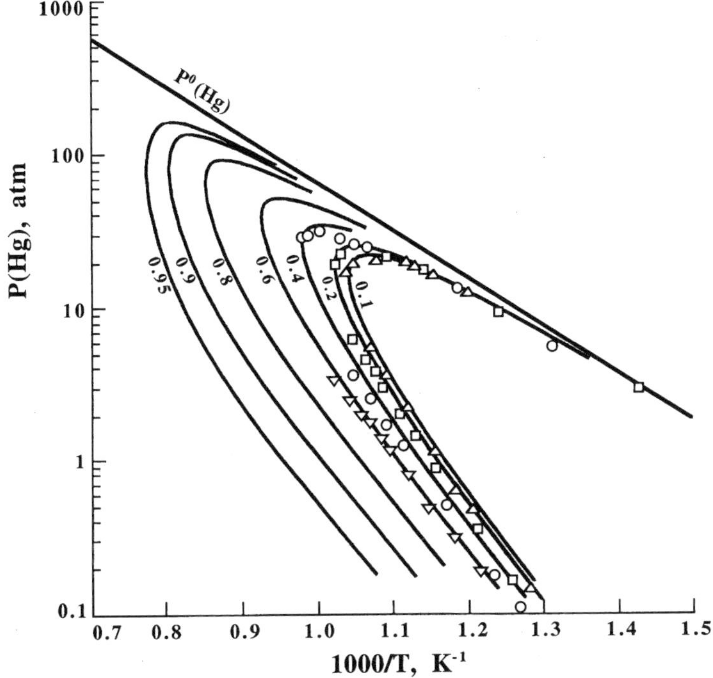

Fig. 76. Partial pressures $P(\mathrm{Hg})$ in three-phase equilibrium for solid solutions $\mathrm{Cd_xHg_{1-x}Te}$. The labels correspond to x values

3.1.2 III–VI compounds

The main applied interest of III–VI compounds, according to review [22], is their potential usage in optoelectronic and thermoelectric devices. Presently, gallium and indium chalcogenides, such as $\mathrm{CuInSe_2}$, $\mathrm{CuGaSe_2}$ and $\mathrm{Cu(In,Ga)Se_2}$, are primarily used as components of ternary photovoltaic materials.

P–T–X phase diagrams for the systems B–X and Al–X (X = S, Se, and Te) have not been studied. Some information on T–X diagrams can be found in reviews (see for example [22, 131]). In these systems, compounds with the stoichiometry $\mathrm{A^{III}_2B^{VI}_3}$ were identified. In addition, boron dichalcogenides $\mathrm{BS_2}$ and $\mathrm{BSe_2}$ were described by Medvedeva et al. [157] and Borjakova et al. [158]. Indium, gallium, and thallium chalcogenide systems have been studied in more detail. In all of these systems, compounds $\mathrm{A^{III}B^{VI}}$ and $\mathrm{A^{III}_2B^{VI}_3}$ were reported. Some general trends for the III–VI systems were described by Zlomanov and Novosyolova [22] when going from gallium down to thallium. (1) The stability of the $\mathrm{A^{III}B^{VI}}$ and $\mathrm{A^{III}_2B^{VI}_3}$ compounds in this series reduces: the melting temperature decreases, and the mode of melting changes from congruent for gallium and indium to incongruent for thallium. (2) The stability of the $\mathrm{A^{III}_2B^{VI}}$ solid compounds increases: this type of compound is not known for gallium and indium, whereas for thallium,

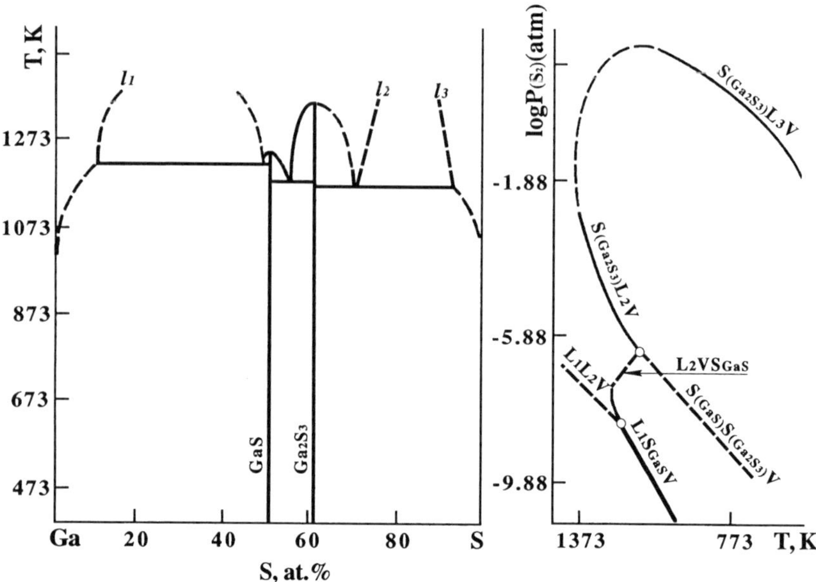

Fig. 77. *T–X* and *P–T* projections of the Ga–S diagram

these compounds reportedly melt congruently. (3) In the composition range 0–50 at.% a miscibility gap has been observed. *P–T* projections of the *P–T–X* phase diagrams have been reported only for Ga–S, Ga–Te and In–Se systems.

The *P–T–X* phase diagram for the gallium–sulfur system is shown in *T–X* and *P–T* projections in Fig. 77 [22]. Two gallium sulfides were identified: GaS with the congruent melting temperature of 1235 K and Ga_2S_3 with the congruent melting point of 1363 K. The eutectic mixture between them is 55 at.% in composition with the melting temperature of 1166 K. The miscibility gap in the liquid results in two invariant equilibria: $L_1L_2S(GaS)V$ at 1231 K and 10–48 at.% S, and $S(Ga_2S_3)L_2L_3V$ at 1160 K and 65–90 at.% S. The main vapor phase species in Ga–S are Ga_2S and S_2 [159,160]. Congruent sublimation of Ga_2S_3 below 1228 K was observed, whereas GaS sublimes incongruently. The *P–T* projection was constructed from DTA data with fixed vapor pressures of sulfur and investigation of the composition of the vapor, formed when samples were annealed in hydrogen [159]. The *P–T* projection (Fig. 77) is in need of more detailed studies.

The *T–X* diagram for the gallium–selenium system was studied in [161,162]. Two solid compounds were reported: GaSe melts congruently at 937±1°C, and the congruent melting temperature of Ga_2Se_3 is 1007±1°C. The eutectic coordinates were given as T_e=889±1°C, X_e=55.1±0.1 at.% Se. In the composition range 10 to 45.5 at.% Se, a miscibility gap in the liquid resulted in an invariant equilibrium $L_1L_2S(GaSe)V$ at 917°C, and the critical temperature of the miscibility gap was found at 994°C. The *P–T* projection for this diagram has not been reported. The

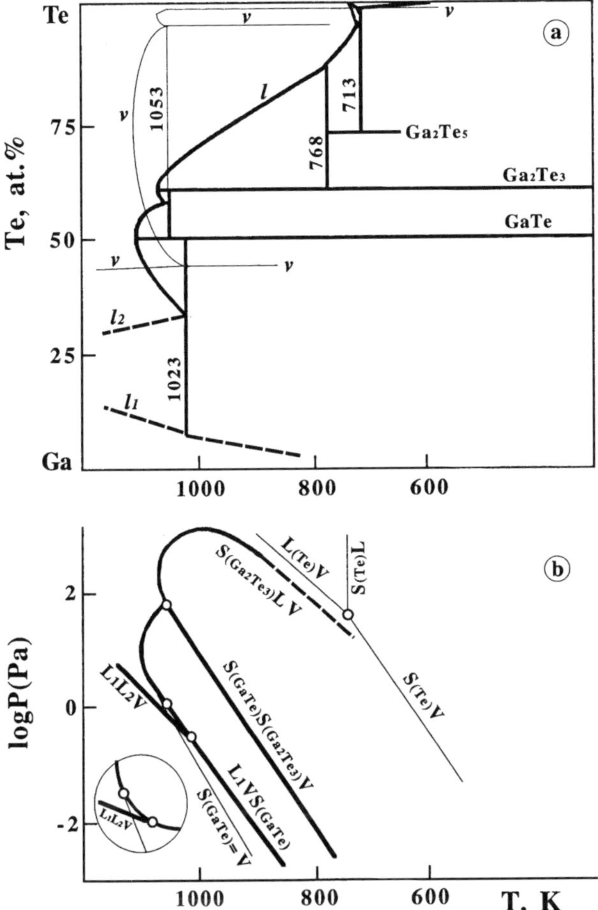

Fig. 78. *T–X* and *P–T* projections of the Ga–Te diagram

principal vapor species in this system are Ga_2Se and Se_2 [163]. Solid Ga_2Se_3 sublimes congruently, and GaSe sublimes incongruently.

The *P–T–X* phase diagram of the gallium–tellurium system is presented in *T–X* and *P–T* projections in Fig. 78 according to [164]. Reliable data were reported for GaTe and Ga_2Te_3: the maximum melting temperatures are 1123 K and 1083 K, respectively. The existence of the third gallium telluride, Ga_2Te_5, with an incongruent melting temperature of 768 K needs confirmation. Two eutectic equilibria were observed in this system: at T_e=1053 K, X_e=56 at.% Te and T_e=713 K, X_e=90 at.% Te. The miscibility gap in the liquid is seen in the *P–T–X* diagram (Fig. 78) as an invariant equilibrium $L_1L_2VS(GaTe)$ in the *T–X* projection, and the corresponding three-phase equilibrium curves in the *P–T* projection. The range of non-stoichiometry for

Ga_2Te_3 at 1053 K is 59.5–60.0 at.% Te, and it decreases at lower temperatures. The solidus and liquidus curves intersect at the congruent melting point: $T_{cm}(Ga_2Te_3)$=1071 K, X_{cm}=59.85 at.% Te. The DTA pattern of the sample containing 50 at.% Te revealed two endothermal effects at 1053 and 1099 K. The first is the eutectic melting temperature, and the second corresponds to the liquidus. Consequently, the single-phase region of existence of GaTe is on the Ga-side of the stoichiometric plane and does not include the stoichiometric composition. The vaporus curve (the thin line in the T–X projection) is drawn from mass spectrometric data [22].

The P–T projection (Fig. 78) was also constructed in [22] on the basis of mass spectrometric results. The major vapor-phase species were Ga_2Te and Te_2, and traces of $GaTe_2$ and Ga_2Te_2 were in the vapors. The minimum vapor pressure in the system S(GaTe) = V corresponds to the congruent sublimation of GaTe. The other gallium telluride, Ga_2Te_3, sublimes incongruently. The thin lines in the P–T projection are for the two-phase equilibria of pure tellurium. The phases in the three-phase equilibria are shown on the corresponding univariant curves in the P–T projection.

Indium–selenium system. The P–T–X diagram of the In–Se system was studied by Greenberg et al. [165] and Lazarev et al. [95] in the range of existence of In_2Se_3 by measuring the total vapor pressure for the three-phase equilibria of In_2Se_3 with liquid and vapor and the two-phase equilibrium of In_2Se_3 with vapor. The P–T projection of the sublimation region of In_2Se_3 is presented in Fig. 79 as a field within the three-phase equilibrium curve ABCD. The upper part of it (ABC) corresponds to the three-phase equilibrium between the selenium-saturated solid In_2Se_3, liquid (60 to 100 at.% Se in composition), and vapor. The lower portion of the loop (CD) is for indium-saturated In_2Se_3 in equilibrium with the melt (X_L < 60 at.% Se) and vapor. Both parts merge at the maximum melting point C, T_{max}, of the $In_2Se_{3\pm\delta}$ phase with specific deviation from stoichiometry δ and a fixed vapor pressure. The maximum melting temperature derived from the vapor pressure measurement, T_{max} = 888°C, coincided with the melting point obtained by DTA and, as will be shown in the following discussion, did not correspond to the exact stoichiometry of In_2Se_3. The vapor pressure at point C was $P(T_{max})$ = ~35 mmHg.

The three-phase equilibrium curve divides two two-phase equilibrium regions: solid–vapor (within the three-phase boundary) and liquid–vapor (outside of it). The maximum vapor pressure P_{max} in the three-phase equilibrium is at point B; its coordinates are $P \cong 1000$ mmHg, $T \cong 750$°C. The three-phase curve ABCD was obtained in [165] with samples of various compositions: 60 at.% to 63 at.% Se for the upper part and 59 at.% Se for the lower branch.

The P–T projection in Fig. 79 represents the total vapor pressure. To determine the molecular composition of the vapors in equilibrium with In_2Se_3, optical absorption spectra for the vapor phase were studied in the visible and IR regions [165]. At 850°C, the electronic spectra exhibited continuous absorption at 3200–3400 Å and bands of the Se_2 molecule in the region 3400–3800 Å. With increasing temperature (up to 950°C), continuous absorption intensified and extended from 2300 to 3800 Å; furthermore, lines of atomic In (4102 and 4511 Å) appeared in the spectrum. At 1200°C, only continuous absorption was recorded, which ex-

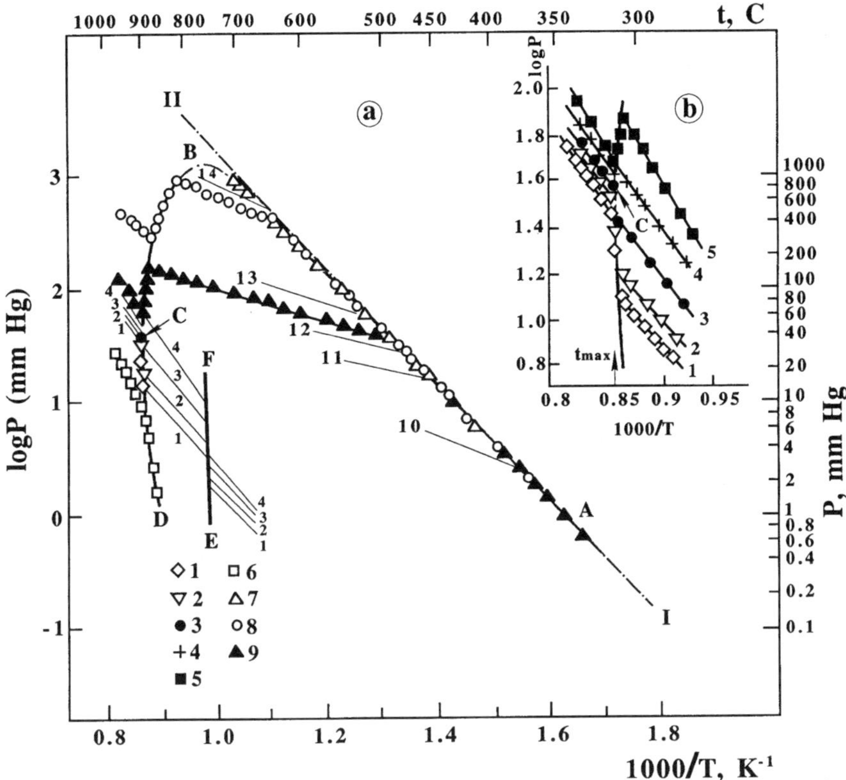

Fig. 79. *P–T* projection of the sublimation range of In_2Se_3. Composition of the samples (at.% Se): 60(1–4), 59.99(5), 59.0(6), 63.0(7), 62.0(8), 60.3(9). Concentration of $In_{32}Se_3$ (g/ml): 0.018(1), 0.025(2), 0.05(3), 0.10(4)

tended up to 4700 Å. The IR spectrum of the vapors above In_2Se_3 exhibits two bands: a strong one whith absorption maximum at about 235 cm⁻¹ and a weaker band with the maximum at about 385 cm⁻¹. These bands [165] correspond to the frequencies v_3 and $(v_3 + 2v_2)$ of the $In_2Se(g)$ molecule, respectively. The presence of this molecule is apparently responsible for the continuous absorption in the electronic spectra of the vapors.

Thus, spectroscopic analysis [165] revealed three species in the saturated vapor above solid In_2Se_3: In_2Se, Se_2, and In. In agreement with [166], according to the equilibrium

$$2In(g) + 0.5Se_2(g) = In_2Se(g),$$

the partial pressure of indium at 950°C due to this reaction cannot exceed ~10^{-3} mmHg. This vapor pressure could be detected by the method used in [165], as was shown by checking against the saturated vapor pressure of pure metallic indium. Same as in the sublimation of In_2O_3 and In_2S_3 [166,168], the partial vapor pressure of In(g) above In_2Se_3(s) was negligibly small compared with those of In_2Se(g) and Se_2(g).

These results, along with the known temperature dependences of the saturated vapor pressure above In_2Se(s) [167], showed that the total vapor pressure along almost entire three-phase curve ABCD is practically equal to the partial vapor pressures of the selenium species, since even at $T = 900°C$, the partial pressure of In_2Se(g) cannot be higher than a few mmHg.

Below ~550°C, the upper branch ABC of the three-phase equilibrium line does not differ noticeably from the saturated vapor pressure of pure selenium shown in Fig. 79 by the straight line I–II according to [142]. This agrees with the *T–X* diagram for the In–Se system [169], in which the (In_2Se_3 + Se) eutectic is degenerate ($T_{eut} = 220°C$) and the liquidus line in the temperature range 220–550°C is almost vertical. An assessment in the approximation of ideal solution showed that the solubility of indium in liquid selenium below 500°C is not expected to exceed ~3 at.%.

Non-stoichiometry of In_2Se_3.   The maximum selenium non-stoichiometry of In_2Se_3 was determined in [165] at temperatures of the phase-transition between the three-phase equilibrium $S(In_2Se_3)LV$ and the two-phase equilibrium $S(In_2Se_3)V$. The phase-transition points are registered as breaking points in the vapor pressure curves 1 to 14 in Fig. 79. The composition of the crystal at these temperatures was calculated from Eq. (25), which for this case is

$$X_S \text{ (at.\% Se)} = \left[N(\text{Se}) - n(\text{Se})\right]/\left[N(\text{In}) + N(\text{Se}) - n(\text{Se})\right] \times 100\%,$$

because the vapor may be taken as pure selenium. The validity of this approximation, along with the previous arguments, is seen directly in Fig. 79: the vapor pressure on the three-phase curve ABC is about three orders of magnitude higher than for the stoichiometric In_2Se_3 (curves 1 to 4) at corresponding temperatures. Therefore, to determine the composition of Se-saturated In_2Se_3, the mass of evaporated selenium $n(\text{Se})$ must be calculated from the total vapor pressure and the equilibrium constants of the polymerization reactions:

$$Se_n(g) = 0.5nSe_2(g).$$

The polymerization of selenium in the vapors was studied in [170–173]. According to [170,171], n = 2, 4, 6, and 8 in these reactions. Mass spectrometric study [172,173] showed that n = 2, 3, 5, 6, 7, and 8. Equilibrium constants for these two models of selenium vapors were used to calculate the partial pressures for six phase-transition points (curves 9 through 14 in Fig. 79) and, ultimately, the maximum Se non-stoichiometry of In_2Se_3 at these temperatures. The corresponding results are presented in Table 8 for the vapor-phase species and in Table 9 for the composition of the solid.

An important conclusion about the composition of the selenium vapor may be drawn from Table 9: calculations based on the data [172,173] show that the stoichiometric In_2Se_3 is single-phase throughout the entire temperature range presented, whereas using the [170,171] data results in the composition of the solid $X_S < 60$ at.% Se at 722, 748, and 791 K implying that the stoichiometric composition is a two-phase mixture ($In_2Se_{3-\delta}$ + Se) at these temperatures. This, however, is inconsistent with the experimental results of Fig. 79: as it is seen in Fig. 79, curves 1–4 for the stoichiometric composition lie within the two-phase sublimation range, and not on the three-phase equilibrium line. Hence, stoichiometric In_2Se_3 is single-phase throughout the entire temperature interval of Fig. 79. Thus, the experimental data of Fig. 79 confirm the composition of the selenium vapor reported by Berkowitz et al. [172,173], according to which n = 2,3,5,6,7,and 8 in Se_n.

Table 8. Partial vapor pressures (mmHg) of Se_n species in the three-phase equilibrium $S(In_2Se_3)LV$ according to different models of the vapor [170–173]

Curve#, Fig. 79	T (K)	Se_2	Se_3	Se_4	Se_5	Se_6	Se_7	Se_8	Model
10	648	0.304	0.011	—	1.466	0.474	0.206	0.039	[172,173]
		0.184	—	0.435	—	0.976	—	0.905	[171]
		0.241	—	0.647	—	1.067	—	0.544	[170]
11	722	2.736	0.106	—	9.675	1.876	0.849	0.159	[172,173]
		2.031	—	2.639	—	7.455	—	3.275	[171]
		2.387	—	3.359	—	6.824	—	2.830	[170]
12	748	5.219	0.206	—	16.699	2.774	1.268	0.235	[172,173]
		4.100	—	4.414	—	13.257	—	4.628	[171]
		4.678	—	5.434	—	11.720	—	4.567	[170]
9	755	6.227	0.246	—	19.369	3.084	1.412	0.261	[172,173]
		4.965	—	5.071	—	15.480	—	5.085	[171]
		5.626	—	6.194	—	13.582	—	5.198	[170]
13	791	13.941	0.562	—	38.041	4.981	2.304	0.420	[172,173]
		11.878	—	9.514	—	31.214	—	7.644	[171]
		13.071	—	11.284	—	26.540	—	9.355	[170]
14	915	145.30	6.117	—	259.32	18.87	8.860	1.531	[172,173]
		147.10	—	54.80	—	216.26	—	21.84	[171]
		150.08	—	61.75	—	175.53	—	47.64	[170]

Table 9. Solubility of selenium in β-In_2Se_3 (at.% Se) according to different vapor phase models [170–173]

T (K)	[172,173]	[171]	[170]
648	60.01	60.00	60.01
722	60.01	59.96	59.97
748	60.03	59.95	59.97
755	60.05	60.02	60.03
791	60.05	59.98	59.99
915	60.11	60.02	60.01

It can be seen in Table 9 that β-In_2Se_3 dissolves 0.01 to 0.11 at.% Se at temperatures 648 to 915 K and that the solubility increases with temperature. Rigorous calculation of the errors in the composition of In_2Se_3 was not done because of uncertainties in the composition of the vapor. However, even a qualitative examination of the temperature dependence of the vapor pressure for the samples with 60 at.% Se (curves 1–4 in Fig. 79) and 60.05 at.% Se (curve 10) confirms the results of Table 9: the maximum non-stoichiometry δ in $In_2Se_{3+\delta}$, according to Fig. 79, is within $0 < \delta < 0.05$ at.% because the vapor pressure for the former sample comes within the two-phase sublimation range solid–vapor, whereas for the latter sample, it is on the three-phase curve below 648 K. The solubility of indium in In_2Se_3 was estimated from curve 5: below 877°C it is not less than 0.01 at.%.

The vapor pressure curves 1–4, Fig. 79, for stoichiometric In_2Se_3 with different concentrations of the solid in the reaction volume pass through the γ–δ phase-transition range at temperatures 743–748°C. Part of the three-phase equilibrium $\gamma\delta V$ is seen in Fig. 79 as the curve EF. Calculation of the composition of the solid at the $\gamma\delta V \rightarrow \delta V$ phase-transition temperatures showed that the single-phase volume of δ-In_2Se_3 is on the In-side of the stoichiometric plane and the composition of the solid at the maximum melting temperature is 59.99 at.% Se.

3.1.3 IV–VI compounds

IV–VI compounds have a great variety of applications [22]. The most important are radiation detectors and emitters, such as high power lasers in "atmospheric windows" at 4–14μ and photodetectors with high radiation and thermal stability. These devices can work at temperatures above 77 K and are retunable due to dependence of the energy gap on the composition of the material. Other areas of application of IV–VI compounds are thermoelectric devices working at temperatures up to 800–900 K, optical filters, switches, etc. [22].

3.1.3.1 Silicon–chalcogen systems

P–T–X phase diagrams for silicon–sulfur and silicon–selenium are unknown. According to Abrikosov and Shelimova [174], the only silicon sulfide is SiS_2 whose congruent melting temperature is 1363 K; in the Si–Se system, Si_2Se_3 has been reported. In the silicon–tellurium system, Si_2Te_3 was observed [175] with incongruent melting point 1165 K. The composition of the melt at the invariant point of incongruent melting of Si_2Te_3 is $X_L = 62$ at.% Te. The coordinates of the eutectic in Si–Te are $T_e = 682$ K, $X_e = 80$–90 at.% Te. The range of the single-phase existence for Si_2Te_3 is $0.594 \leq X_S \leq 0.605$ [175].

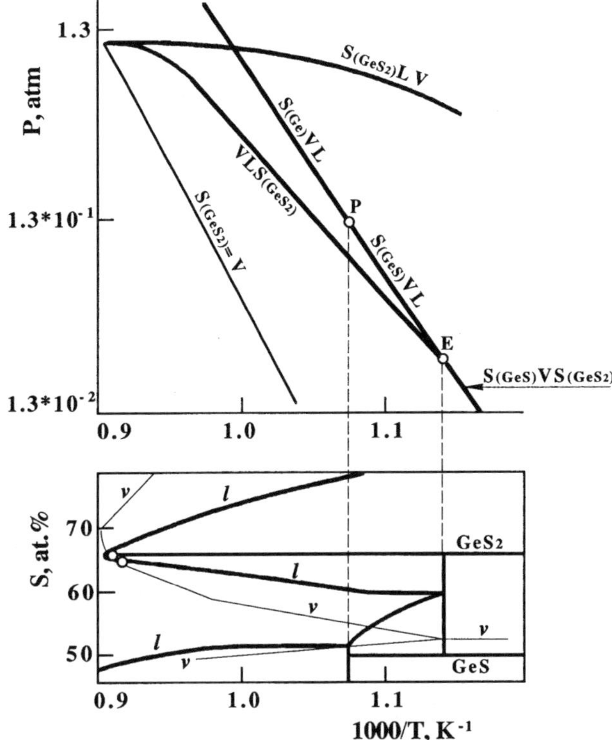

Fig. 80. *P–T* and *T–X* projections of the Ge–S diagram

3.1.3.2 Germanium–chalcogen systems

The *P–T–X* diagram of the **germanium–sulfur** system is presented in Fig. 80 in *P–T* and *T–X* projections according to [176,177]. Two compounds were found in the Ge–S system: monosulfide GeS and disulfide GeS_2. GeS melts incongruently at 931 K (point P in the *P–T* projection) into solid germanium and liquid containing 53 at.% S. GeS_2 melts congruently at 1113 K. The eutectic (GeS+GeS_2) melts at 870 K and corresponds to 58 at.% S. In the compositional range 3–45 at.% S (outside the limits of the *T–X* projection, Fig. 80), a miscibility gap in the liquid appears as an invariant equilibrium at 1193 K. The *P–T* projection and the vaporus curve in *T–X* (the thin line *v* in Fig. 80) are constructed from total vapor pressure measurements [178]. Vapor pressures in the three-phase equilibria S(Ge)VS(GeS), S(GeS)VS(GeS_2), and S(GeS)VL coincide within the experimental errors with the saturated vapor pressure of GeS(s) at temperatures 827 to 930 K. Because the vapor pressures for the heterogeneous regions (Ge + GeS), (GeS + GeS_2) and for GeS(s) are the same, it means that germanium monosulfide sublimes incongruently. It is consistent with the analysis of the condensed and vapor phases: metallic germanium and GeS were identified in the condensed phase along with GeS_2 in the vapors. Above the incongruent melting point of GeS (931 K), vapor pressure for compositions Ge_xS_{1-x}, x = 0.6–0.8, fall within the same line, which confirms the incongruent mode of melting for GeS(s). The three-phase equilibrium curves $VLS(GeS_2)$ and $S(GeS_2)LV$ were constructed from the vapor pressure measurements for samples $0.62 \leq x \leq 0.73$.

The saturated vapor pressure for $GeS_2(s)$ was measured at temperatures of 925 to 1044 K. It is lower than in the neighboring heterogeneous regions $VLS(GeS_2)$ and $S(GeS_2)LV$, (Fig. 80, *P–T* projection), meaning that germanium disulfide sublimes congruently, $S(GeS_2)$ = V. This is confirmed by the intersection of the vaporus line and the solidus of GeS_2 (Fig. 80, *T–X* projection). Comparison of the *P–T–X* arrangement for the $L(GeS_2)V$, $S(Ge)LV$, and $L(S)V$ curves suggests congruent vaporization of the GeS_2 melt.

Germanium–selenium system. The *T–X* projection of the germanium–selenium phase diagram is shown in Fig. 81 according to the results reported in [179–187]. Germanium monoselenide melts incongruently at 943 ± 3 K into solid Ge and liquid containing 50.7 ± 0.3 at.% Se. The non-stoichiometry of GeSe is within 0.2 at.% Se [179,182]. In the temperature range 923 to 937 K, an α–β phase-transition, most probably of an incongruent type (inset in the *T–X* projection, Fig. 81) was observed in GeSe [188]. At 860 K, GeSe and $GeSe_2$ form a eutectic mixture whose melting temperature is 860 K and composition is 56.5 ± 0.5 at.% Se. Vapor pressures of the samples in the composition range $0.2 \leq X \leq 0.55$ fall on the same curve at 583–831 K, meaning that GeSe sublimes incongruently. The $VLS(GeSe_2)$ equilibrium in the *P–T* projection, constructed from vapor pressure measurements of samples $X=0.62$ and $X=0.65$, was the same for both compositions up to the respective liquidus temperature. This confirms the previously cited composition of the $(GeSe+GeSe_2)$ eutectic. The vapor pressure for the three-phase equilibrium $S(GeSe_2)LV$ was measured with sample $X=70$ at.% Se. No breaking points were observed in this curve, which suggests the absence of a miscibility gap in the liquid for the $GeSe_2$ – Se part of the system.

The saturated vapor pressure for solid $GeSe_2$ is lower than in the adjoining heterogeneous fields $(GeSe + GeSe_2)$ and $(GeSe_2 + Se)$, meaning that $GeSe_2$ sublimes congruently (minimum vapor pressure $S(GeSe_2)$ = V in the *P–T* projection, Fig. 81). An unusual, "beak-like" shape of the three-phase equilibria $VLS(GeSe_2)$ and $S(GeSe_2)LV$ is a result of the partial pressures $P(Se_2)$ and $P(GeSe)$ being much higher than $P(GeSe_2)$ [22]. The vapor pressure for the liquid–vapor equilibrium presented in the *P–T* projection, Fig. 81, is for the liquid $GeSe_2$. It is lower than the vapor pressures of pure liquid selenium and the melt X_L = 50 at.% Se in composition, suggesting a congruent mode of vaporization (L = V curve in the *P–T* projection). The relative position of the three congruent lines in the near-melting region of $GeSe_2$ is shown in the insets on the *P–T* and *T–X* projections, Fig. 81 (the single-phase ranges of existence for the solids in the *T–X* insets are shaded). The near-vertical lines in the *P–T* projection correspond to condensed phase equilibria.

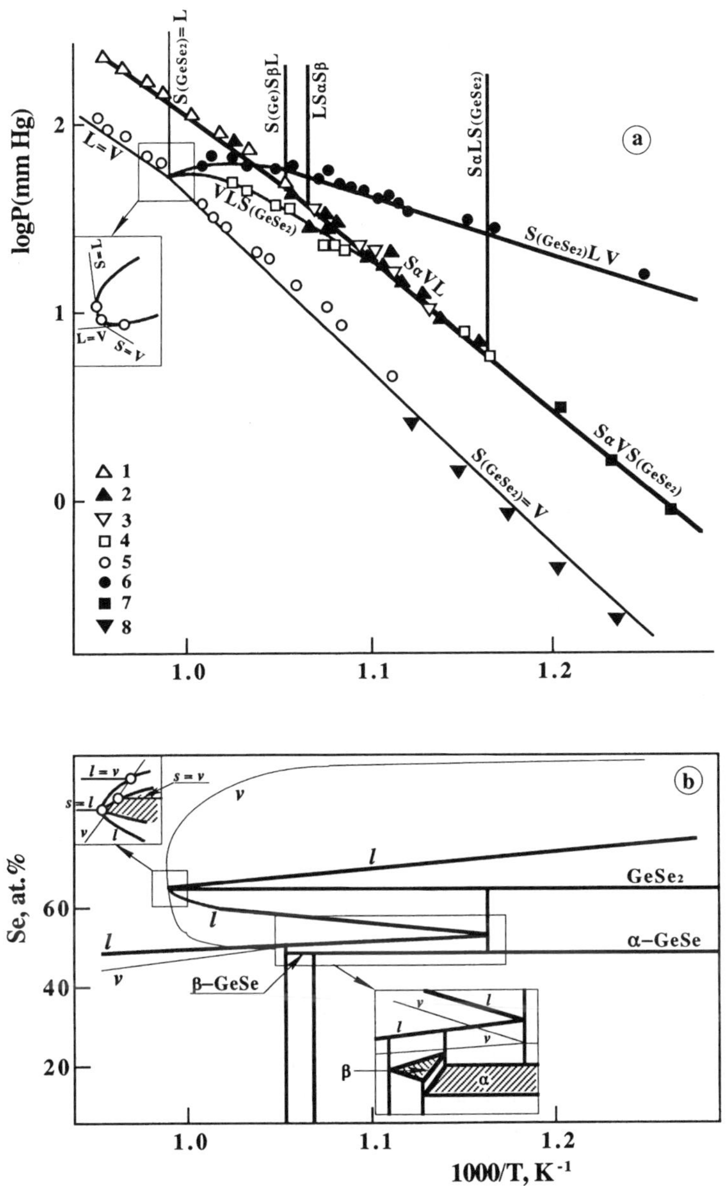

Fig. 81. *P–T* (**a**) and *T–X* (**b**) projections of the Ge–Se diagram. Composition of the samples (in at.% Se): 20–47(1), 50(2), 53–55(3), 62–65(4), GeSe₂(5), 70(6). 7,8 - [22]

Germanium–tellurium system. The only solid compound in this system is germanium monotelluride, GeTe. It melts congruently at 998 K and participates in two invariant equilibria, at $T = 648$ K, $X_e = 85$ at.% Tc and at $T = 996$ K, $X_e = 49.85$ at.% Te [174,179,189]. According to mass spectrometric studies [190], the predominant vapor phase species in Ge–Te are GeTe, GeTe$_2$ and Te$_2$. Projections of the P–T–X phase diagram Ge–Te are shown in Fig. 82 according to the results of [50,99,191]. The vapor pressure at the congruent melting point is 10^{-2} atm. Germanium monotelluride sublimes incongruently, although the vaporus curve (the thin line in the T–X projection, Fig. 82) is very close to the GeTe solidus. The maximum in the S(Ge)LV equilibrium can be explained by the temperature dependence of the condensed phase compositions [22]: $X_L \rightarrow X_S$ when $T \rightarrow T_{max}$. The small positive slopes of the condensed phase equilibrium curves in the P–T projection is connected with the increase in volume associated with the melting of GeTe and tellurium [22]. The details of the melting region of GeTe can be seen in the insets in Fig. 82.

3.1.3.3 Tin–chalcogen systems

Tin–sulfur system. In this system, two compounds were identified, SnS and SnS$_2$. Both of them melt congruently at 1154 and 1143 K, respectively [174,189] and give a eutectic mixture whose melting temperature is 1013 K and composition is $X_e = 55$ at.% S. The eutectics on both extremities of the system are degenerate, and the melting temperatures are 505 K (on the Sn-side) and 392 K (on the S-side). A miscibility gap in the melt appears in the diagram at 1133 K in the composition range $0.10 \leq X \leq 0.48$ and at 1143 K in the range $0.7 \leq X \leq 0.9$. The single-phase region of existence for tin monosulfide is on the S-side of the stoichiometric plane [192]. The vapor pressure and composition of the vapors in the Sn–S system were studied by total vapor pressure measurement [192] and mass spectrometry [193]. At the maximum melting temperature of SnS$_2$, the vapor pressure $P(T_{max}) = 40$ atm and for SnS(s) it is about 1.3×10^{-3} atm. In the eutectic equilibrium, $T_e = 1013$ K, $P_e = 2 \times 10^{-2}$ atm. The principal vapor phase species are SnS, S$_2$, Sn$_2$S$_2$, Sn$_3$S$_3$, and Sn$_4$S$_4$ [193].

Tin–selenium system. According to [174,189], two tin selenides, SnSe and SnSe$_2$, have been identified (Fig. 83). Tin monoselenide melts congruently at 1148±5 K. The non-stoichiometric range of SnSe, according to Dumon et al. [194], is less than 10^{-4} at.% and is on the Se-side of the stoichiometric composition. Tin diselenide has a congruent melting point of 929 K [189]. A miscibility gap in the liquid is seen in Fig. 83 at 1093 K in the composition range 18 to 49 at.% Se. The eutectic point between SnSe and SnSe$_2$ is at $T_e = 898$ K, $X_e = 61$ at.% Se. The P–T projection of the P–T–X phase diagram (Fig. 83) was constructed by Kulyukhina et al. [195]. The vapor pressure minimum in the sublimation range of SnSe$_2$ corresponds to the congruent sublimation of this compound, S(SnSe$_2$) = V, because the saturated vapor pressure for SnSe$_2$(s) is lower than that in the neighboring heterogeneous regions

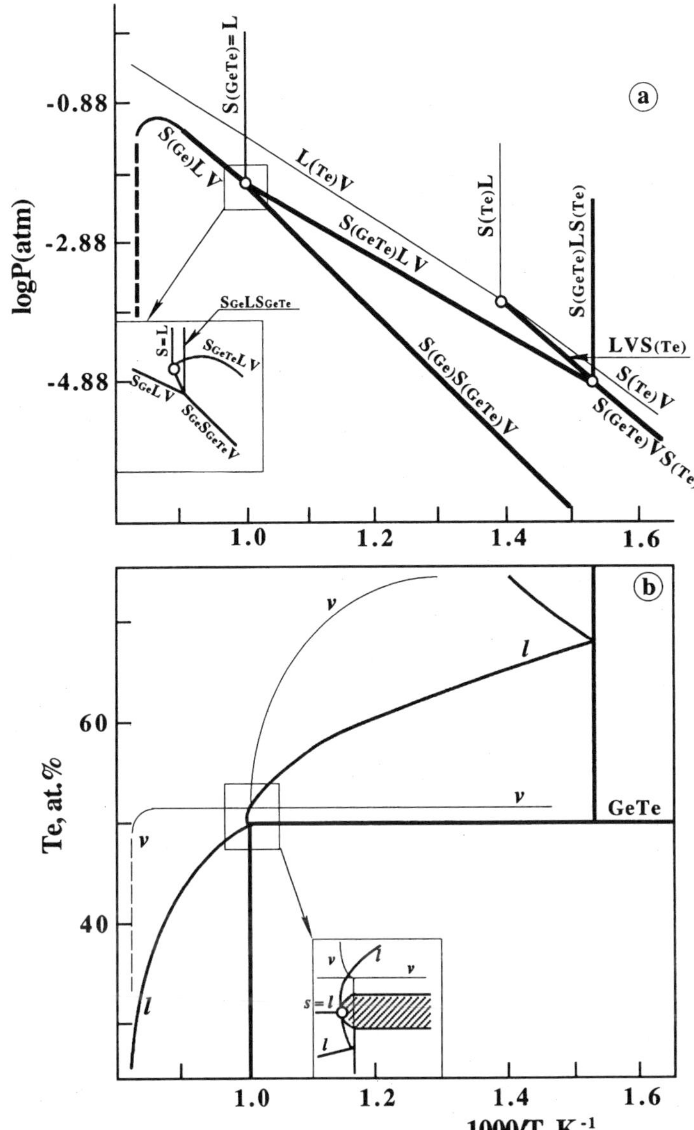

Fig. 82. *P–T* (**a**) and *T–X* (**b**) projections of the Ge–Te diagram

(SnSe$_2$+Se) and (SnSe+SnSe$_2$). The congruent sublimation line S(SnSe$_2$) = V touches the three-phase equilibrium curve S(SnSe$_2$)LV at T = 913 K; above this temperature the vapor pressure increases sharply, because the system passes through the S(SnSe$_2$) = V equilibrium toward LS(SnSe$_2$)V (Fig. 83). In the *T–X* projection,

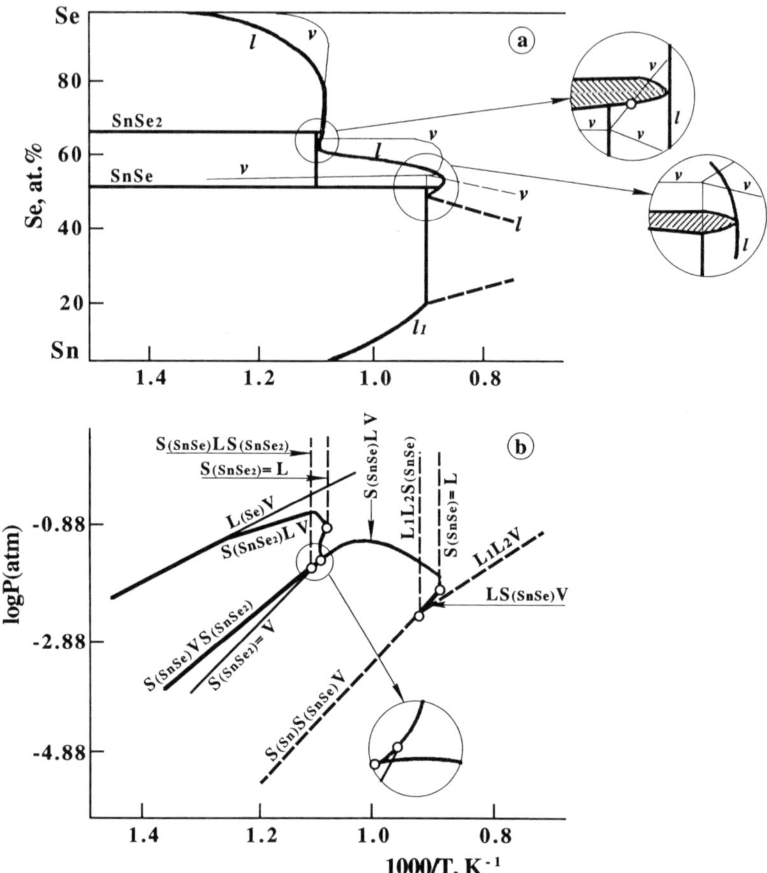

Fig. 83. *T–X* (**a**) and *P–T* (**b**) projections of the Sn–Se diagram

the maximum congruent sublimation point is seen in the upper inset as an intersection point of the vaporus curve v and the $SnSe_2$ solidus (the shaded area in the inset).

Tin monoselenide sublimes incongruently. This was proved [195] by the identification of metallic tin in the residue, analyzed after the sublimation of SnSe. According to mass spectrometric data [195], the saturated vapor above SnSe(s) consists mainly of SnSe(g) and Se_2(g) molecules, and consequently, the vapor is enriched in selenium compared to solid SnSe. In the lower inset in the *T–X* projection, Fig. 83, it can be seen that the vaporus curve v does not cross the shaded region of the single-phase existence of SnSe. The vaporus curve in the *T–X* projection was constructed mainly from mass spectrometric results. The near vertical lines in the *P–T* projection describe the pressure dependence of the melting temperatures for the invariant equilibria. Details of the phase equilibrium in the melting regions of tin selenides are seen in the insets in the *P–T* and *T–X* projections, Fig. 83.

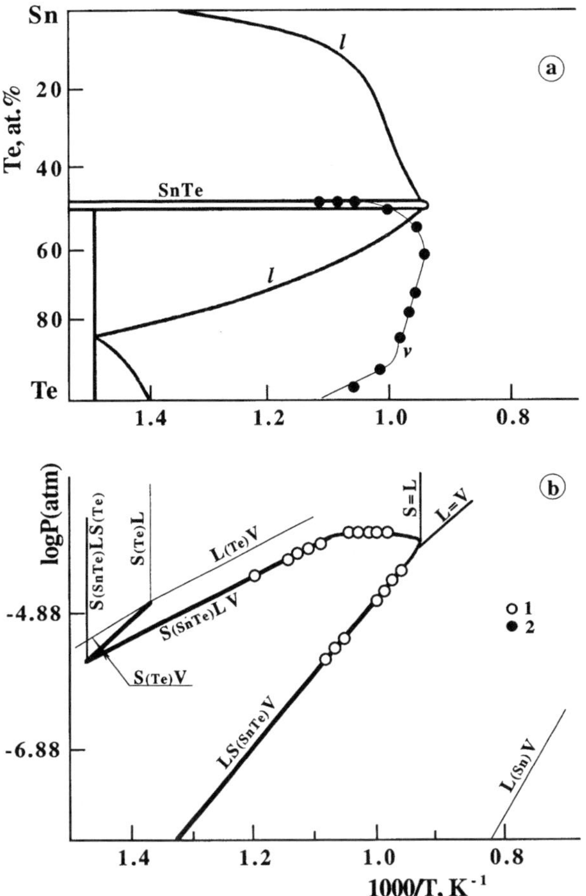

Fig. 84. T–X (**a**) and P–T (**b**) projections of the Sn–Te diagram. 1-calculated composition, 2-calculated pressure

Tin–tellurium system. The T–X projection of the P–T–X phase diagram for tin–tellurium is presented in Fig. 84 according to the data compiled in [174,189,196]. The only chemical compound in this system is tin monotelluride. The liquidus curve was constructed from DTA and calorimetric data [197,198]. The congruent melting temperature for SnTe is 1079 K. Two eutectic invariant equilibria were observed in Sn–Te at 678 K on the Te-side with $X_e = 85$ at.% Te, and a degenerate eutectic on the Sn-side. The single-phase range of existence for tin monotelluride is on the Te-side of the stoichiometric composition; at $T = 873$ K, it extends from 50.1 at.% Te to 51.1 at.% Te. It can be seen in the P–T projection, Fig. 84, that SnTe sublimes incongruently because no vapor pressure minimum was observed in the sublimation region of this phase. The principal vapor phase species in this system are SnTe, Te_2, $SnTe_2$, Sn_2Te_2, and Sn_3Te_3 [22]. Because the density of the solid SnTe is greater than that of the liquid, the congruent melting curve S(SnTe) = L has a positive slope dP/dT. Hence, the congruent melting temperature for SnTe is lower than the maximum melting point, and the S(SnTe) = L curve touches the three-phase equilibrium line on the LS(SnTe)V side. The saturated vapor pressure for the liquid SnTe, shown in the P–T projection, Fig. 84, corresponds to congruent vaporization L = V.

3.1.3.4 Lead–chalcogen systems

Lead–sulfur system. The P–T–X phase diagram of this system [199] was among the first reported for semiconductor systems. It has been extensively used by Kroger [20] in his discussion of the defect structure of lead sulfide. P–T and T–X projections of the P–T–X diagram for lead–sulfur are presented in Fig. 85 according to the results compiled in [174,189,199]. The only compound in this system is lead sulfide (PbS) whose congruent melting temperature is 1400 K. In the composition interval $0.705 \leq X \leq 0.991$ and $T = 1072$ K, a miscibility gap was registered in the liquid phase. The range of single-phase existence of lead sulfide is on both sides of the stoichiometric plane and goes up to 3×10^{-4} atomic fractions of both components. It can be seen in the T–X projection, Fig. 85, where the compositional axis is given on the logarithmic scale in terms of the deviation from the stoichiometric composition a. The solubility of both sulfur and lead in PbS is retrograde. At the maximum melting temperature T_{max}(PbS) = 1400 K the compositions of the solid S, liquid L, and vapor V are different, which is clearly seen in the inset in the T–X projection: the vapor is almost pure sulfur, whereas the solid and melt are both on the Pb-side of the stoichiometric composition with about 10^{-4} atomic fraction of Pb excess. Three congruent processes were observed in the Pb–S system: congruent melting S(PbS) = L, congruent sublimation S(PbS) = V, and congruent vaporization L = V. The spatial arrangement of the corresponding curves in the melting region of PbS is shown on an enlarged scale in insets in the P–T and T–X projections, Fig. 85. It can be seen in the P–T projection that the congruent sublimation curve does not correspond to the stoichiometric composition PbS. Sublimation of the latter is inside the three-phase loop and intersects the VLS(PbS) three-phase equilibrium at $T = 1350$ K, whereas the congruent sublimation composition proved to have an excess of lead. This is the reason that PbS crystals and films grown from the vapor phase always have n-type conductivity. The principal vapor phase species in the lead–sulfur system are Pb(g) and S_2(g), along with sulfur polymers.

Lead–selenium system. The P–T–X phase diagram for this system is shown in projections in Fig. 86 according to [174,179,189]. The only compound in the system is lead selenide whose congruent melting temperature is 1353 K. At $T = 953$ K, a miscibility gap in the liquid was observed in the composition interval $0.76 \leq X \leq 0.998$ (Fig. 86, T-X projection). Along with congruent melting, congruent sublimation was observed for PbSe, which is the vapor pressure minimum S(PbSe) = V in the sublimation region of the PbSe phase (Fig. 86, P–T projection). This two-phase sublimation region is within the three-phase equilibrium curves L_1VS(PbSe),

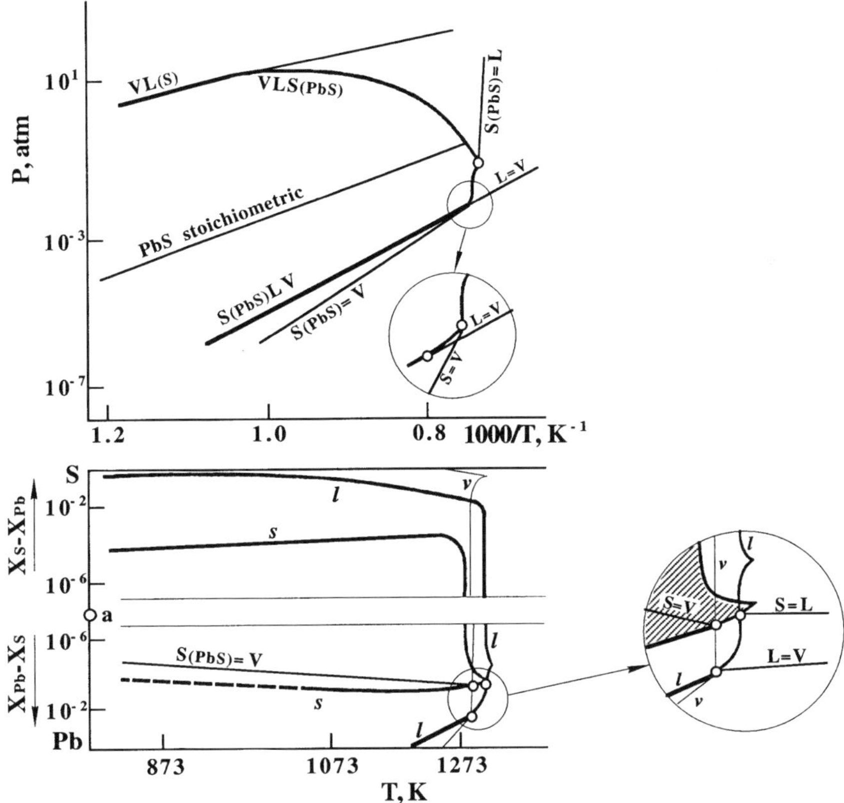

Fig. 85. P–T and T–X projections of the Pb–S diagram

S(PbSe)L$_2$V, S(PbSe)L$_3$V, and S(PbSe) = V (below the maximum sublimation point). Because the density of the solid PbSe is higher than that of the liquid, the congruent melting curve S(PbSe) = L has a positive P–T slope. Consequently, the congruent melting point is lower than the maximum melting temperature and touches the L$_1$ S(PbSe)V branch of the three-phase equilibrium curve. The L(Pb)V line in the P–T projection is for vaporization of pure liquid lead. The vaporization of pure selenium below the monotectic temperature coincides with the three-phase equilibrium curve S(PbSe)L$_3$V, and above this temperature, it is the same as for the L$_2$L$_3$V equilibrium. Mass spectrometric analysis of the vapors of Pb–Se [200] showed that, along with Pb(g) and Se$_2$(g), several heteroatomic species were present in the vapor phase: PbSe, PbSe$_2$, Pb$_2$Se$_2$, Pb$_2$Se, and Pb$_2$Se$_3$.

Lead–tellurium system. The shape of this system (Fig. 87 [22]) is similar to that of Pb–Se. The only compound in the system is lead telluride PbTe, whose congruent melting temperature is 1197 K. The composition of the solid at the congruent melting

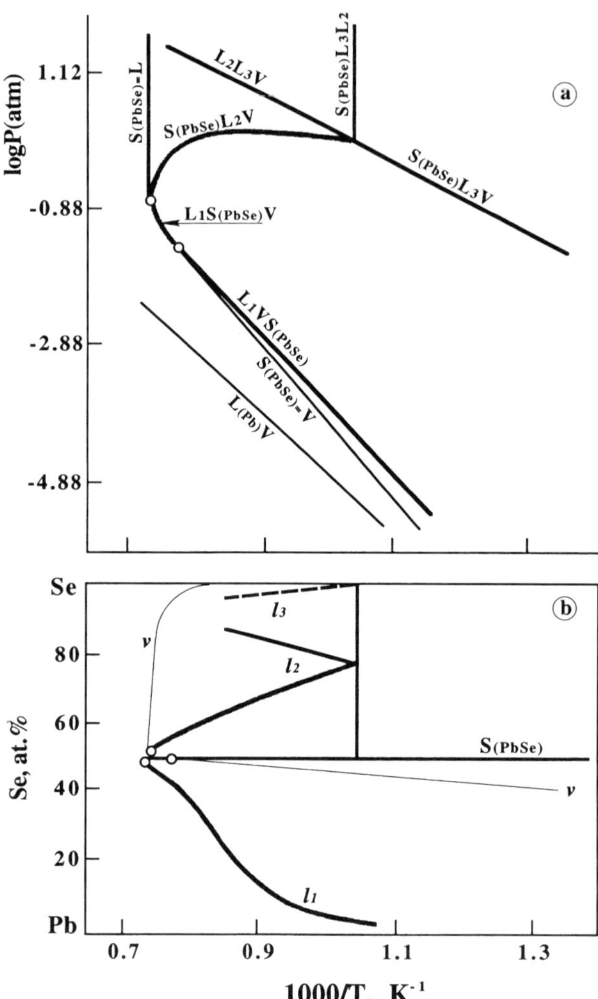

Fig. 86. *P–T* (**a**) and *T–X* (**b**) projections of the Pb–Se diagram

point is $X_{cm} = 50.012$ at.% Te, i.e., on the Te-side of the stoichiometric plane. Two invariant eutectic equilibria correspond to $T_e = 600$ K, $X_e = 0.08–0.16$ and $T_e = 686$ K, $X_e = 0.893$. The defect structure of PbTe is discussed in detail in review papers [179,200]. The *P–T* projection was constructed in Fig. 87 from total vapor pressure measurements, mass spectrometric data, and calculations, based on the associated melt approximation [22]. The coordinates of the invariant points for this system, cited by Zlomanov and Novosyolova [22], are the following. For S(Pb)LVS(PbTe): $T = 599.7$ K, $P = 1.1 \times 10^{-11}$ atm, $X_V = 0.3598$, $X_L = 0.0012$, $X_{S(PbTe)} = 0.49999$; for the

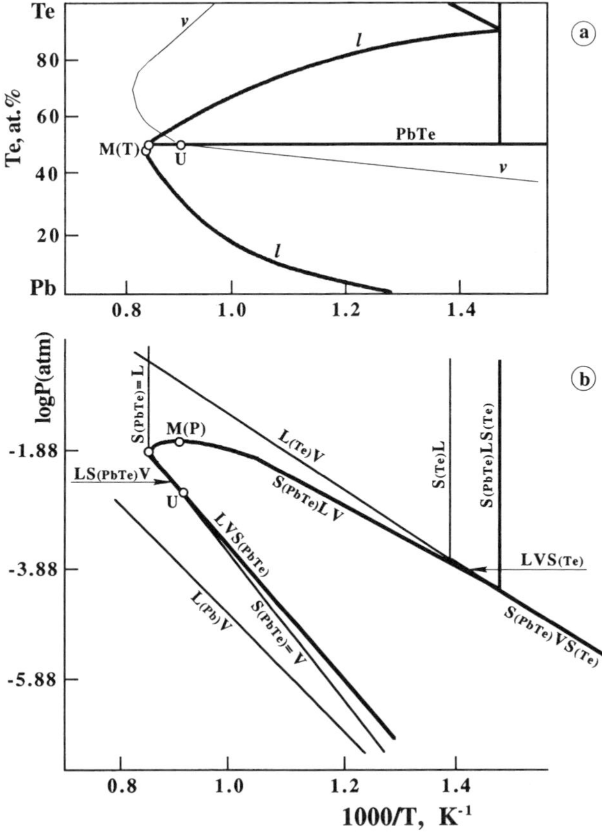

Fig. 87. $T–X$ (**a**) and $P–T$ (**b**) projections of the Pb–Te diagram

maximum sublimation point U: $T = 1113$ K, $P = 2 \times 10^{-3}$ atm, $X_V = X_S = 0.50008$, $X_L = 0.35$; for the congruent melting point of PbTe(s): $T = 1197$ K, $P = 1.7 \times 10^{-2}$ atm, $X_S = X_L = 0.50012$, $X_V = 0.6010$; for S(PbTe)LVS(Te): $T = 679$ K, $P = 7 \times 10^{-5}$ atm, $X_V = 1.0$, $X_L = 0.855$, X_S(PbTe) $= 0.50008$ (all of the compositions are in atomic fractions of tellurium). It should be stressed that the composition of the vapors at the congruent melting point of PbTe is essentially different from that of the condensed phases: the vaporus curve v at the M(T) temperature is strongly shifted towards tellurium (Fig. 87, $T–X$ projection). The pressure dependence of the congruent melting temperature for PbTe has a small negative slope.

The vapor pressure minimum in the region of single-phase existence of PbTe(s) corresponds to the congruent sublimation S(PbTe) = V (Fig. 87, $P–T$ projection). The vapor phase consists of the following species: $Te_2(g)$, $Pb(g)$, $PbTe(g)$, $PbTe_2(g)$, $Pb_2Te(g)$, $Pb_2Te_2(g)$, $Pb_2Te_3(g)$, and $Pb_3Te_2(g)$. In the $P–T$ projection, L(Pb)V is the vaporization curve of pure lead, and L(Te) is that for pure tellurium, which approaches the S(PbTe)LV curve at low temperatures. The congruent sublimation composition for PbTe was shown to have an excess of tellurium. This means that the material grown from the vapor phase is expected to be of p-type conductivity, and to prepare stoichiometric or n-PbTe, it should be annealed in Pb vapor.

3.1.4 V–VI compounds

3.1.4.1 Arsenic chalcogenides

Arsenic–chalcogen phase diagrams have not been studied in sufficient detail because of extended regions of glassy states in these systems. Arsenic reacts with chalcogens to produce two types of solid compounds, $As_2B_3^{VI}$ and AsB^{VI}. The stability of the $As_2B_3^{VI}$ compounds increases in the series B^{VI} = S, Se, Te, whereas for the AsB^{VI} compounds, it decreases in the same sequence [131,189]. $As_4S_4(s)$ melts incongruently at 569 K to form $As_2S_3(s)$ and the liquid $X_L \cong 48$ at.% S in composition. At $T = 576$ K in the composition range $0.48 \leq X \leq 0.58$ (in atomic fractions of sulfur) and at $T = 1048$ K, $X < 0.35$ a miscibility gap in the melt was observed. Arsenic and As_4S_4 at 452 K give a eutectic mixture whose composition is $X \cong 45$ at% S. The third arsenic sulfide, As_2S_5, was prepared at high pressure [22].

3.1.4.2 Antimony chalcogenides

In the antimony–chalcogen systems, the only solid compound is $Sb_2B_3^{VI}$. The other type, $Sb_2B_5^{VI}$, is thermally unstable; these compounds were prepared either by precipitation from aqueous solutions or at high pressure [22]. The stability of antimony chalcogenides increases in the series B^{VI} = S, Se, Te. The miscibility gap formation in the liquid decreases in the same sequence: in the Sb–S system, a miscibility gap was observed in the two-phase fields Sb–Sb_2S_3 and Sb_2S_3–S, in Sb–Se, it is only in Sb–Sb_2Se_3, whereas in Sb–Te, no miscibility gap was found. In the antimony–tellurium system, solid solutions are formed in composition ranges $0.173 \leq X \leq 0.369$ (β-phase has a minimum melting temperature), $0.409 \leq X \leq 0.538$ (γ-phase), and the δ-phase, the Sb_2Te_3 based solid solution with a range of single-phase existence on the Sb-side of the stoichiometric composition. The vapor phase in these systems is made up mainly of $B_2^{VI}(g)$ and $(SbB^{VI})_n(g)$ species with n = 1 to 4 [22].

3.1.4.3 Bismuth chalcogenides

In the **bismuth–sulfur** system (Fig. 88), the only compound is Bi_2S_3 whose congruent melting temperature is 1033 K [131,189,201]. At $T = 1000$ K a miscibility gap in the liquid phase was registered. The homogeneity range for bismuth sulfides includes the stoichiometric composition (Fig. 88c) and extends up to 2×10^{-3} atoms per mole on both sides. The solubility of both components is retrograde. At

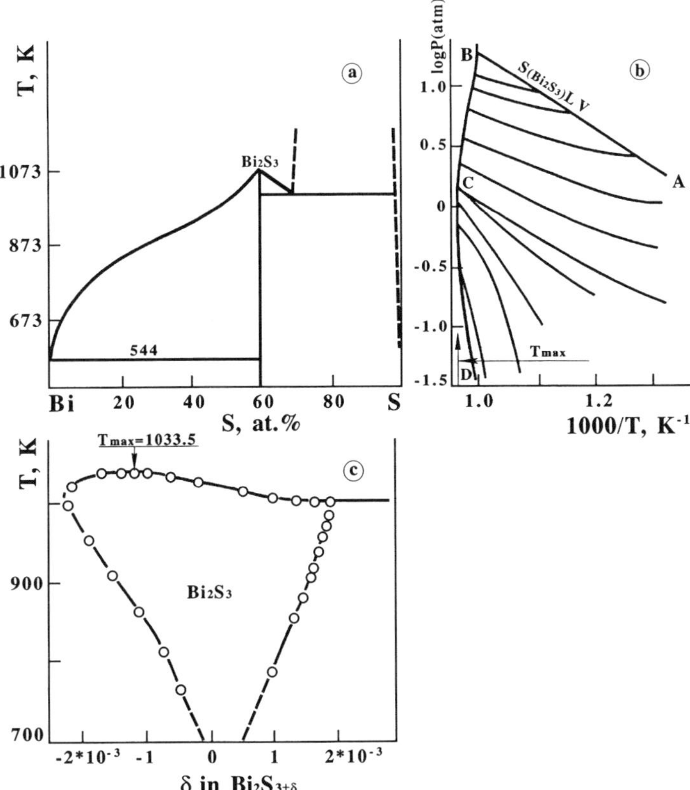

Fig. 88. T–X (**a**), P–T (**b**) projections of the Bi–S diagram and the non-stoichiometry range of Bi_2S_3 (**c**)

the maximum melting temperature, the composition of the solid is on the Bi-side of the stoichiometry (1×10^{-3} at. Bi/mole). The P–T projection, Fig. 88, was constructed from total vapor pressure measurement [201]. The three-phase equilibrium $S(Bi_2S_3)LV$ practically coincides with the saturated vapor pressure for pure sulfur, implying that solubility of Bi in liquid sulfur is negligible. The ABCD curve is for the three-phase equilibrium $S(Bi_2S_3)LV$, and inside of it is the two-phase sublimation region $S(Bi_2S_3)V$. Vapor pressures for Bi–S melts of different composition were measured by Cubicciotti [202]. Bismuth sulfide sublimes incongruently; the principal vapor phase species are $S_2(g)$, $Bi_2(g)$, $BiS(g)$, and $Bi_2S_2(g)$.

The T–X diagram for the **bismuth–selenium** system (Fig. 89, [22]) contains three solid **bismuth selenide** compounds: Bi_2Se, $Bi_{1-x}Se_x$ ($0.413 \leq X \leq 0.555$) and Bi_2Se_3. The first two melt incongruently at 741 and 880 K respectively,

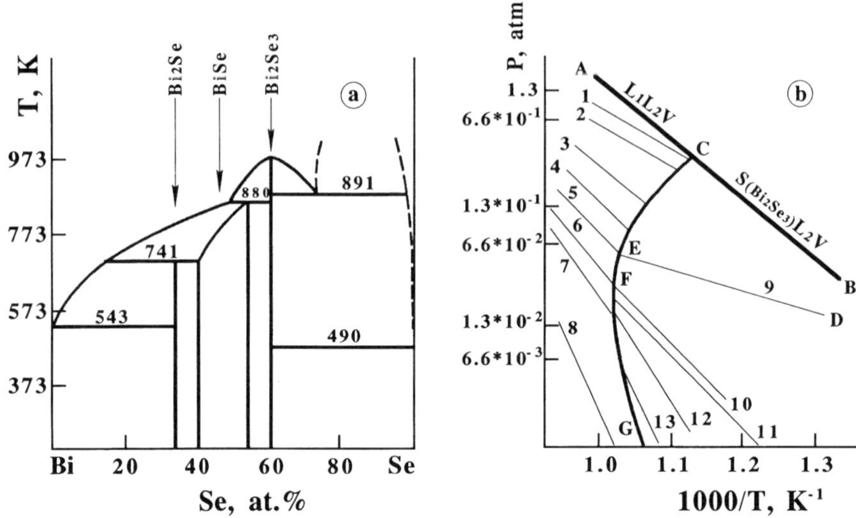

Fig. 89. *T–X* (**a**) and *P–T* (**b**) projections of the Bi–Se diagram. Composition of the samples (at.% Se): 72.6(*1*), 70.2(*2*), 66.2(*3*), 63.8(*4*), 61.8(*5*), 60.0(*6*), 58.8(*7*), 53.1(*8*). Curves *9–13* represent equilibrium S(Bi_2Se_3)V

whereas Bi_2Se_3 has a congruent melting temperature of 979 K. At the maximum melting point, the composition of the solid is 0.02 at.% on the Bi-side of the stoichiometry. Invariant equilibrium at 891 K corresponds to the miscibility gap in the liquid phase in the composition interval $0.71 \leq X_L \leq 0.99$ [131]. The *P–T* projection (Fig. 89) is given according to the "dew point" measurement [203] of the selenium vapor pressure for the three-phase equilibrium S(Bi_2Se_3)LV and two-phase equilibria S(Bi_2Se_3)V and L($Bi_{1-x}Se_x$)V. The former is seen in the *P–T* projection as the BCEFG curve; sublimation is inside and vaporization outside of it. The main vapor phase species in this system are $Bi_2(g)$, $Se_2(g)$, and BiSe(g).

In the **bismuth–tellurium** system (Fig. 90, [22]), three solid phases were reported. Bismuth telluride Bi_2Te_3 (δ-phase) has a congruent melting point at 858 K and forms a eutectic mixture with tellurium at $T_e = 686$ K, $X_e = 0.9$. The composition of the melt at the maximum melting temperature of Bi_2Te_3 is on the Bi-side of the solid, $X_L(T_{max}) = 59.95$ at.% Te, and the melting temperature increases with growing pressure. The single-phase range of existence for Bi_2Te_3 at $T = 733$–793 K is 0.4 at.% in composition, and it is symmetrical relative to the stoichiometric plane. Along with Bi_2Te_3, two more bismuth tellurides are known: Bi_2Te (β-phase with the composition $0.322 \leq X \leq 0.330$) has an incongruent melting point at 713 K, and the peritectic equilibrium at 836 K corresponds to the incongruent melting of BiTe (γ-phase with the composition $0.456 \leq X \leq 0.547$). The vaporus curve *v* in the *T–X* projection, Fig. 90, is shown according to the mass spec-

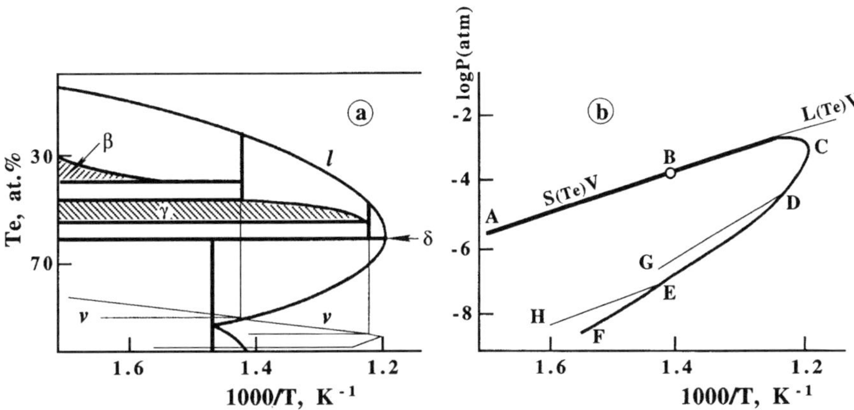

Fig. 90. $T–X$ (**a**) and $P–T$ (**b**) projections of the Bi–Te diagram

trometric data. It does not intersect the solidus lines of the bismuth tellurides, meaning that all of them sublime incongruently.

The $P–T$ projection, Fig. 90, was constructed from mass spectrometric and optical absorption measurements [204,205]. The vapor phase is made up mainly of $Te_2(g)$, $Bi_2(g)$, and $BiTe(g)$. In the $P–T$ projection, the following three-phase equilibrium curves are shown: BCD is for $S(Bi_2Te_3)LV$, DG corresponds to $S(Bi_2Te_3)S(BiTe)V$ $(S_\delta S_\gamma V)$, DE is for $LS(BiTe)V$ $(LS_\gamma V)$, EH represents $S(Bi_2Te)S(BiTe)V$ $(S_\beta S_\gamma V)$, and EF is for $LS(Bi_2Te)V$ $(LS_\beta V)$. The $P–T$ stability regions of $S(Bi_2Te_3)$, $S(BiTe)$, and $S(Bi_2Te)$ are within ABCDG, GDEH and HEF, respectively. The vapor phase for Bi–Te liquid alloys is enriched in tellurium. The partial pressure of $Te_2(g)$ for the melt $X_L = 0.19$ at the beginning of vaporization is greater than that of all of the other species, but it gradually decreases and at the end of vaporization becomes less than $P(BiTe)$ and $P(Bi)$. A similar relation between the partial pressures, $P(BiTe) > P(Bi) > P(Te)$, was observed for the melt $X_L = 0.096$. Consequently, a vapor pressure minimum is anticipated in the composition interval $0.01 \leq X_L \leq 0.19$ [22].

3.1.5 VI–VI compounds

The VI Group transition metals form a number of chalcogenides with various stoichiometry. Application of these materials is in different fields. MX_2 compounds, where M is tungsten or molybdenum, and X is sulfur or selenium, have a 2D layered structure that provides good lubricating properties. It has been shown [206] that nanoparticles of these compounds are unstable in a planar configuration and spontaneously tend to form hollow-cage fullerene-like structures. Numerous applications were predicted for these nanostructures. In particular, recently WS_2 nanotubes were used as tips in scanning probe microscopy [207]. Detailed vapor pressure scanning study of non-stoichiometry in Cr_2Se_3 [208] was triggered by interest in the photoferromagnetic properties of cadmium and mercury seleno-chromites. These properties strongly depend on stoichiometry, and Cr_2Se_3 is a component part of these ternary compounds.

The formation of non-stoichiometric phases in the Cr–Se system is due to the ordering of chromium vacancies in the NiAs structure. The study of phase diagrams for such systems by conventional DTA and XRD methods is difficult because of the high probability of non-equilibrium states in different degrees of ordering and the limited applicability of quenching methods. In such cases, the most reliable way to study the phase equilibria is to use static methods with control of the vapor pressure of the volatile component. For the Cr–Se system, the *T–X* diagram was reported by Haraldsen [209] and Babitzina et al. [210]. Information on chromium selenides, their crystal structure, temperature, and composition boundaries of existence is contradictory. In the single-phase existence region of the compound Cr_2Se_3, two phases were found [210] with the hexagonal structure of the NiAs type: a low-temperature ordered γ-phase with a homogeneity range of about 1 at.%, stable up to 811°C and a high-temperature α-phase with disordered structure, whereas in [209] only the γ-phase was observed. In the same composition region, depending on the preparation conditions (temperature and selenium vapor pressure), phases were reported that had the approximate composition of Cr_2Se_3 with rhombohedral [211,212], monoclinic, and trigonal structures [213], as well as the phase $Cr_{0.68}Se$ [212,213] with a structure identical to that of the γ-phase [209,210].

The *P–T–X* phase diagram for the Cr–Se system was studied by Zhegalina et al. [208] by vapor pressure measurements; single-phase ranges of existence for crystalline phases were determined by vapor pressure scanning of the solidus in the α–γ phase-transition region of Cr_2Se_3. Vapor pressure was measured by a quartz Bourdon gauge at temperatures up to 920°C and vapor pressures up to 1 atm. The *P–T–X* phase diagram for Cr–Se is shown in Fig. 91 in *P–T* and *T–X* projections for the composition range of ~60 to 100 at.% Se. To construct the *P–T* projection, temperature dependences of the vapor pressure for samples with 68 at.% Se (line AC) and 60–62 at.% Se (CDF) were measured. The *T–X* projection is shown on an arbitrary scale. According to Alikhanyan et al. [214], the vapor in this composition range is made up of almost pure selenium. The temperature dependence for the vapor pressure of the γLV equilibrium (Fig. 91, *P–T* projection) is described in the temperature range from 300 to 670°C by the equation

$$\log P \ (\mathrm{mmHg}) = -(4930 \pm 150)/T + (8.01 \pm 0.29).$$

Within experimental errors, this vapor pressure corresponds to that for pure liquid selenium [122]. This suggests that chromium is not appreciably soluble in liquid selenium; the Cr_2Se_3–Se eutectic is degenerate, and the liquidus curve on the *T–X* projection up to ~670°C coincides with the selenium ordinate.

The univariant line CDF is formed by the parts of curves 1 to 10 in the region of the α–γ phase-transition in Cr_2Se_3 and corresponds to the three-phase equilibrium of

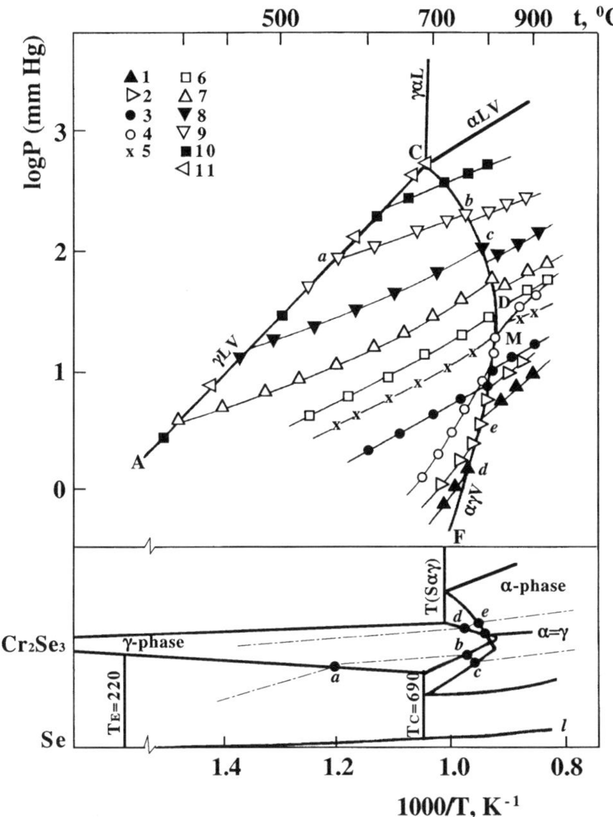

Fig. 91. *P–T* and *T–X* projections of the non-stoichiometrie range of Cr$_2$Se$_3$. Composition/concentration of the samples (at.% Se/g×ml^{-1}): 60.0/0.002(1), 60.0/0.004(2), 60.0/0.025(3), 60.2/0.00289(4), 60.2/0.00761(5), 60.2/0.01515(6), 60.4/0.01468(7), 60.8/0.01534(8), 62.0/0.01534(9), 62.0/0.05681(10), 68.0(11)

two solid phases (γ and α) and vapor. The maximum temperature of the γ–α phase-transition in the *P–T* projection corresponds to point D (T_D = 820°C). The coexisting γ- and α-phases are selenium-saturated on CD and chromium-rich on DF. The invariant equilibrium at T_C is shown in the *T–X* projection according to the coordinates of the point C (T_C = 690°C, P_C = ~1 atm) found from the *P–T* projection. It corresponds to a four-phase equilibrium between two solids (γ and α), liquid, and vapor. The temperatures at points C and D determine the interval (~130°C) of the γ–α phase-transition for Se-saturated Cr$_2$Se$_3$. The sublimation field γV of the low-temperature γ-phase in the *P–T* projection is within ACDF, and for the α-phase, it is to the right of the CDF curve (on the high-temperature side of it). The composition limits of the γ- and α-phases can be seen in the isothermal sections, Fig. 92.

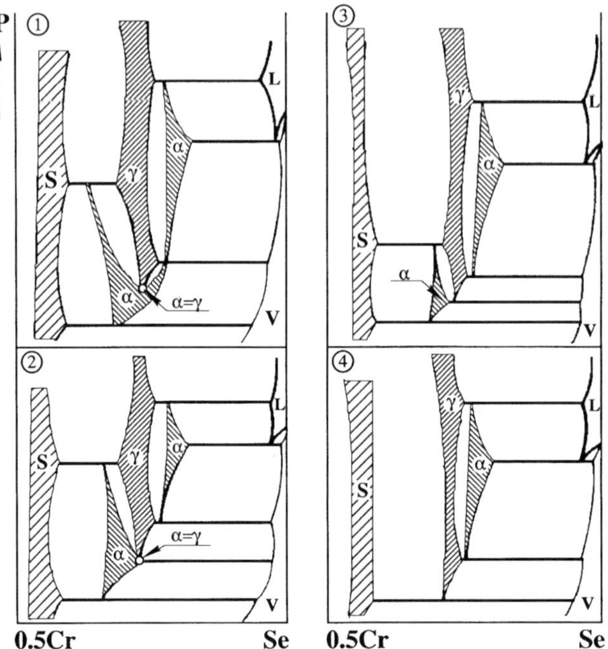

Fig. 92. Isothermal sections of the diagram of Fig. 91 at $T_{max}(1)$, $T_{\alpha=\gamma}(2)$, $T_C < T < T_{\alpha=\gamma}(3)$, $T_{S\alpha\gamma} < T < T_C(4)$

It was experimentally shown that the γ–α phase-transition in Cr_2Se_3 passes through the temperature maximum T_{max} in the three-phase equilibrium $\gamma\alpha V$ (point D, Fig. 91). Therefore, it is of the type shown in Fig. 39; the only difference is that the invariant point C in Fig. 91 is lower in temperature than the second invariant point $T(S\alpha\gamma V) = 811°C$ [210] (S is the adjacent chromium selenide Cr_3Se_4 [210]). This difference appears to be sufficient to prove that the γ–α phase-transition is the congruent type (Fig. 39a,b).

Detailed geometrical analysis of the *P–T–X* diagram shows that an incongruent phase-transition with T_{max} (Fig. 39a,c with $T_2 < T_1$) is possible in two cases:

1. The compositional sequence of phases $X_\alpha < X_\gamma$ (X_i is the composition of the phase *i*, at.% Se, in the phase-transition region) can be unchanged over the whole *P* and *T* interval only if the isothermal relation of pressures in the corresponding three-phase equilibrium is $P(S\alpha\gamma) > P(\alpha\gamma L)$. It is seen in Fig. 91 that this inequality does not hold for the Cr–Se system.

2. The $X_\gamma < X_\alpha$ sequence in the whole phase-transition region requires that $P(S\gamma\alpha) < P(\gamma\alpha V)$. This inequality does not hold at $T > T(S\alpha\gamma)$ for the Cr–Se system (Fig. 91).

Thus, the γ–α phase-transition in Cr_2Se_3 is of the congruent type. Accordingly, in Fig. 91, curve $\alpha = \gamma$ appears (it is not shown in the *P–T* projection to avoid further complicating the figure). This curve has a positive $P–T$ slope because according to the structural data [210], the γ-phase is more compact than α.

The $P–T$ slope of the $\gamma=\alpha$ line determines the coordinates of the invariant point of the phase-transition in relation to point $D(T_{max})$: $P_{ct} < P(T_{max})$; $T_{ct} < T_{max}$; $X_{ct} < X_{\gamma,\alpha}$ (T_{max}), i.e., the $\gamma–\alpha$ phase-transition involves P_{min}, and the congruent point $\gamma=\alpha$ is on the Cr side of both γ- and α-phases. It is clearly seen in the isothermal section of the $P–T–X$ diagram at $T = T_{max}$, which is schematically shown in Fig. 92.1 in the pressure limits of the phase-transition and for $X = 50$ to 100 at.% Se. Only narrow parts of the liquid and vapor spaces, immediately adjacent to pure selenium are seen in Fig. 92.1. The point $\gamma=\alpha$ in Fig. 92.1 is on the Cr-side of γ and α in the $\gamma\alpha V$ equilibrium. At T_{max}, only the α-form is involved in the two-phase equilibrium with the vapor; the γV space is not cut by this isotherm. Upon lowering the temperature to T_{ct} (Fig.92.2), the γV space is crossed by the $T=$const plane in a narrow pressure interval, from $P(\alpha=\gamma V)$ to $P(\gamma\alpha V)$. The point $\gamma=\alpha$ appears on the $\alpha\gamma V$ horizontal, and on account of this, the compositional sequence of phases in the lower part of CDF is changed from $\gamma\alpha V$ at $T > T_{ct}$ to $\alpha\gamma V$ at $T < T_{ct}$ (Fig. 92.3).

Thus, in the three-phase equilibrium of γ-Cr_2Se_3, α-Cr_2Se_3, and vapor, the γ-phase may be enriched in both Cr and Se compared to the α-phase. At $T < T_C$, the α-form is on the Se-side of γ at any pressure (Fig. 92.4).

Homogeneity range of Cr_2Se_3. Cr_2Se_3 sublimes incongruently [214]; up to $\sim 1000°C$ the vapor phase in the Cr–Se system consists almost entirely of selenium. When the substance is heated in a closed volume, the composition of the condensed phase gradually alters, and in certain experimental conditions (initial composition of the sample, the ratio of the mass to the reaction volume), the number of phases in the system may change. This change in the phase state is seen in the vapor pressure curves as points of discontinuity (Fig. 91, curves 1 to 10). For example, on curve 9, point a corresponds to a transition from the three-phase equilibrium γLV to the two-phase equilibrium γV, point b to a transition from the equilibrium γV to the $\gamma\alpha V$, and point c to the $\gamma\alpha V \rightarrow \alpha V$ transition. On curve 1, the change from γV to the three-phase equilibrium $\alpha\gamma V$ is at point d, and $\alpha\gamma V \rightarrow \alpha V$ is at e. The corresponding points on the $T–X$ projection are labeled by the same letters; the dot-and-dash lines show the shift in composition of the condensed phases for curves 1 and 9.

When only two phases (vapor and solid) are involved in an equilibrium, the composition of the latter can be determined from the difference in mass between the components (chromium and selenium) in the initial sample and in the vapors. In this way, the composition of the γ-phase can be found at every point on curves 1–10 in the γV field, whereas the composition of α-Cr_2Se_3 is determined from these curves in αV. At the limiting points a, b, and d of the γV equilibrium and at c and e of αV, the composition of the coexisting phases (γ or α) corresponds to the boundary of the homogeneity region of these phases at the respective temperatures.

Equation (25) can be applied at every transition point to determine the composition of the solid. For Cr–Se, Eq. (25) transforms into

$$X_S(\text{at.\% Se}) = \left\{ N_{Se} - [v\textstyle\sum nP(Se_n)]/RT \right\} / \left\{ N_{Se} + N_{Cr} - [v\textstyle\sum nP(Se_n)]/RT \right\} \times 100\%.$$

Here N_{Cr} and N_{Se} are the numbers of gram-atoms of the components in the initial sample, $P(Se_n)$ the partial pressures of the selenium polymers in the vapors, and v the volume of the vapor phase. This equation can be used to determine the temperature dependence of the solubility of selenium in γ-Cr_2Se_3 (calculation of X_S at points *a*) and the composition limits of the phases in the γ–α phase-transition region (at points *b* and *d* for the γ-phase, *c* and *e* for α).

The partial pressures of selenium $P(Se_n)$ were calculated from the measured total vapor pressure,

$$P = \sum P(Se_n),$$

and equilibrium constants for the polymerization of selenium in the vapor phase [172,173]:

$$K_n = P(Se_2)^{n/2}/P(Se_n).$$

Calculated partial pressures $P(Se_n)$ and compositions of the solid X_S are given in Tables 10 and 11. It can be seen in Table 11 that the solubility of selenium in γ-Cr_2Se_3 does not exceed 1 at.% and varies appreciably with temperature.

The results of calculating the compositions of the solid phase at points *b, c, d,* and *e* of the three-phase line CDF were used to construct the *T–X* projection of the diagram in the region of the γ–α phase-transition (Fig. 93). The composition boundaries of both coexisting phases, γ and α, were calculated from the vapor pressure experiment. In the left-hand part of the $(\alpha+\gamma)$-field the γ-phase is on the selenium side of α, whereas on the right, the sequence is the reverse: $X_\alpha > X_\gamma$. This means that in the $\alpha\gamma V$ equilibrium, the compositional sequence of phases changes, which is a consequence of the congruent phase-transition point in $\alpha\gamma V$. It can be seen in Fig. 93 that the non-stoichiometric range of the γ-phase in the phase-transition region is about 59.8 at.% to 60.2 at.% Se, and the region of coexistence of γ- and α-phases does not exceed 0.15 at.% Se.

Special attention was paid in [208] to uncertainties in the composition of the solids determined by vapor pressure scanning of the solidus surfaces of the γ- and α-phases. The errors in X_S were calculated by applying the error accumulation law, and the results were compared with those obtained at the intersections of the vapor pressure curves in both αV and γV two-phase equilibria. In Fig. 91 curves 3 and 4 intersect in the sublimation range of the γ-phase and 4 and 5 intersect in αV. It has been shown that the composition of the solid X_S was reproducible to within 0.002 at.% and the maximum uncertainty in the X_S values did not exceed 0.03 at.% over the entire phase-transition range.

Table 10. Partial pressures $P(Se_n)$ on the three-phase curves γLV (at points a) and $\gamma \alpha V$ (at points b, c, d, and e)

Point, (Fig.91)	T (°C)	P(mmHg)	$P(Se_n)$ (mmHg)					
			Se_2	Se_3	Se_5	Se_6	Se_7	Se_8
a_7[a]	393	4.0	0.554	0.020	2.406	0.671	0.293	0.055
a_8	467	22.0	4338	0.169	13.850	2.074	1.072	0.197
a_9	543	95.2	24.390	0.990	59.418	6.732	3.113	0.557
a_{10}	626	345.0	111.106	4.634	204.768	15.832	7.383	1.276
b_9	770	269.2	225.280	4.970	37.821	0.890	0.214	0.018
c_9	773	254.1	216.189	4.603	32.395	0.729	0.169	0.014
b_8	795	128.5	121.816	1.709	4.897	0.067	0.010	5.10^{-4}
c_8	802	101.9	98.215	1.186	2.466	0.029	0.004	2.10^{-4}
b_7	813	70.8	69.287	0.663	0.841	0.007	8.10^{-4}	3.10^{-5}
c_7	818	63.2	62.067	0.547	0.581	0.004	4.10^{-4}	1.10^{-5}
d_5	816	22.3	22.137	0.117	0.045	2.10^{-4}	1.10^{-5}	2.10^{-7}
e_5	818	26.4	26.184	0.149	0.066	3.10^{-4}	2.10^{-5}	5.10^{-7}
d_4	809	12.7	12.630	0.053	0.013	5.10^{-5}	2.10^{-6}	3.10^{-8}
e_4	817	24.8	24.604	0.137	0.057	3.10^{-4}	2.10^{-5}	4.10^{-7}
d_3	806	9.4	9.360	0.034	0.006	2.10^{-5}	9.10^{-7}	1.10^{-8}
e_3	808	11.2	11.150	0.004	0.010	3.10^{-5}	1.10^{-6}	2.10^{-8}
d_2	794	4.6	4.586	0.012	0.001	4.10^{-6}	1.10^{-7}	1.10^{-9}
e_2	801	7.2	7.173	0.024	0.004	1.10^{-5}	4.10^{-7}	5.10^{-9}
d_1	784	2.0	1.996	0.004	2.10^{-4}	4.10^{-7}	8.10^{-9}	6.10^{-10}
e_1	795	4.7	4.685	0.013	0.001	4.10^{-6}	1.10^{-7}	1.10^{-9}

[a]The numerical subscript corresponds to the number of the experimental curve in Fig. 91

Table 11. Non-stoichiometry in Cr_2Se_3

Solubility of chromium			Solubility of selenium		
Point (Fig.91)	T (°C)	X_S (at.% Se)	Point (Fig.91)	T (°C)	X_S (at.% Se)
		γ-Cr_2Se_3			
d_1	784	59.90	a_7	393	60.31
d_2	794	59.91	a_8	467	60.41
d_3	806	59.93	a_9	543	60.58
d_4	809	59.96	a_{10}	626	60.80
d_5	816	59.96	b_9	770	60.21
			b_8	796	60.07
			b_7	813	60.00
		α-Cr_2Se_3			
e_1	795	59.81	c_9	772	60.35
e_2	801	59.85	c_8	802	60.24
e_3	808	59.88	c_7	818	60.05
e_4	817	59.92			
e_5	818	59.92			

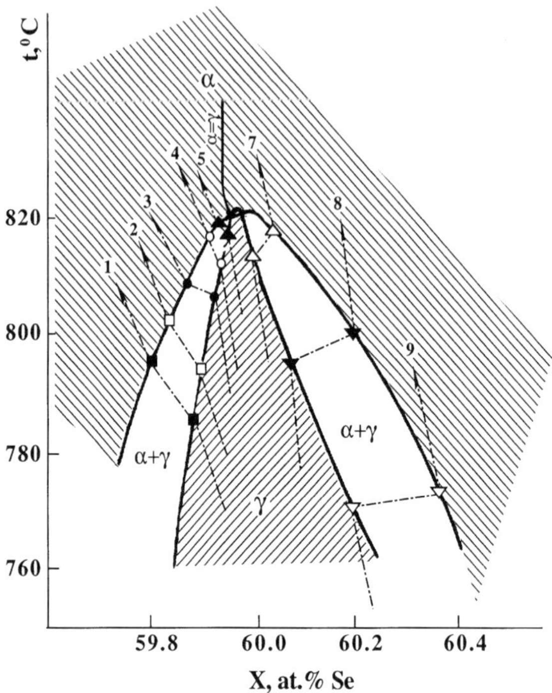

Fig. 93. Non-stoichiometry ranges of α-Cr$_2$Se$_3$ and γ-Cr$_2$Se$_3$. The numbers correspond to Fig. 91

3.1.6 III–V compounds

III–V compounds have numerous and ever growing technical applications in various fields of modern electronics: lasers, microwave and digital devices, and optoelectronics to name a few [215]. Today gallium arsenide is a well-established number one semiconductor compound with two major domains of application — high frequency microelectronics and optoelectronics. In spite of tremendous achievements in crystal growth of large (up to 150 mm in diameter) semi-insulating single crystals of GaAs by the Liquid Encapsulated Czochralski (LEC) method, in particular [216], this technology still relies heavily on an empirical approach to the key problem of non-stoichiometry and tailoring of the defect structure during the crystal growth process and in after-growth treatment. Bulk crystal growth is essentially a quasi-equilibrium process [215]. Hence, information on phase equilibrium is crucial for this purpose.

The *P–T–X* phase diagram for the gallium–arsenic system, presented in *P–T* and *T–X* projections in Figs. 94 and 95, according to Wenzl et al. [215], cannot be considered the ultimate version because most of it was constructed on the basis of model considerations of phase equilibrium in this system rather than on experimental results. The maximum melting temperature of GaAs is 1513 K. Eutectic

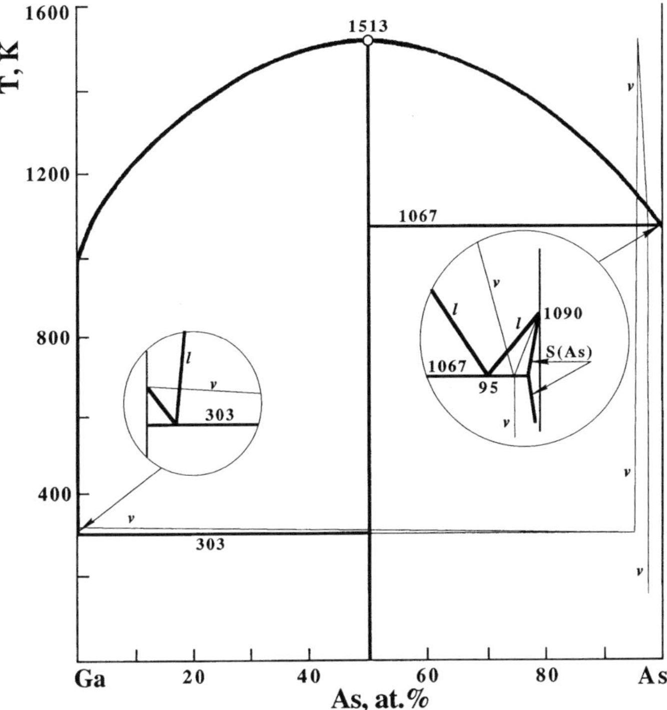

Fig. 94. *T–X* projection of the Ga–As phase diagram

temperatures are 303 K (for the degenerate gallium eutectic) and 1067 K (for the arsenic eutectic, 95 at.% As in composition). Details of the space arrangement of the solidus, liquidus, and vaporus curves near the eutectic points are given in the insets in the *T–X* projection, Fig. 94. The vaporus curve in the *T–X* projection (the thin *v* line) is very close to the As ordinate because the vapor phase in all of the phase equilibria was proved to be made up almost completely of arsenic. The *P–T* projection, Fig. 95 [215], is presented in partial pressures of the principal species, $As_4(g)$ and $As_2(g)$, as well as those of the minor vapor phase components, $As(g)$ and $Ga(g)$, which are several orders of magnitude less than the predominant molecules. The left-hand sides of the arsenic loops and the right-hand side of the gallium loop correspond to the three-phase equilibrium between Ga-saturated GaAs(s), liquid, and vapor, whereas the opposite branches of these curves are for the As-saturated GaAs(s) in the three-phase equilibrium with an As-rich melt and almost pure arsenic vapor. Every pair of these arms converges at the maximum melting point of GaAs. According to the data compiled by Wenzl et al. [215] from different sources, the total vapor pressure at the maximum melting temperature is between 0.7 and 2.0 atm. Experimental vapor pressure measurements for this system are unknown.

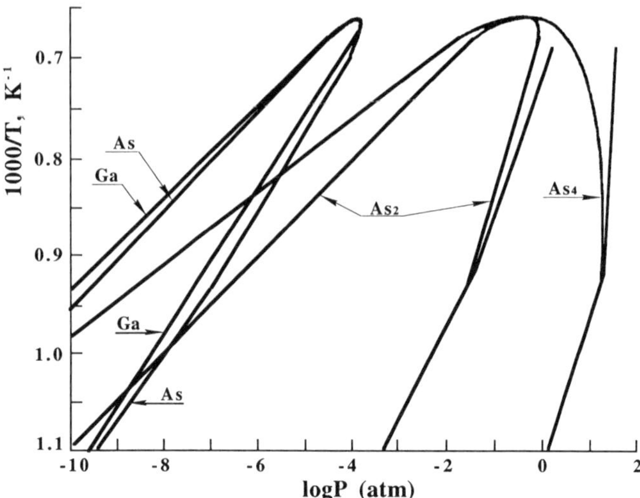

Fig. 95. Partial pressures in the three-phase equilibria VLS and SLV

The straight lines for As_4(g) and As_2(g) in Fig. 95 describe the partial vapor pressures of these species in the saturated vapor of pure arsenic below and above its melting point.

Experimental studies of non-stoichiometry in GaAs proved to be a formidable task. Even the best modern chemical analysis techniques are not precise enough to generate reliable results on the *in situ* composition of GaAs(s) in equilibrium with vapor or (vapor + melt) [215]. The single-phase range of existence for GaAs(s), shown in Fig. 96, was constructed on the basis of thermodynamic modeling of the defect structure of GaAs. The composition of the solidus was calculated for the three-phase equilibrium with a conjugated liquid of fixed composition and vapor at pressures listed in Fig. 96. Hence, these compositions correspond to the maximum non-stoichiometry of GaAs(s) at corresponding temperatures. The main composition axis in Fig. 96 is for the liquidus; the solidus is shown on an enlarged scale, given just below the principal abscissa. The shape of the asymmetrical solidus "bubble" in Fig. 96 is such that the maximum melting temperature of GaAs is on the arsenic side of the stoichiometric plane $X = 50$ at.%, implying, in particular, that stoichiometric GaAs is expected to be grown from a Ga-rich melt. Calculations based on various defect models [217–220] lead to substantially different shapes and arrangement of the GaAs solidus, as can be seen in Fig. 97. Unfortunately, these results cannot be verified against direct experimental measurements of non-stoichiometry, such as vapor pressure scanning, because of the lack of relevant experimental data. This means that the functional dependences of a crystal composition on the vapor pressure and composition of the conjugated melt are yet to be studied. That is why modern technology of GaAs with controlled composition is based essentially on a trial-and-error approach.

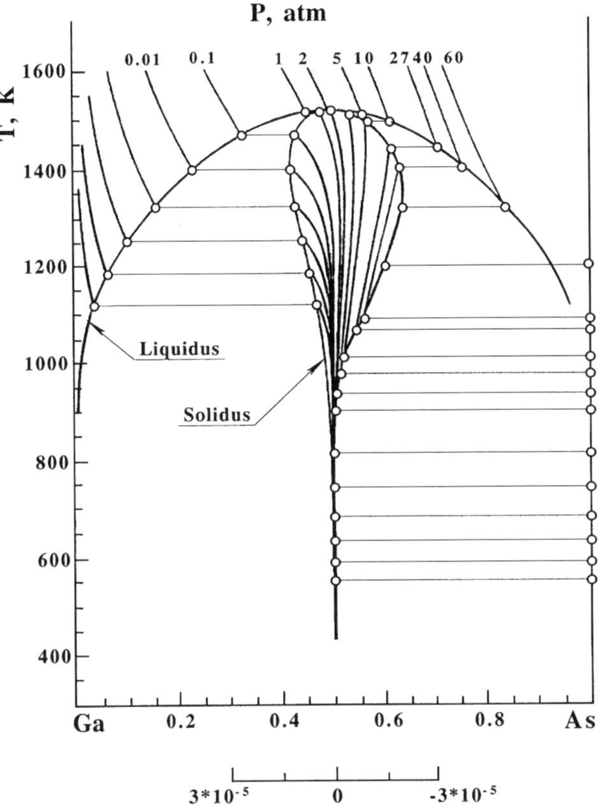

Fig. 96. T–X diagram Ga–As and a close-up of the non-stoichiometry range of GaAs. The secondary composition axis is for the solidus

3.1.7 II–V compounds

The applications of II–V compounds are due to their semiconductor properties [221]. At present for application purposes, probably, the best studied among the II–V compounds is Zn_3P_2. It meets the major requirements for terrestrial photo-voltaic devices [222]; its direct band gap of 1.5 eV is the theoretical optimum for solar power conversion efficiency in air, and several Zn_3P_2-based solar cells have already been constructed [222]. Among other applications of II–V compounds, infrared and ultraviolet sensors were also mentioned.

But the main interest in these compounds presently is for the student of P–T–X phase diagrams because in these systems one comes across almost all of the most complicated phenomena of phase equilibrium: formation of several compounds in a binary system (for example, five cadmium phosphides have been reported); congru-ent and incongruent melting; eutectoid decomposition and miscibility gaps in the liquid phase (Cd–P system); negative pressure dependences of the melting tempera-

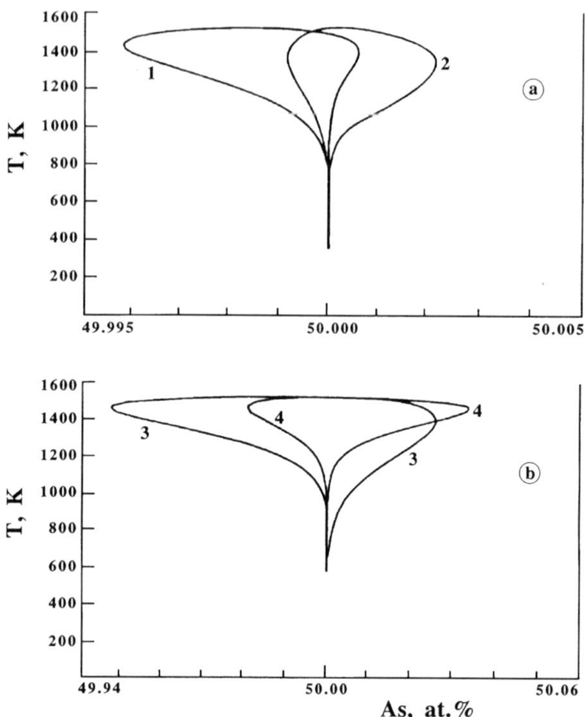

Fig. 97. Non-stoichiometry range for GaAs. 1-[217], 2-[218], 3-[219], 4-[220]

ture (Zn and Cd arsenides) ; two congruent sublimation curves in one system (also Cd–P); metastable states (Cd–As system); congruent and incongruent solid-state phase-transitions, including a unique incongruent first-order phase-transition in Zn_3As_2 involving multiple extrema in the three-phase equilibrium, etc. That is why it seems appropriate for a book on phase equilibrium to discuss *P–T–X* diagrams of II–V systems in detail.

3.1.7.1 Zinc phosphides

The *P–T–X* phase diagram for the zinc–phosphorus system was reported by Lazarev et al. [223]. The *T–X* projection of this diagram was constructed for the region $Zn–Zn_3P_2$ in [224] and for $Zn–ZnP_2$ in [225]. Several authors reported results for the sublimation of Zn_3P_2 [226–228], but only Alikhanyan et al. [229] studied the evaporation of the second phosphide, ZnP_2. At present, it is established that zinc and phosphorus form two compounds, Zn_3P_2 (melting point 1193°C) and ZnP_2 (melting point 1040°C). Both zinc phosphides melt congruently and undergo a first order phase-transition, Zn_3P_2 at 880°C and ZnP_2 at ~ 1000°C [225]. Deviation from stoichiometry in both of them proved too small to be recorded by conventional analytical methods.

Sublimation of Zn_3P_2 was shown to be a congruent process,

$$Zn_3P_2(s) = 3Zn(g) + 0.5P_4(g),$$
$$P_4(g) = 2P_2(g),$$

described by a minimum vapor pressure in the phase diagram. The vapor pressure for the congruent sublimation $S(Zn_3P_2) = V$ was studied in [226], where the total vapor pressure of stoichiometric Zn_3P_2 was measured by the static method using a Bourdon gauge. Extrapolation of the [226] data to the congruent melting point of the high temperature β-Zn_3P_2 showed that Zn_3P_2 melted under a vapor pressure of about 7–8 atm.

The sublimation of ZnP_2 proved to be an incongruent process; the saturated vapor contains more phosphorus than the condensed phase [229]. As a result, the composition of the condensed phase shifted toward Zn in the process of heating. It was found that even at high ZnP_2 concentrations in the reaction bulb with a minimum volume of the vapor, the measured saturated vapor pressure corresponded to the three-phase equilibrium between crystalline Zn_3P_2, ZnP_2, and vapor. This means that, even at low vapor pressures, which shifted the initial ZnP_2 composition only slightly toward Zn_3P_2, the second solid phase, Zn_3P_2, is immediately formed and recorded by the vapor pressure measurement. The sublimation curves of ZnP_2 did not exhibit any points of discontinuity, which proved the absence of Zn non-stoichiometry in ZnP_2 (within the limits of the vapor pressure scanning method).

In a number of experiments, breaking points were observed in the vapor pressure curves of the three-phase equilibrium $S(Zn_3P_2)S(ZnP_2)V$, which corresponded to the change in the phase state of the system due to complete evaporation of the more volatile crystalline phase, ZnP_2. As a result of the phase-transition $S(Zn_3P_2)S(ZnP_2)V$ → $S(Zn_3P_2)V$, crystalline Zn_3P_2 was formed with composition corresponding to the maximum phosphorus non-stoichiometry of the Zn_3P_2 phase at the phase-transition temperature. It has been shown that the maximum solubility of phosphorus in Zn_3P_2 was 0.009 to 0.012 at.% at temperatures of 785 to 820°C and tended to decrease at lower temperatures.

3.1.7.2 Cadmium phosphides

The cadmium–phosphorus system was studied in [224,230] by thermal analysis, X-ray diffraction, and metallography in the composition range 0–66.7 at.% P. The following cadmium phosphides were identified: Cd_3P_2, Cd_6P_7, Cd_2P_3, and CdP_2. The only congruently melting compound is CdP_2 (melting point 782°C). The other compounds melt incongruently at 740°C (Cd_3P_2), 734°C (Cd_6P_7), and 746°C (Cd_2P_3). One more cadmium phosphide, CdP_4, was reported in [231,232]. The P–T–X phase diagram for the cadmium–phosphorus system in the composition range 66.7–100 at.% P was reported by Lazarev et al. [233]. Along with DTA and XRD, vapor pressure was measured and a complete P–T–X phase diagram was constructed.

According to [233], CdP_4 is involved in an incongruent melting process $S(CdP_4)$ → $S(CdP_2)$ + L at ~755°C, and Cd_6P_7 is stable in a closed temperature interval [230]. Peritectic melting is described as $S(Cd_6P_7)$ → $S(Cd_2P_3)$ + L, and peritectoid decomposition is the phase reaction $S(Cd_6P_7)$ → $S(Cd_3P_2)$ + $S(Cd_2P_3)$. Two congruent sublimation curves were found in the cadmium–phosphorus system, $S(Cd_3P_2)$ = V and $S(CdP_2)$ = V, which corresponded to vapor pressure minima in $S(Cd_3P_2)V$ and $S(CdP_2)V$ two-phase equilibria. An additional vapor pressure minimum, in the liquid–vapor equilibrium, corresponds to the congruent vaporization L = V. A rare mode of peritectic melting was observed for Cd_3P_2: this phosphide melted at 740°C into the miscibility gap in the liquid phase Cd_3P_2 → L_1 + L_2.

The vapor pressure scanning procedure was used in [233] to measure the maximum phosphorus non-stoichiometry of CdP_2. Points of discontinuity were registered on the vapor pressure curves and attributed to the $S(CdP_2)S(CdP_4)V$ → $S(CdP_2)V$ phase-transition. At these points, complete evaporation of the more volatile phosphide (CdP_4) was observed, and the composition of the second solid, CdP_2, corresponded to the phosphorus saturation at the phase-transition temperatures. The boundary composition of the solid varied from 67.04 at.% P to 67.34 at.% P at temperatures of 805 to 837 K, i.e., it was proved that CdP_2 dissolved up to 0.7 at.% of phosphorus.

3.1.7.3 Zinc arsenides

According to Hansen [189], the zinc–arsenic system comprises two arsenides, Zn_3As_2 and $ZnAs_2$. The *T–X* projection of the zinc–arsenic phase diagram, given by Hansen [189], is in quite good agreement with subsequent investigations reviewed by Lazarev et al. [221]. According to [221], Zn_3As_2 non-stoichiometry is very small, and $ZnAs_2$ can dissolve as much as several at.% of arsenic. More recent thermal analysis, X-ray powder diffraction, and metallographic investigations [234,235] showed that the solubilities of both zinc and arsenic in $ZnAs_2$ are about 0.5 at.%. High-temperature sublimation studies of zinc arsenides [236–239] led to a general agreement that Zn_3As_2 sublimes congruently, whereas sublimation of $ZnAs_2$ is an incongruent process. As a result of extensive vapor pressure studies [240], the *P–T–X* phase diagram for the Zn–As system was constructed, and detailed vapor pressure scanning of the Zn_3As_2 and $ZnAs_2$ solidus surfaces led to high precision determination of the arrangement of single-phase volumes for these solids in the *P–T–X* phase space. It was proved that the phase relationships in the Zn–As system were very complicated, and it would be useful for those who aim to master a knowledge of phase equilibrium to follow carefully the subsequent description of the [240–244] results. Also, because of the exceptionally high accuracy of the results claimed in [240–244], detailed description of the experimental procedure [240–244] is relevant here.

Composition and DTA data for the Zn–As samples used for constructing the *P–T–X* phase diagram are given in Table 12 together with the vapor pressure results for the α–β phase-transition in Zn_3As_2. The samples were prepared by direct synthesis from high purity elements in quartz tubes with vacuum jackets.

Table 12. Composition of the samples, DTA, and vapor pressure data

# according to Fig. 98	Composition (at.% As)	α–β phase trans. from vapor pressure (K)	DTA (K) α–β trans.	Solidus	Liquidus
1	40.0000	932.0–941.5	945		1288
2	40.0000	934.0–944.5	945		1288
3	40.0000	935.0–944.5	945		1288
4	40.0000	944.0–953.5	945		1293
5	35.9961		945	698	1288
6	30.4518		945	693	1258
7	42.5544		945	1033	1283
8	46.6379		945	1028	1233
9	79.8961			993	1033
10	88.6396			993	1087

The inner surface of the tube was covered with pyrolitic carbon film. The starting elements (semiconductor purity grade) were additionally purified. Zinc was etched in nitric acid, carefully washed in distilled water, and dried in vacuum. Arsenic was purified by vacuum distillation. The samples were weighed directly in the reaction tubes. The total mass was about 30 g, and the precision of the balance was ($\pm$ 5×10^{-5}) g. The quartz tube with the sample was pumped, sealed, and placed in a rotating furnace to obtain a homogeneous alloy. The temperature was increased up to the melting point of the sample, and after a two-hour exposure, the rotation of the furnace was stopped. The sample was slowly cooled down to 720 K and held at this temperature for annealing. The isothermal annealing time for the two-phase samples varied from three hours to two weeks and did not influence the measured saturated vapor pressure. Special attention was paid to avoiding partial sublimation of the reagents. The samples were characterized by X-ray powder diffraction and differential thermal analyses.

The vapor pressure was measured with a quartz Bourdon gauge. The reaction bulb of the gauge was coated with pyrolitic carbon and held for several hours in vacuum (10^{-5} mmHg) at 1200 K. After cooling to room temperature, the sample was introduced into the reaction chamber and heated in vacuum for several hours at 400 to 600 K, depending on the composition of the alloy. The temperature was measured with Pt-Pt/Rh thermocouples placed in special pockets at both ends and in the middle of the reaction vessel. The thermocouples were calibrated in the temperature range 450–1200 K against the melting points of several metals. During the vapor pressure experiments, the reaction chamber of the Bourdon gauge was held in isothermal conditions; the temperature of the top was somewhat (0.2–0.6 K) higher than that at the bottom to prevent condensation on the membrane. Temperature uncertainties were believed to be within 0.5 K. The vapor pressure apparatus was calibrated against tabulated values of the saturated vapor pressures of cadmium and selenium. Heating and cooling vapor pressure experiments, as well as observation at constant temperatures for up to 48 hours, showed that the equilibrium state in the Zn–As system was attained in 30 to 60 min after the temperature became stationary.

P–T–X phase diagram. Projections of the *P–T–X* phase diagram for the Zn–As system are given in Fig. 98. The *T–X* projection is constructed mainly from the data compiled in [221], which is in reasonable agreement with the DTA results [240,241] also presented partially in Fig. 98 and Table 12. The *P–T* projection represents the saturated vapor pressure measurements for alloys with different compositions. Three-phase equilibria were studied in the vapor pressure range up to 800 mmHg.

The vaporus curve shown schematically in the *T–X* projection, Fig. 98, is based on Knudsen cell sublimation data with mass spectrometric analysis. These were obtained in special experiments on complete evaporation of the condensates formed by abrupt quenching in liquid nitrogen of the vapor in $VL(Zn)S(Zn_3As_2)$ and $S(Zn_3As_2)S(ZnAs_2)V$ equilibria.

The *P–T* projection comprises the following univariant curves. The three-phase equilibrium $S(\alpha\text{-}Zn_3As_2)S(\beta\text{-}Zn_3As_2)V$ is represented by two lines, KO and MO. These are the vapor pressure curves of samples whose composition was nearly 40 at.% As. The curve AOM is shown in Fig. 98 on an enlarged scale. Its shape is quite unusual and is the result of a complicated phase arrangement in the α–β phase-transition area, which is discussed in the next section. Line BKD corresponds to the three-phase equilibrium of the liquid phase (the composition is given in the *T–X* projection by the liquidus in the concentration range 0 to 40 at.% As), the Zn-rich crystalline Zn_3As_2, and the vapor. Curve CME_1 represents the equilibrium $S(Zn_3As_2)$ $S(ZnAs_2)V$ of two crystalline phases (As-rich Zn_3As_2 and Zn-rich $ZnAs_2$) with the saturated vapor. Two curves originate from the eutectic point E_1: BE_1 corresponds to the three-phase equilibrium $LS(Zn_3As_2)V$, and HE_1 results from the equilibrium $LS(ZnAs_2)V$. In these equilibria, L is the liquidus in the composition range 40–66.7 at.% As. Point E_1 is a four-phase invariant eutectic involving two crystalline phases (β-Zn_3As_2 and $ZnAs_2$), liquid of the eutectic composition, and the saturated vapor. The *P-T* coordinates of the point E_1 are P = 754 mmHg, T = 1026 K. The space, limited by the curve $BKD(AOB)CME_1$, is the P-T projection of the three-dimensional range of stability of crystalline Zn_3As_2 in equilibrium with vapor. BKD corresponds to maximum Zn non-stoichiometry of Zn_3As_2, whereas CME_1QB is for As-rich Zn_3As_2. The two-phase equilibrium S_1V for Zn-rich Zn_3As_2 is within AOBKD, whereas that for As-rich Zn_3As_2 is within $AOBQE_1MC$. It will be seen from Fig. 98 that these two-phase fields are partially juxtaposed in the *P–T* projection.

As-rich $ZnAs_2$ is involved in two three-phase equilibria: $S(ZnAs_2)S(As)V$ (represented by FE_2) and $LS(ZnAs_2)V$ (the E_2H line). Point E_2 is the invariant eutectic point, which involves the following four phases: crystalline $ZnAs_2$ and As, the liquid, and the saturated vapor. Point H is the maximum melting point in the range of existence CME_1HE_2F of the crystalline $ZnAs_2$ phase. The absence of a minimum vapor pressure in the range of stability of $ZnAs_2$ shows that sublimation of this compound is an incongruent process. Unary equilibria are not shown in Fig. 98. Evaporation of zinc is within KD, and sublimation of arsenic is within E_2F.

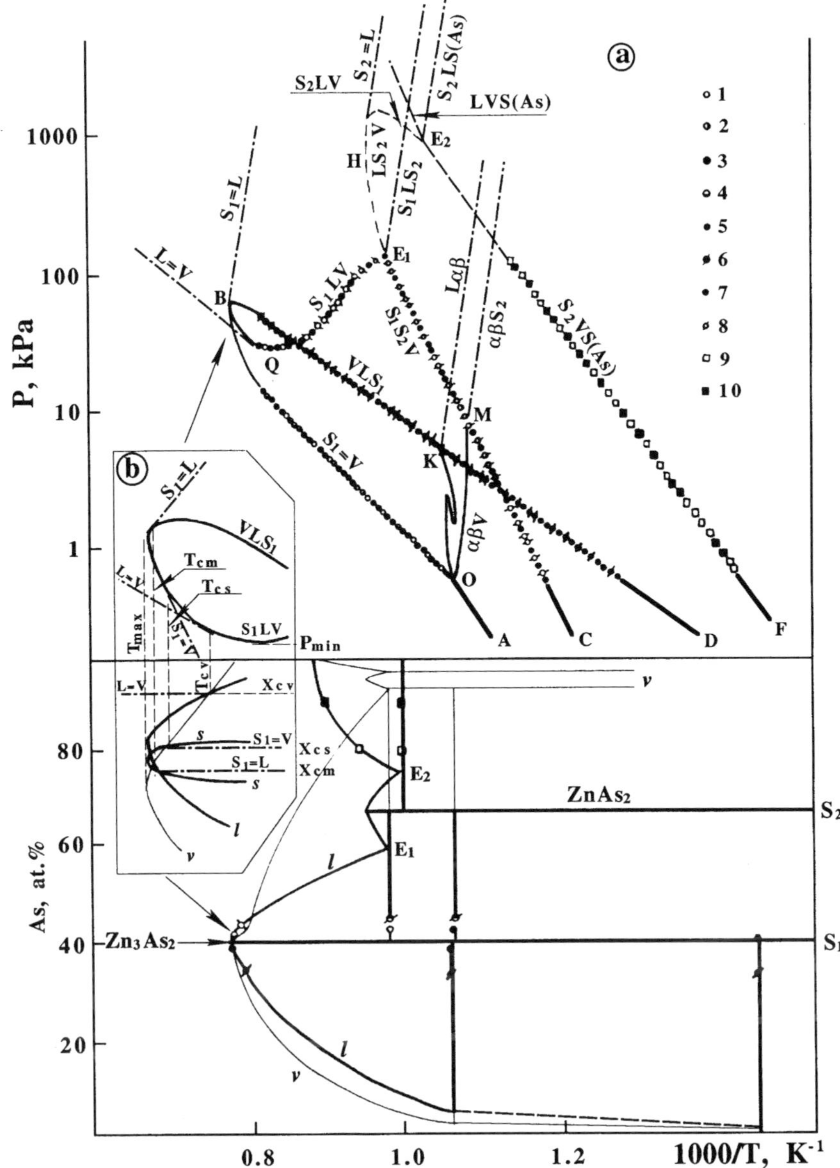

Fig. 98. *P–T* and *T–X* projections of the Zn–As diagram. Points are numbered according to Table 12

Four congruent curves are observed in Fig. 98: the minimum vapor pressure describes the congruent sublimation $S(Zn_3As_2) = V$, two congruent melting curves are $S(Zn_3As_2) = L$ and $S(ZnAs_2) = L$, and $L = V$ is the azeotrope. According to [245], the pressure dependences of the melting temperatures are negative for both zinc arsenides. Furthermore, β-Zn_3As_2 is known to be of more compact crystal structure than the low-temperature α-form [221]. Therefore $LS(\alpha$-$Zn_3As_2)S(\beta$-$Zn_3As_2)$ and $S(\alpha$-$Zn_3As_2)S(\beta$-$Zn_3As_2)S(ZnAs_2)$ also have negative P–T slopes. These experimental results together with detailed examination of the phase arrangement for this system show that $S(Zn_3As_2)LS(ZnAs_2)$ and $S(ZnAs_2)LS(As)$ most probably also have negative P–T slopes.

A peculiar feature of the phase diagram, Fig. 98, is the drop in vapor pressure in the $S(Zn_3As_2)LV$ equilibrium down to P_{min} with rising temperature. The approximate coordinates of P_{min} are $T = {\sim}1200$ K, $P = 248$ mmHg. The vapor pressure in three-phase equilibrium $S(Zn_3As_2)LV$ at T_{max} is lower than that for both sublimation of arsenic and evaporation of zinc. Such relative arrangement of $S(Zn_3As_2)LV$, $L(Zn)V$ and $S(As)V$ together with $dP/dT < 0$ for the univariant condensed phase equilibria leads necessarily to the appearance of an azeotropic point in $S(Zn_3As_2)LV$ (point Q in Fig. 98a where $T = T_{cv}$). The compositions of the phases at this point are seen in the T–X projection as intersections of the liquidus and vaporus and the corresponding composition of S_1 at $T = T_{cv}$. In Fig. 98b, enlarged parts of P–T and T–X projections are shown for the temperature limits between T_{cv} (point Q) and T_{max}. Examination of the arrangement of the three-phase curves in the vicinity of the maximum melting point of Zn_3As_2 shows the relative positions of the three congruent points in the $S(Zn_3As_2)LV$ equilibrium. Because $dP/dT < 0$ for $S(Zn_3As_2) = L$, the point of tangency of this curve with $VLS(Zn_3As_2)$ is on the Zn-side of T_{max} (Fig. 98). This means that for the congruent melting point, $T_{cm} < T_{max}$ and $X_{cm} < X_{L,S}(T_{max})$ (in at.% As). It also leads to $X_S(T_{max}) < X_L(T_{max})$, i.e., at T_{max}, the solid is on the Zn-side of the liquid.

Experimental data, Fig. 98, show that the $S(Zn_3As_2) = V$ line is tangent to $S(Zn_3As_2)LV$ and Zn_3As_2 is enriched in arsenic. Therefore, for the congruent sublimation point, $T_{cs} < T_{max}$ and $X_{cs} > X_{S,V}(T_{max})$. Moreover, $X_V(T_{max}) < X_S(T_{max})$ becomes necessary, i.e., at T_{max}, the solid is on the As-side of the vapor. Thus, vapor pressure investigation shows that at the maximum melting point of the Zn_3As_2 phase, T_{max}, compositions of the conjugated phases, $X(T_{max})$, do not coincide: $X_V(T_{max}) < X_S(T_{max}) < X_L(T_{max})$. The minimum vapor pressure in the liquid–vapor equilibrium is a result of the negative deviation from ideal behavior. For the azeotrope, $X_{cv} < X_L(T_{max})$, and consequently, $T_{cv} < T_{max}$.

Thus, it was seen that the phase arrangement in the melting region of Zn_3As_2 is rather complicated. It is elucidated in isothermal sections of the space model presented in Fig. 99 for temperatures from $T_1 = T_{max}$ down to T_6, which is below the melting temperature of $ZnAs_2$. In P–X isotherms at $T > T_1$, a liquid–vapor loop is seen with P_{min} at azeotropic points $L = V$. Upon lowering the temperature down to $T_1 = T_{max}$ the solid phase Zn_3As_2 emerges with composition $X_S(T_{max})$ (Fig. 98). As a result, a three-phase horizontal appears in the left-hand part of the LV loop for the VS_1L equilibrium (Fig. 99.1). Upon further lowering the cross-section temperature, point S spreads to a narrow field S, which is the single-phase region for Zn_3As_2 and is limited by two VS_1L horizontals (in Fig. 99 single-phase spaces of crystalline phases are shaded). When the cross-section at temperature is lowered to T_{cm}, the S field touches the liquidus surface, and the congruent melting point $L = S_1$ appears on

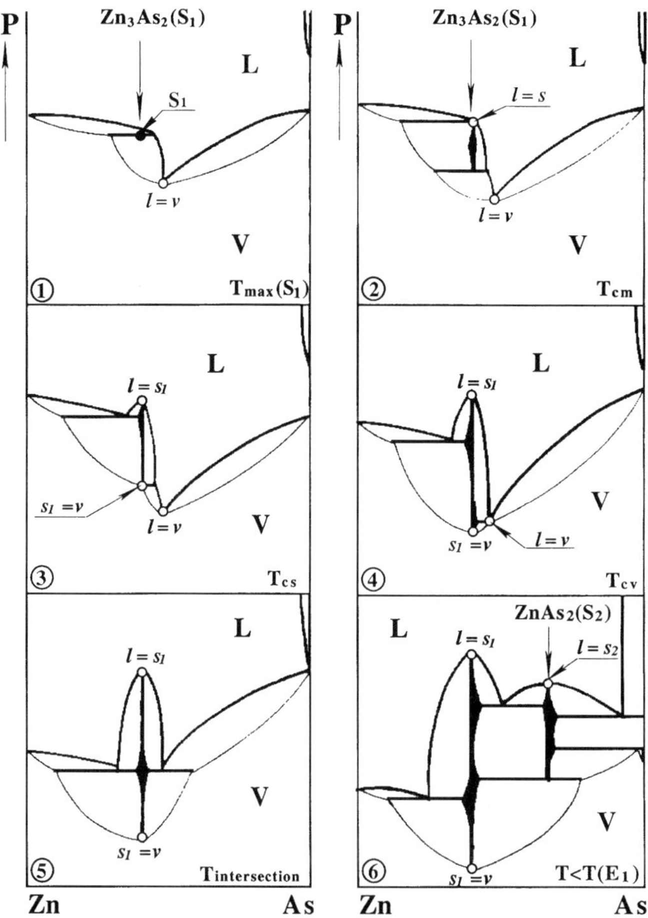

Fig. 99. Isothermal section of the diagram Fig. 98.

the VS_1L horizontal (Fig. 99.2). On still further decreasing the temperature, the S field contacts the vaporus surface at $T = T_{cs}$ to produce another invariant point, $S_1 = V$ (Fig. 99.3). At this temperature the azeotropic pressure ($L = V$) is lower than that for $S_1 = V$. When the temperature is decreased, the vapor pressure in $S_1 = V$ becomes lower than for $L = V$, and eventually at T_{cv}, the $L = V$ point appears on the S_1LV horizontal (Fig. 99.4). At $T < T_{cv}$, evaporation becomes incongruent. A characteristic feature of this system is that the VLS_1 and S_1LV curves intersect in the P–T projection (but not in the P–T–X space). A P–X section made at this intersection temperature (Fig. 99.5) clearly shows that, although the total vapor pressures for both equilibria is the same, the compositions of all three coexisting phases in different equilibria are different.

The second zinc arsenide, $ZnAs_2$, emerges in the isothermal section at $T_{max}(ZnAs_2)$ in a way, analogous to that of Zn_3As_2. Because of this, only one section at $T < T_{E1}$ is shown in Fig. 99.6.

The two-phase field L+S(As) is seen in all of the isotherms above the triple point of pure arsenic because of the positive *P–T* slope of the melting temperature of pure arsenic [31]. Solidification of zinc is not shown in the sections (Fig. 99) because it occurs at high pressures, outside the limits of the figures. The following temperature sequence is assumed in Fig. 99: $T_{cv} > T_{cs} > T_{cv}$. As seen in Fig. 99, the Zn–As system is characterized by maximum temperatures in all two-phase equilibria involving liquid and crystalline phases.

Vapor pressure scanning of the phase-transition in Zn_3As_2. To study the phase-transition in Zn_3As_2, saturated vapor pressure was measured for a total of about twenty alloys of Zn and As with different compositions, which deviated slightly from the stoichiometry of 40 at.% As; the samples contained a small excess of either Zn or As. The measurements were made in a wide range of concentrations $m/v = 10^{-3}$ to 1 g/ml (m is the mass of the sample and v the volume of the vapor phase). Experimental points, corresponding to the most interesting vapor pressure measurements, are shown in the *P–T* projection, Fig. 100, where α is α-Zn_3As_2, β is β-Zn_3As_2, S_2 is $ZnAs_2$, L is liquid, and V vapor. The three-phase equilibrium curves VLα, βS_2V and the coordinates of the quadruple points N_1 and N_2 are taken from [240]. Univariant lines Lαβ and αβS_2 are drawn schematically. In the *P–T* projection, only those parts of the vapor pressure curves are shown, that immediately adjoin the univariant equilibrium line αβV. This latter is the limit of the shaded space in the *P–T* projection and together with Lαβ and αβS_2 curves, constitutes the phase-transition region.

Because of the congruent sublimation of both α- and β-Zn_3As_2 (α = V and β = V curves), the *P–T* projection of the phase-transition region is not single-valued. The shaded area restricted by the univariant lines βS_2V, αβV, and β = V is the *P–T* projection of the solid–vapor bivariant equilibria βV of As-saturated β-Zn_3As_2, whereas the part of this area within the VLα, αβV, and α = V lines is the projection of the bivariant αV equilibrium of Zn-saturated α-Zn_3As_2. All of the experimental curves shown in Fig. 100 exhibit points of discontinuity, which correspond to phase-transition in the system during the vapor pressure experiments. These phase-transition points are due to the formation or disappearance of a condensed phase in the process of heating (or cooling) the initial sample.

Vapor pressure measurements for the samples with the initial phase composition Zn_3As_2 + $ZnAs_2$ are represented in Fig. 100 by solid points, and the hollow points correspond to the Zn + Zn_3As_2 samples. The former vapor pressure curves originate in the three-phase equilibrium αS_2V (these initial parts of the curves are outside Fig. 100), then with increasing temperature, they cross the bivariant field αV consecutively, follow the three-phase line αβV, and pass through the bivariant βV area. Each of these curves has two points of discontinuity, corresponding to the phase-transition αV → αβV (formation of β-Zn_3As_2) and αβV → βV (disappearance of the α-phase). It is seen from the *P–T* projection that all of these curves are charac-

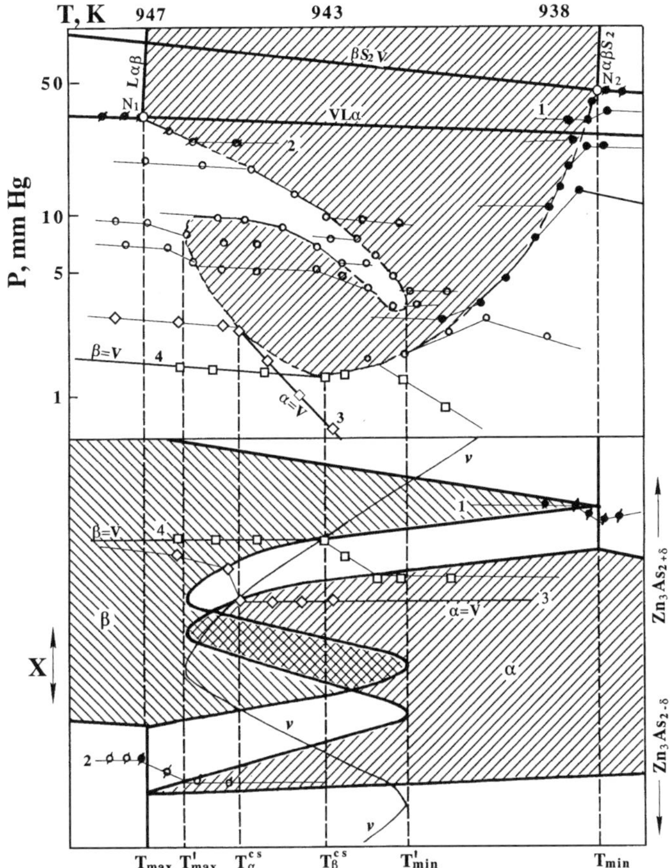

Fig. 100. $P–T$ and $T–X$ projections of the α-β phase-transition region

terized by a decrease of vapor pressure in the $\alpha\beta V$ equilibrium with rising temperature. Experimental curves of the second group (hollow points, Fig. 100) originate in the $VL\alpha$ equilibrium (outside of Fig. 100) and proceed through αV, $V\alpha\beta$, and βV equilibria. The heating process for these samples is accompanied by an increase of vapor pressure throughout all of the phase reactions.

The resulting $P–T$ curve of the three-phase equilibrium involving α-Zn_3As_2, β-Zn_3As_2, and the saturated vapor is drawn in Fig. 100 through the corresponding $\alpha\beta V$ parts of the experimental vapor pressure curves. It extends between two invariant quadruple points N_1 and N_2 and has an unusually complicated form with two temperature extrema at T'_{min} and T'_{max} and three vapor pressure extrema.

The compositions of the α-Zn_3As_2 and β-Zn_3As_2 phases in the α-β transition region were estimated from special experiments. In these, the vapor phase formed in the VLβ and $\beta S_2 V$ equilibria at elevated temperature was quenched in liquid nitrogen, and the resulting condensates were completely evaporated from Knudsen cells with mass spectrometric analysis. It has been shown that the vapor phase in VLβ equilibrium consisted of almost pure zinc, whereas the $\beta S_2 V$ vapor was pure arsenic. Consequently, the saturated vapor is continuously enriched in zinc through the entire $\alpha\beta V$ equilibrium range as the temperature rises from point N_2 up to point N_1. This is true for both the decreasing vapor pressure interval at $T > T(N_2)$ and the increasing vapor pressure at $T < T(N_1)$. It means that at temperatures $T_{min} < T < T(P_{min})$, the vapor pressure drop is due to arsenic condensation and the increases in vapor pressure at $T_{max} > T > T_{min}$ are due to the preferential evaporation of zinc. It is clear that in both cases, i.e., through the whole phase-transition region, the composition of the condensed phases is invariably shifted toward arsenic.

In this connection, it is interesting to trace the $P–T$ and $T–X$ projections of the heating curves for samples 1–4 (Fig. 100). Curve 1 passes from the $\alpha S_2 V$ equilibrium through $\alpha\beta V$ to βV. Because the composition of the crystalline phases in $\alpha\beta V$ shifts toward arsenic in the process of heating ($T–X$ projection, Fig. 100), the composition sequence of phases near the invariant point $T_{min} = T(N_2)$ is $X_\beta > X_\alpha$, the composition X_i of the phase i was measured in at.% As. Similar consideration for the $P–T$ and $T–X$ projections of curve 2 implies that the compositional sequence is also $X_\beta > X_\alpha$ near the second invariant $T_{max} = T(N_1)$. Thus, the compositional order of the crystalline phases is the same at both extremities of the solid-state phase-transition region.

Examination of the $P–T$ and $T–X$ projections of the heating curves for samples 3 and 4 shows that the arrangement of the invariant points of the congruent sublimation $\alpha = V$ and $\beta = V$ in the three-phase equilibrium of α-Zn_3As_2, β-Zn_3As_2, and saturated vapor is such that $X_\beta^{cs} > X_\alpha^{cs}$ and $T_\alpha^{cs} > T_\beta^{cs}$. The second inequality can also be derived from thermodynamic considerations. Because the enthalpy of the phase transformation is positive, $\Delta H_{tr} > 0$, the slope of the $\alpha = V$ line is greater than that of $\beta = V$, $\partial P/\partial T|_{\alpha=V} > \partial P/\partial T|_{\beta=V}$. Therefore the point of tangency of the $\alpha = V$ line to V$\alpha\beta$ is higher in temperature than that for the $\beta = V$ line. Of course, in the $P–T–X$ space the $\alpha = V$ and $\beta = V$ lines do not intersect.

The following coordinates of the invariant points were found from vapor pressure measurements: $T(N_2) = 937$ K, $P(N_2) = 68.9$ mmHg; $T(N_1) = 948$ K, $P(N_1) = 34.6$ mmHg. These are the extremities of the phase-transition in Zn_3As_2 at the vapor pressure of the system. The precision of the measurement (± 0.5 K, ± 0.15 mmHg) made it possible to scan the whole of the three-phase V$\alpha\beta$ curve. As a result, three separate nonoverlapping parts of the vapor pressure increase were located on the zinc-side of the V$\alpha\beta$ curve (Fig. 100, hollow points) and two local temperature extrema: $T'_{min} = 941$ K, and $T'_{max} = 946$ K.

An unusual geometrical configuration of the three-phase equilibrium of α-Zn_3As_2, β-Zn_3As_2, and vapor is the result of the complicated phase relations in the α-β transformation region. To elucidate the $P–T–X$ space model of the phase-transition area, isothermal sections of this region were made. Schematic $P–X$ isotherms are seen in Fig. 101. Figure 101.1 corresponds to the $T > T_{max}$ interval, and

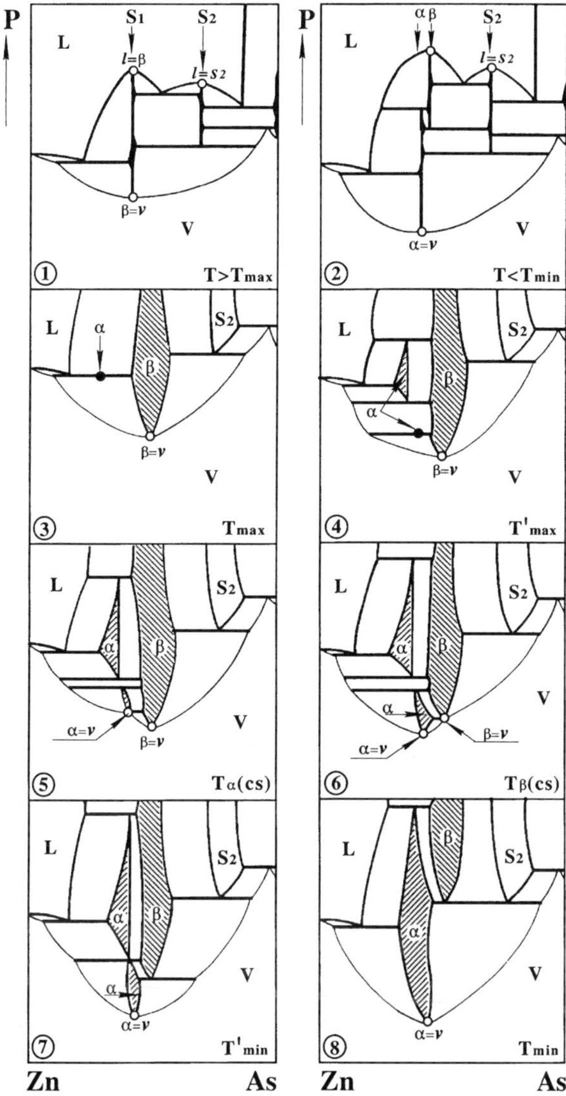

Fig. 101. Isothermal section of the α–β phase-transition region

Fig. 101.2 sections represent $T < T_{min}$. The homogeneity ranges of the crystalline phases are shown by narrow vertical shaded fields. The shape of the isotherms, Fig. 101.1 and Fig. 101.2, is determined by the negative slopes dP/dT of the univariant condensed phase equilibrium curves, as well as the congruent sublimation ($\alpha = V$, $\beta = V$) and congruent melting lines $\beta = L$, $S(ZnAs_2) = L$. The isotherms at temperatures gradually decreasing from $T_1 = T_{max}$ down to $T_6 = T_{min}$ are given in Fig. 101.3 through Fig. 101.8 for the pressure range of the α-β phase-transition. For obvious reasons, the homogenous ranges of the crystalline phases in Fig. 101 are greatly ex-aggerated.

At $T_1 = T_{max}$, the isothermal plane meets the phase transformation area at the invariant point N_1 of the four-phase equilibrium VLαβ. Accordingly, point α appears on the VLβ horizontal (Fig. 101.3) at $T = T_1$ (Fig. 101.3), and the composition coordinate $X_α$ corresponds to the maximum solubility of Zn in Zn_3As_2 (q.v. *T–X* projection, Fig. 100). In the temperature interval $T_{max} - T'_{max}$, cross-sections of the α-phase area are restricted by the univariant equilibrium lines Vαβ, VLα, and Lαβ (in the order of rising pressure). The $T_2 = T'_{max}$ isothermal plane crosses the Vαβ equilibrium twice, giving rise to two univariant horizontals Vαβ in Fig. 101.4. Figures 101.5 and 101.6 represent $T_α^{cs}$ and $T_β^{cs}$ isotherms and clearly show that the α = V line originates in the three-phase Vαβ equilibrium, whereas the β = V line touches the αβV branch of this curve. It is also seen in Figs. 101.5 and 101.6 that $X_β^{cs} > X_α^{cs}$ (in at.% As). $T_7 = T'_{min}$ is the lowest temperature of existence of the zinc-side of the α–β transformation in equilibrium with the vapor. Accordingly, Fig. 101.7 is the last isotherm to cross the Vαβ equilibrium. At $T < T'_{min}$, the saturated vapor in equilibrium with α-Zn_3As_2 and β-Zn_3As_2 is enriched in arsenic, and the corresponding *P–X* section contains αβV horizontals. When the temperature is reduced to $T_8 = T_{min}$, the third crystalline phase, S($ZnAs_2$) appears on the horizontal at point N_2 (Fig. 100) to participate in the invariant equilibrium αβS_2V (Fig. 101.8). An important characteristic of Fig. 101 is the invariable compositional sequence of phases $X_α < X_β$ throughout the whole $T_1 - T_8$ temperature interval of the phase-transition. Together with the absence of the $X_α = X_β$ point, it proves that the phase transformation in Zn_3As_2 is of the incongruent type at the vapor pressure of the system.

In conclusion, it should be stressed that such an unusual solid-state phase-transition was resolved by the vapor pressure scanning method for a compound with only about a 0.01 at.% non-stoichiometry region.

Vapor pressure scanning of non-stoichiometry in β-Zn₃As₂. Non-stoichiometry in Zn_3As_2 was studied only for the β-modification because the vapor pressure in the sublimation range of the α-form was not high enough for precise determination of the composition of this solid (the maximum vapor pressure in the range of existence of α-Zn_3As_2 is 68.9 mmHg at point N_2, Fig. 100). The congruent sublimation curve divides the two-phase equilibrium field of sublimation into two parts. In the first, the vapor is enriched in zinc compared to the solid (the sequence of phases is VS), and in the second (SV), the vapor is enriched in arsenic. To scan the non-stoichiometry of β-Zn_3As_2, a total of about 60 vapor pressure curves were obtained and analyzed in [244]. The composition of the solid was determined from Eq. (31), which can be written for the Zn–As system as

$$X_S = [N_{As1}v_2 - N_{As2}v_1]/\left\{(N_{Zn1} + N_{As1})v_2 - (N_{Zn2} + N_{As2})v_1\right\}. \tag{50}$$

The composition of the conjugated vapor is given by the equation

$$X_V = [2P(As_2) + 4P(As_4)]/[P(Zn) + 2P(As_2) + 4P(As_4)].$$ (51)

The experiments showed that vapor pressure curves in the fields VS and SV are essentially different. Hence, it would be appropriate to discuss the relevant results for these two fields separately.

Phase equilibrium VS. A total of 96 intersection points of the vapor pressure curves were obtained in VS. To fix their (P,T) coordinates, each of the curves was approximated in the polynomial form $\log P = \sum a_i T^i$, and the corresponding system of equations

$$P_i = f_i(T),$$
$$P_j = f_j(T),$$

was solved for every pair of vapor pressure curves $i \neq j$. At each of the intersection points the composition of the solid was determined from Eq. (50). The partial pressures of $Zn(g)$, $As_2(g)$, and $As_4(g)$ were calculated at the intersection points from the corresponding total vapor pressure and the equilibrium constant for the reaction $As_4(g) = 2As_2(g)$. After that, the composition of the vapor was calculated from Eq. (51). In this way, two conjugated data sets, $\{P,T,X_S\}$ and $\{P,T,X_V\}$, were obtained, which described the solidus and vaporus surfaces of the equilibrium VS.

To obtain the isotherms $P_i(X_S)|_T$ and isopleths $P_i(T)|_{X_S}$ of the partial pressures ($i = Zn$ or As_4), temperature dependences of $\log P_i$ and X_S were approximated in the polynomial form. Relevant calculations showed that in the interval of temperatures and vapor pressures not less than 3.3 kPa away from the minimum vapor pressure curve $S(Zn_3As_2) = V$, the vapor is made up of at least 99 at.% of zinc. That is why zinc partial pressures were computed from Eq. (50), whereas $As_4(g)$ partial pressures were calculated from the Gibbs–Duhem equation. Very strong dependence $X_V = \varphi(X_S)$ was observed in the vicinity of the $S(Zn_3As_2) = V$ curve. It was shown that $\partial X_V/\partial X_S$ at $T = \text{const}$ was as high as 10^5–10^6, meaning that even very small changes in the composition of $Zn_3As_{2\pm\delta}$ resulted in the vapor composition change from almost pure zinc to almost pure arsenic.

Isopleths ($X_S = \text{const}$) of the partial pressures for $Zn(g)$ and $As_4(g)$ are presented in Table 13. An interesting feature of Zn_3As_2 can be seen in Table 13: equilibrium VS at low temperatures and vapor pressures is on the zinc-side of the stoichiometric composition, $X < 40$ at.% As, whereas at high temperatures, it is on the $X > 40$ at.% As side. This means that the congruent composition is not constant and changes with the temperature along the $S(Zn_3As_2) = V$ curve.

Table 13. Partial pressures $P(Zn)$(kPa) and $P(As_4)$(kPa), activities a, and partial thermodynamic functions ΔH^M (kJ/mole) and ΔS^M (J/mole×K) for $Zn_3As_{2\pm\delta}$

X(at. %As)	Zinc						Arsenic					
	$logP$=A–B/T		$loga$=A–B/T		ΔH^M	ΔS^M	$logP$=A–B/T		$loga$=A–B/T		ΔH^M	ΔS^M
	A	B	A	B			A	B	A	B		
					Equilibrium VS							
39.995	7.3	6500	−0.79	−260	5.0	15.1	11.1	17500	0.280	2500	−48.1	−5.50
39.996	7.4	6700	−0.70	−90	1.7	13.4	10.5	16400	0.140	2250	−43.1	−2.50
39.997	7.5	6900	−0.61	70	−1.3	11.7	10.0	15500	0.002	2000	−38.5	−0.04
39.998	7.7	7200	−0.35	400	−7.5	6.7	8.4	13500	−0.380	1500	−29.1	7.50
39.999	8.0	7500	−0.10	700	−13.8	1.7	6.9	11600	−0.770	1050	−19.7	14.60
40.000	8.2	7700	0.07	940	−18.0	−1.3	5.9	10300	−1.010	700	−13.4	19.20
40.001	8.5	8200	0.44	1400	−26.4	−8.4	3.7	7600	−1.560	50	−0.8	30.10
40.002	8.6	8300	0.53	1500	−28.9	−10.0	3.2	6800	−1.700	−150	2.9	32.60
40.003	9.5	9300	1.40	2500	−48.5	−26.8	−2.1	750	−3.010	−1700	32.2	57.30
					Equilibrium SV							
40.022	8.9	9400	0.77	2600	−49.4	−14.6	1.7	420	−2.070	−1750	33.6	39.50
40.045	8.8	9400	0.73	2600	−49.4	−13.8	2.0	410	−2.000	−1750	33.6	39.50
40.052	8.8	9400	0.71	2600	−49.4	−13.4	2.1	420	−1.970	−1750	33.7	37.70
40.055	8.8	9400	0.69	2600	−49.4	−13.4	2.2	410	−1.950	−1750	33.7	37.40
40.062	8.8	9400	0.68	2600	−49.4	−13.0	2.3	420	−1.930	−1750	33.7	36.90

Phase equilibrium SV. Vapor pressure curves of this field do not intersect even with essential variation in the initial composition of the samples (40.02 to 46.58 at.% As) and m/v values (2.5 orders of magnitude). The reason is that the composition of the condensed phase in the process of heating in a closed volume in the SV equilibrium does not change, at least at $X_S > 40.02$ at.% As. Also unchanged (within experimental errors) is the Gibbs energy of the crystalline phase at T=const. The vapor composition in the SV equilibrium (Table 13) was calculated using this approximation. It can be seen in Table 13, that there is less than 1 at.% Zn in the vapors.

The calculated partial pressures of $Zn(g)$ and $As_4(g)$ as a function of the temperature and composition of the solid phase in equilibria SV and VS made it possible to calculate the activities of the components in the non-stoichiometric $Zn_3As_{2\pm\delta}$ in the investigated interval of the parameters. In these calculations, the saturated vapor pressures of the pure components were taken from Standard Thermodynamic Tables [141] for zinc and [246] for $As_4(g)$. From the activities, partial thermodynamic functions of the components in the single-phase region of existence of $Zn_3As_{2\pm\delta}$ were computed using standard thermodynamic methods. Numerical values are presented in Table 13.

Uncertainties in the composition of the solid X_S and vapor X_V were calculated by applying the error accumulation law to Eqs. (50 and 51). The accuracy in the crystal composition proved to be within 10^{-3}–10^{-4} at.%, and for the vapor δX_V was up to 1 at.%.

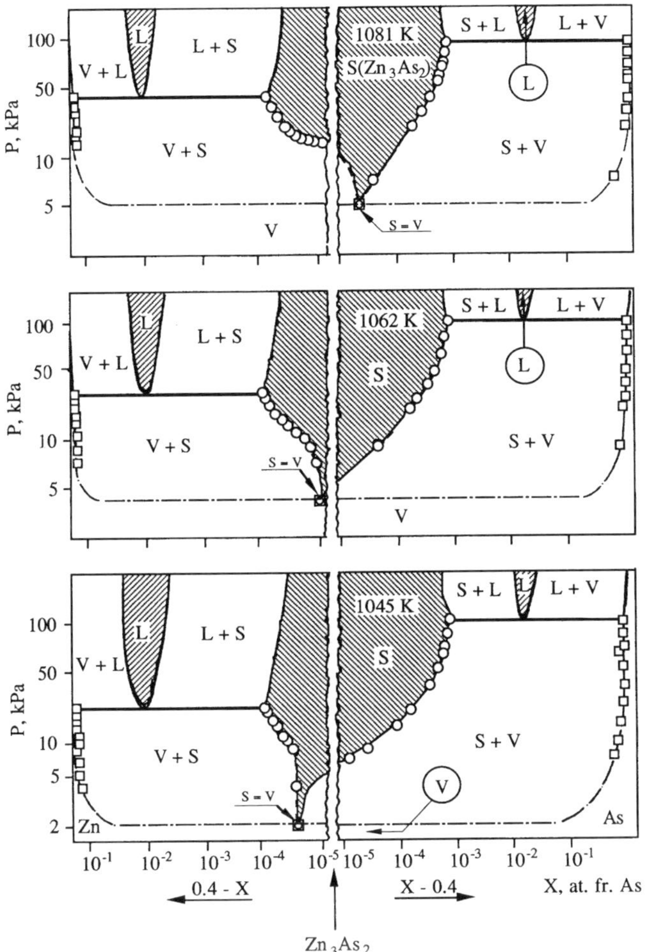

Fig. 102. Isothermal sections of the non-stoichiometry range of Zn_3As_2

P–T–X range of existence of Zn_3As_2. The key factor in outlining the single-phase volume of Zn_3As_2 is the P–T–X arrangement of the congruent sublimation curve $S(Zn_3As_2) = V$. To determine this, experimental vapor pressures were examined in the vicinity of the vapor pressure minimum in the system. The corresponding experimental data are cited in the first two columns of Table 14, and isothermal sections for some of these temperatures are shown in Fig. 102. The circles on the solidus surface, which accommodates the single-phase solidus volume S (shaded in Fig. 102), and the squares on the vaporus surface were calculated from the data, Table 13, for the SV and VS equilibria at the corresponding temperatures. These two branches, SV on the right and VS on the left, converge at the vapor pressure minimum. In Fig. 102, the single-phase range of existence of vapor V is below the vaporus surface, and the compositional axis is drawn on a logarithmic scale in terms of the deviation from the stoichiometric plane. Columns 3 and 4 in Table 14 list compositions of the conjugated solid and vapor phases at the corresponding experimental points. These were calculated by assuming that the Gibbs energy ΔG_S for $Zn_3As_{2\pm\delta}$ does not change appreciably at T=const within the single-phase range of existence of the solid. The ΔG_S values for each temperature listed in Table 14 were calculated from the sublimation data [242]. It is clear that the congruent sublimation composition X_{cs} at the minimum vapor pressure falls within the range $X_S(VS) < X_{cs} < X_S(SV)$ between the compositions of the crystal in the VS and SV equilibria. This is reflected in Table 14 and Fig. 102. An interesting consequence follows from the shape of the $Zn_3As_{2\pm\delta}$ solidus: in a specific temperature range, the over-stoichiometric $Zn_3As_{2+\delta}$ is in equilibrium with almost pure zinc vapor, whereas at other temperatures (see the first two rows in Table 14) the substoichiometric $Zn_3As_{2-\delta}$ is in equilibrium with arsenic-rich vapor.

To determine the maximum non-stoichiometry of Zn_3As_2 (zinc and arsenic saturation), isotherms of the partial pressures, $P(Zn) = f_1(X_S)$ and $P(As_4) = f_2(X_S)$, were approximated in the polynomial form, and intersections of these isotherms with the corresponding three-phase equilibrium curves were found. Zinc-saturation corresponds to the intersection of the Zn-isotherm with $VLS(Zn_3As_2)$ and arsenic-saturation to that of the As-isotherm and $S(Zn_3As_2)S(ZnAs_2)V$ (below the eutectic temperature) or $S(Zn_3As_2)LV$ (above the eutectic). The results are presented in the $T–X$ projection of the single-phase range of existence of Zn_3As_2 (Fig. 103) and are summarized in numerical form in Table 15. The following temperature dependences of the vapor pressure in the three-phase equilibria were obtained from the experimental data. For $VLS(Zn_3As_2)$ in the temperature range 740–1185 K,

$$\log P(\text{kPa}) = 6.675 - 5630/T \pm t\sigma,$$
$$\sigma = 0.0049 - 9.2/T + 4.3 \times 10^3/T^2; \quad r = 0.998.$$

Table 14. Composition of conjugated phases at some experimental (P,T) points and on the congruent sublimation curve $S(Zn_3As_2) = V$

T (K)	P (kPa)	X_S(at.% As)	X_V(at.% As)	$S(Zn_3As_2) = V$	
				P(kPa) [242]	$X_{S=V}$(at.% As)
1045	4.52	<39.998	>79.7	2.07	39.997 – 39.998
1062	5.05	<40.000	>75.4	2.77	39.998 – 40.000
1081	5.83	<40.002	>70.5	3.75	40.000–40.002
1098	6.81	<40.004	>65.4	4.93	—
1115	8.04	<40.005	>61.6	6.35	40.003 – 40.005
1135	10.16	<40.005	>58.2	8.59	40.004 – 40.005
1152	12.51	<40.005	>55.6	11.00	40.004 – 40.005

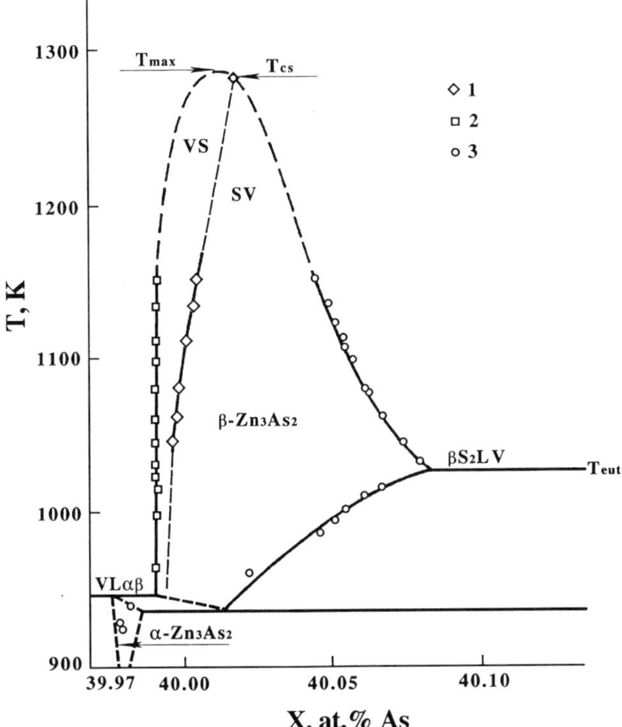

Fig. 103. Non-stoichiometric range of Zn_3As_2. 1 - congruent sublimation, 2 - maximum Zn solubility, 3 - maximum As solubility

For the equilibrium $S(Zn_3As_2)LV$ at $T = 1060–1185$ K,

$$\log P(kPa) = -1.984 + 4210/T \pm t\sigma,$$
$$\sigma = 0.0146 - 32/T + 1.8 \times 10^5/T^2; \quad r = 0.994.$$

In these equations, t is the Student criterion, and r is the correlation coefficient. Also in Table 15 are the boundary compositions of the $S(Zn_3As_{2\pm\delta})$ phase calculated at the experimental points of the phase-transition between the two- and three-phase equilibria. Although the vapor pressure for sublimation of the low-temperature α-Zn_3As_2 is too low (up to 6.7 kPa) to localize the single-phase volume of this phase in the P–T–X space precisely, nevertheless, it can be seen in Fig. 103 that the range of existence of this phase is completely on the zinc-side of the stoichiometric plane.

The single-phase volume of β-Zn_3As_2 (Fig. 103) is strongly asymmetric. The maximum solubility of zinc is about 10^{-3} at.% As and almost independent of the temperature, whereas the arsenic saturation reaches ~0.08 at.% As at the eutectic temperature. The uncertainties in the compositions of the solid phase, presented in

Table 15 and Fig. 103, were estimated using the error accumulation law. They were within $\pm(5\times10^{-5} - 5\times10^{-4})$ at.%.

3.1.7.4 Cadmium arsenides

Cadmium–arsenic was the first II–V system to be studied following the discovery of semiconductor properties in II–V compounds [246]. The *T–X* projection of this system was subsequently studied in [247–249] and the *P–T* projection was constructed by Nipan et al. [250,251] at vapor pressures up to 200 kPa and by Clark and Pistorius [252–254] at high pressures. The *P–T–X* space model of the phase equilibrium in Cd–As was constructed by Nipan et al. [255] for the entire available range of experimental parameters.

Table 15. Maximum non-stoichiometry of β-Zn_3As_2

T (K)	Zinc solubility				Arsenic solubility	
	P(VLS) (kPa)	X_S(VLS) (at.% As)	P(SLV) (kPa)	X_S(SLV) (at.% As)	$P(S_1S_2V)^a$ (kPa)	$X_S(S_1S_2V)$ (at.% As)
					Phase-transition $S_1S_2V \rightarrow S_1V$	
960					19.93	40.022
985					38.60	40.045
994					49.26	40.052
1002					59.92	40.055
1012					76.79	40.062
1045	19.57	39.9914	111.58	40.0748		
1062	23.98	39.9915	95.83	40.0696		
1081	29.58	39.9917	81.90	40.0643		
1098	35.69	39.9919	71.17	40.0595		
1115	42.50	39.9920	62.45	40.0551		
1135	52.33	39.9922	53.44	40.0491		
1152	62.11	39.9924	47.01	40.0454		
	Phase-transition VLS $\rightarrow$ VS		Phase-transition SV $\rightarrow$ SLV			
964	6.39	39.9925				
998	10.16	39.9924				
1024	14.25	39.9929				
1078			83.19	40.062		
1108			68.53	40.055		
1125			60.00	40.052		

^{a} S_1 is Zn_3As_2; S_2 is $ZnAs_2$.

***P–T* projection.** Nipan et al. [250,251] reported results of total vapor pressure measurements in Cd–As using quartz Bourdon gauge in the temperature range 650–1000 K and pressures up to 200 kPa. The following univariant equilibria are presented in Fig. 104: VLS_1, S_1S_2V, $S_2S_{As}V$, S_1LV, and $S_1 = V$, where S_1, S_2, and S_{As} correspond to the crystalline phases Cd_3As_2, $CdAs_2$, and As; L is the melt, and V is vapor. The minimum vapor pressure in the sublimation field of $Cd_3As_2(s)$ proves that this phase sublimes congruently, whereas the other cadmium arsenide, $CdAs_2$, sublimes incongruently. Both compounds melt congruently. The maximum melting temperature of Cd_3As_2 is 988 K at vapor pressure of 20 kPa; for $CdAs_2$ the corresponding parameters are $T_{max} = 900.5$ K and $P(T_{max}) = 60$ kPa. Invariant eutectic points appear in Fig. 104 at the following coordinates: $E_1(S_1LS_2V)$ at $T_{E_1} = 883$ K, $P(T_{E_1}) = 14.7$ kPa; $E_2(S_2LS_{As}V)$ at $T_{E_2} = 892.5$ K, $P(T_{E_2}) = 105$ kPa. The vapor pressure minimum in the S_1LV equilibrium appears at $T_{min} = 941.0$ K, $P(T_{min}) = 13.3$ kPa. According to Clark and Pistorius [252], Cd_3As_2 melts congruently, and the congruent melting temperature has a negative pressure dependence at pressures up to 2.45 GPa. A similar $P–T–X$ phase arrangement of the congruent sublimation $S_1 = V$, congruent melting $S_1 = L$, and three-phase univariant curves S_1S_2V, S_1LV, and VLS_1 in the zinc–arsenic system (Fig. 98 [256]) resulted in the compositional inversion L $\leftrightarrow$ V of the liquid and vapor phases at the congruent vaporization point, where the S_1LV is tangent to the congruent vaporization curve L = V. According to [253], $CdAs_2$ is also a congruent melting phase. The $S(CdAs_2) = L$ curve has a negative $P–T$ slope and touches the LS_2V three-phase equilibrium curve at the congruent melting point, which is lower than the maximum melting temperature and on the E_2 side of it [251].

$P–T$ projection of high-pressure phase equilibria were reported by Clark and Pistorius [253] from DTA measurements at 750–900 K and pressures up to 5 GPa. In Fig. 104, these equilibria are shown as bold curves E_2E_3, E_3E_5, E_3E_4, E_3E_6, and E_1E_4. The congruent melting curve $S(Cd_3As_2) = L$ in Fig. 104 is taken from [252]. The E_2E_3, E_3E_5 and E_3E_6 lines correspond to the decomposition of $CdAs_2(s)$ into crystalline arsenic S_{As} and cadmium monoarsenide CdAs(s) labeled S_3 in Fig. 104. The two-phase equilibrium $S(CdAs_2)S(CdAs)$ is below E_3E_6, and $S(CdAs)S(As)$ is above E_3E_6. Hence, E_3E_6 is the univariant three-phase equilibrium $S_3S_2S_{As}$ between CdAs(s), $CdAs_2(s)$ and crystalline arsenic. This curve is a border-line between two two-phase regions, $S(CdAs)S(CdAs_2)$, and $S(CdAs)S(As)$. Curve E_3E_5 is the high-temperature limit of the $S(CdAs)S(As)$ equilibrium, corresponding to the appearance of the liquid phase L. Consequently, it is the three-phase equilibrium curve $S(CdAs)S(As)L$, and E_3 is the invariant point of the four-phase equilibrium $S(CdAs)LS(CdAs_2)S(As)$. From the Phase Rule it follows that E_3 should be an intersection point of four univariant equilibrium curves: the previously mentioned $S(CdAs)S(CdAs_2)S(As)$ and $S(CdAs)LS(As)$ and two other, $LS(CdAs_2)S(As)$ and $S(CdAs)LS(CdAs_2)$. The former, E_2E_3, originates in the eutectic point E_2 and passes through the congruent melting point of $CdAs_2(s)$, at which the congruent melting curve $S(CdAs_2) = L$ touches the three-phase equilibrium line resulting in the change in the compositional sequence of phases L and $CdAs_2(s)$. The experimental curve E_4E_3 [253] describes the equilibrium $S(CdAs)LS(CdAs_2)$, and E_1E_4, which corresponds to the three-phase equilibrium $S(Cd_3As_2)LS(CdAs_2)$, originates in the invariant point E_4. According to [250], this is the only condensed phase equilibrium originating in the E_1 eutectic. Thus, from high pressure experimental data [252–254], the coordinates of two invariant equilibria were obtained: $E_3(S_3LS_2S_{As})$ at $T = 853$ K and $P = 2$ GPa, resulting from the intersection of experimental curves $LS(CdAs_2)S(As)$, $S(CdAs)LS(As)$, $S(CdAs)S(CdAs_2)S(As)$ and $S(CdAs)LS(CdAs_2)$, and $E_4(S_1S_3LS_2)$ at $T = 863$ K and $P = 800$ MPa, resulting from intersection of the univariant curves $S(Cd_3As_2)LS(CdAs_2)$ and $S(CdAs)LS(CdAs_2)$.

The *P–T* projection, constructed from experimental data [250–254] (Fig. 104, bold lines), is supplemented by the projections of four additional univariant equilibria, not obtained experimentally. These are $S(Cd_3As_2)S(CdAs)L$, $S(Cd_3As_2)S(CdAs)S(CdAs_2)$, $S(Cd_3As_2)S(CdAs)S(As)$, and $S(Cd_3As_2)LS(As)$. The existence and location of these equilibria (thin lines in Fig. 104) follow from the phase rule. Extrapolation of the univariant curves $S(Cd_3As_2)S(CdAs)L$ and $S(CdAs)LS(As)$ gives the invariant point $E_5(S_1S_3LS_{As})$ with temperature close to $T(E_3)$ and $T(E_4)$. The coordinates of the E_6 eutectic point ($S_1S_3S_2S_{As}$ equilibrium) were not obtained experimentally. This point was tentatively positioned in Fig. 104 by applying the Schreinemakers rule. Finally, it follows from the differential thermal analysis data [247–250] that there is a eutectic equilibrium between solid cadmium, $Cd_3As_2(s)$, liquid, and vapor, $E_0(S_{Cd}VLS_1)$, which is the intersection point of the univariant lines $S(Cd)VL$, $S(Cd)LS(Cd_3As_2)$, $S(Cd)VS(Cd_3As_2)$, and $VLS(Cd_3As_2)$. The vapor pressure at E_0 was below the experimental limit of measurement ($\sim$10 Pa) [250].

***T–X* projection.** In Fig. 104e the *T–X* projection of the *P–T–X* phase diagram is constructed in accord with the *P–T* projection. It comprises the following features:

1. The triple point temperatures of components $T(Cd, s, l, v)$ and $T(As, s, l, v)$.
2. The temperatures of seven invariant points corresponding to four-phase equilibria $T_{E_0}(S_{Cd}VLS_1)$, $T_{E_1}(S_1LS_2V)$, $T_{E_2}(S_2LS_{As}V)$, $T_{E_3}(S_3LS_2S_{As})$, $T_{E_4}(S_1S_3LS_2)$, $T_{E_5}(S_1S_3LS_{As})$, $T_{E_6}(S_1S_3S_2S_{As})$, and the conodes connecting the compositions of all four phases involved in the appropriate invariant equilibria at these temperatures.
3. The solidus of the crystalline phases $S(Cd)$, $S(Cd_3As_2)$, $S(CdAs)$, $S(CdAs_2)$, $S(As)$, and the maximum melting temperatures of these phases.
4. The vaporus curve (thin lines v), which describes the temperature dependence of the composition of vapors in the three-phase equilibria. The vaporus consists of the following branches. The part going down from the $S(Cd)VLS_1$ conode corresponds to the $S(Cd)VS(Cd_3As_2)$ equilibrium; up toward $T(Cd, s, l, v)$ from this conode the $S(Cd)VL$ equilibrium rises; $VLS(Cd_3As_2)$ goes from $T(E_0)$ to $T_{max}(Cd_3As_2)$; $S(Cd_3As_2)LV$ is from $T_{max}(Cd_3As_2)$ down to S_1LS_2V; $LS(CdAs_2)V$ equilibrium goes from the S_1LS_2V conode up toward $T_{max}(CdAs_2)$; the $S(Cd_3As_2)S(CdAs_2)V$ equilibrium is down from that conode; from $T_{max}(CdAs_2)$ to the eutectic $S_2LS_{As}V$ is $S(CdAs_2)LV$; from the $S_2LS_{As}V$ conode down is the $S(CdAs_2)S(As)V$ equilibrium; and from that conode up toward $T(As, s, l, v)$ is the vaporus of the $LS(As)V$ equilibrium. It is evident from the *T–X* projection that in the entire range of *P–T* parameters, except the region

adjacent to the minimum vapor pressure curve $S(Cd_3As_2) = V$, the saturated vapor is made up practically of one component, either cadmium or arsenic.

5. Liquidus l, which describes the temperature dependence of the melt composition in three-phase equilibria with saturated vapor, is composed of several parts. $E_0T(Cd, s, l, v)$ corresponds to the $S(Cd)VL$ equilibrium; $E_0LT_{max}(Cd_3As_2)$ corresponds to the $VLS(Cd_3As_2)$ equilibrium; $T_{max}(Cd_3As_2)E_1$ is for $S(Cd_3As_2)LV$; $E_1T_{max}(CdAs_2)$ corresponds to $LS(CdAs_2)V$; $T_{max}(CdAs_2)E_2$ is for $S(CdAs_2)LV$; and $E_2T(As, s, l, v)$ for the $LS(As)V$ equilibrium.

6. Liquidus l in the condensed phase equilibria describes the temperature dependence of the liquid composition in the following three-phase equilibria. E_1E_4 is the melt in the $S(Cd_3As_2)LS(CdAs_2)$ equilibrium, E_4E_3 is for $S(CdAs)LS(CdAs_2)$, E_4E_5 is the liquid in $S(Cd_3As_2)S(CdAs)L$, E_5E_3 describes $S(CdAs)LS(As)$, E_5l is for $S(Cd_3As_2)LS(As)$, and E_3E_2 is for $LS(CdAs_2)S(As)$ and $S(CdAs_2)LS(As)$. In the last three-phase equilibrium, the change in the compositional sequence of phases $L \leftrightarrow S(CdAs_2)$ is observed where the liquidus E_3E_2 intersects the $CdAs_2$ solidus. To simplify the T–X projection somewhat, congruent curves $S(Cd_3As_2) = V$, $S(Cd_3As_2) = L$, $L = V$ and $S(CdAs_2) = L$ are left out of Fig. 104e. They are shown only in the P–T projection, Fig. 104a. It is to be noted that at the congruent points the corresponding compositional sequence of the congruent phases in the three-phase equilibrium changes.

***P–T–X* space model of phase equilibrium.** It was pointed out in Part 1 that the phase space in a binary system is made up of two types of volumes, single-phase and two-phase. The latter correspond to two-phase equilibria, and the compositions of the conjugated phases are to be found on the surfaces that restrict the single-phase volumes. These surfaces are projected onto the P–T plane as fields within the univariant curves. Consequently, if the surfaces of all of the two-phase equilibria, in which a certain phase participates, are determined, then the single-phase volume of this phase is also determined. Moreover, if the P–T projections of these surfaces are outlined, then the result is that the P–T projection of the single-phase range of existence of this phase is also outlined. For example, $CdAs(s)$ (S_3 in Fig. 104b) participates in four two-phase equilibria: with the liquid, S_3L, which is projected onto the P–T plane as the $E_3E_4E_5$ field bounded by the three-phase curves S_1S_3L, S_3LS_2, and S_3LS_{As}; with crystalline arsenic (S_3S_{As} within $E_3E_5E_6$); with $CdAs_2(s)$ (the S_3S_2 field $E_3E_4E_6$ confined within S_3LS_2, $S_3S_2S_{As}$, and $S_1S_3S_2$); and with $Cd_3As_2(s)$ (the S_1S_3 field within $E_4E_3E_6$ restricted by S_1S_3L, $S_1S_3S_{As}$, and $S_1S_3S_2$). Consequently, $E_3E_4E_5E_5$ is the P–T projection of the single-phase range of existence of $CdAs(s)$, which is stable in the temperature range $T(E_6)$–$T(E_4)$ and the pressure range $P(E_4)$–$P(E_5)$. In Fig. 104b, the P–T projection of the single-phase volume S_3 is plotted separately, as well as that for two other cadmium arsenides, $Cd_3As_2(s)$ (Fig. 104c) and $CdAs_2(s)$ (Fig. 104d). Along with the T–X projection of these single-phase volumes (seen on the scale of Fig. 104e as vertical lines S_1, S_2, and S_3), they define the shape and space arrangement of the single-phase volumes of cadmium arsenides.

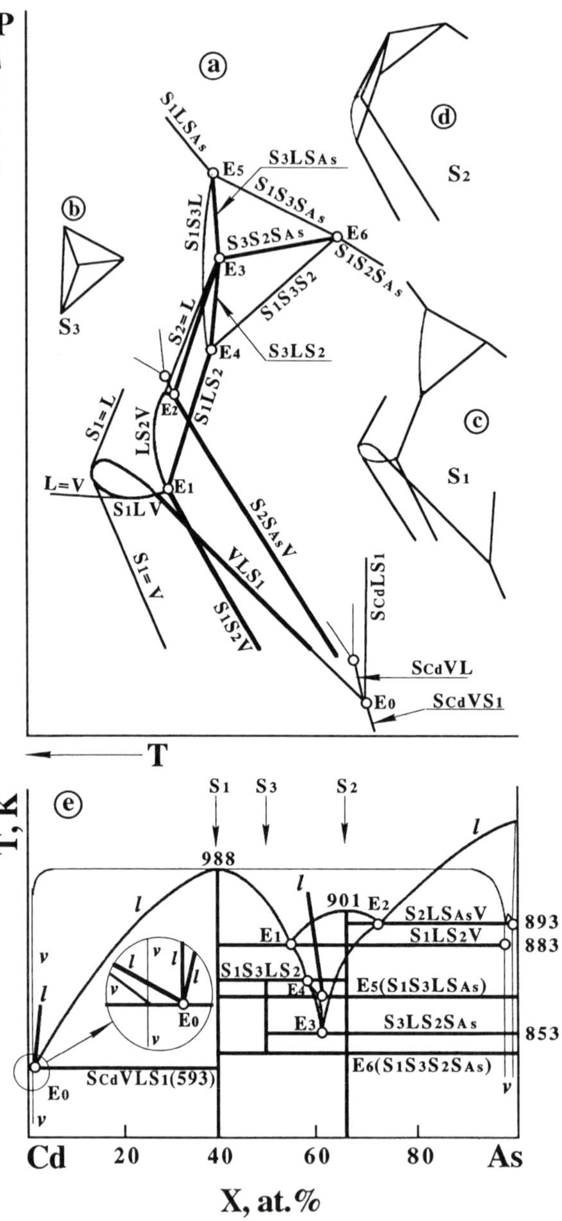

Fig. 104. *P–T* and *T–X* projections of the Cd–As diagram. *P–T* projections of the existence ranges of cadmium arsenides are shown separately

To reconstruct the three-dimensional model of heterogeneous equilibria for the Cd–As system in the P–T–X phase space, it is sufficient to have two projections, P–T (Fig. 104a) and T–X (Fig. 104e). However, because of the large number and the complicated shapes of the surfaces, that contain the single-phase volumes of individual phases, it is helpful to have cross-sections of the diagram as well. Figure 105 is a succession of isobaric sections of the P–T–X diagram for the Cd–As system. The selected representative pressures illustrate the relative arrangement of individual phases in the Cd–As system and show the possible ways of preparing cadmium arsenides from different matrices. The sections are given in the order of pressures increasing from E_0 up to E_5.

Sections of the single-phase volumes in Fig. 105 are all labeled, and those of cadmium arsenides are shaded. The open spaces between the single-phase fields correspond to two-phase equilibria, and the horizontal lines are for the three-phase equilibria. The crystalline phases are shown on an enhanced scale to facilitate understanding crystallization processes. It is apparent from Figs. 104 and 105 that CdAs (S_3) is the most difficult cadmium arsenide to prepare because it exists in a rather narrow range of temperature $T(E_6) < T < T(E_4)$ and pressure $P(E_4) < P < P(E_5)$. The only way to obtain single-phase CdAs(s) is to crystallize it from an arsenic-rich melt at high pressures (Fig. 105,13–15). The resulting CdAs is As-saturated. Preparation of Cd-rich CdAs is a very complicated task because the only two-phase equilibrium on the Cd-side of CdAs is with Cd_3As_2(s). $CdAs_2$(s) occupies a much larger part of the P–T–X space. It can be crystallized from the melt (Fig. 105,5–13), vapor (Fig. 105,1–8), or in the three-phase equilibrium S_2LV (the so-called VLS-method, Fig. 105,5–9). The variety of preparative routes makes it possible to obtain $CdAs_2$(s) with different non-stoichiometries, enriched either in Cd or As and consequently with different properties. The third cadmium arsenide, Cd_3As_2(s), occupies an even larger portion of the P–T–X space. It may be prepared in a single-phase form from the melt (Fig. 105,4–16), vapor (Fig. 105,1–6), or (liquid+vapor) (Fig. 105,4–7). Depending on the growth conditions, Cd_3As_2(s) can be enriched in either As or Cd up to different concentration levels.

Metastable states in the Cd–As system. A characteristic feature of the Cd–As system is the tendency of the alloys with the composition $X > 40$ at.% As to form metastable states. The metastable eutectic (Cd_3As_2 + As) was observed for the first time in an early study of the Cd–As system [246]. The eutectic temperature was given as 799 K and the composition $X_E = 62.5$ at.% As. While studying the $CdAs_2$–As portion of the diagram, Gukov et al. [247] observed the metastable eutectic (Cd_3As_2 + As) at $T_E = 803$ K, and $T_E = 781$ K was reported in [249]. In subsequent investigations of the Cd–As system [248,249], the metastable phase was reported with a composition between $CdAs_2$ and $CdAs_4$, which gave eutectics with arsenic at $T_E = 853$ K and with Cd_3As_2 at $T_E = 821$ K. Metastable modification of $CdAs_2$ with the composition Cd_2As_5 reportedly have a maximum temperature of existence of 823 K and gave two metastable eutectics, with Cd_3As_2 at 808 K and with arsenic at 778 K. The metastable $\alpha \rightarrow \beta$ phase-transition was observed for $CdAs_2$ at 683 K [248]. Metastable crystallization of cadmium arsenides was also observed at high pressures [252]. All of the reported metastable states in the Cd–As system characteristically had no crystallization field of $CdAs_2$. Evaluation of the thermodynamic stability of $CdAs_2(s)$ using thermodynamic data for cadmium arsenides [221, 246, 252, 257, 258] showed that there was a possibility for crystallization of $Cd_3As_2(s)$, $CdAs(s)$, and $As(s)$ or their mixtures in the *P–T–X* field of the equilibrium single-phase existence of $CdAs_2(s)$. Detailed analysis of the experimental data on metastable crystallization in Cd–As in connection with the *P–T–X* space model of this system was reported by Nipan et al. [259].

Metastable diagrams. Metastable diagrams, Figs. 106–108, were constructed using fragments of *P–T* and *T–X* projections of the *P–T–X* phase diagram, Fig. 104. Solid lines in Figs. 106–108 describe univariant equilibria in the *P–T* projection and the corresponding liquidus and solidus curves in the *T–X* projection. The dashed lines indicate their metastable extensions with labels in brackets. Four-phase metastable states are labeled M, and the indices correspond to the type of diagram to be specified in the later discussion. On *T–X* projections, presented for the composition interval 40–100 at.% As (the Cd_3As_2–As portion of the diagram), the vaporus curve is left out to avoid overcrowding the diagrams. As mentioned earlier, the vapors in the composition range $X > 40$ at.% As are made up of almost pure arsenic.

Thermodynamic analysis showed that metastable states of two types are to be expected in the Cd–As system: (1) crystallization of Cd_3As_2 and arsenic instead of the equilibrium crystallization of $CdAs_2$; (2) crystallization of $CdAs(s)$ and arsenic instead of $CdAs_2$. Detailed deliberation of these possibilities was reported in [259].

Any three-phase equilibrium in a binary system can be reached from three different two-phase equilibria by changing the (P,T) parameters (cooling or heating, increasing or decreasing the pressure). For example, the three-phase equilibrium between $Cd_3As_2(s)$, liquid, and vapor (S_1LV in Fig. 104) can be attained in three ways: (1) crystallization of Cd_3As_2 from the liquid–vapor equilibrium by decreasing the temperature or pressure; (2) from the two-phase equilibrium of $Cd_3As_2(s)$ with vapor by partial melting of Cd_3As_2 (heating) or condensation of the vapor (increasing the pressure); (3) by partial evaporation of the melt in the two-phase equilibrium of $Cd_3As_2(s)$ with the liquid (heating or reducing the pressure). When two-phase mixtures are supercooled or superheated (or the pressure is changed correspondingly), the third, metastable phase can be formed instead of the equilibrium phase. The nature of this metastable phase is determined by the metastable extension of the corresponding three-phase equilibrium. The metastable three-phase state, formed in this way, can also be attained by three different (P,T) routes, from three two-phase mixtures. The possibility of forming metastable states depends on the kinetic characteristics of the specific experiment. In the following discussions, the most probable way of crystallization is chosen for every type of metastable state.

Metastable states, Type I. These states originate from the relatively low thermodynamic stability of $CdAs_2(s)$ with respect to decomposition into $Cd_3As_2(s)$ and crystalline arsenic. They may result from supercooling the melt from the (liquid + vapor) or (Cd_3As_2 + liquid) equilibrium or a non-equilibrium decrease of pressure from the S(As)V equilibrium. Metastable crystallization of the melt in the Cd–As system results in the following changes of the phase state.

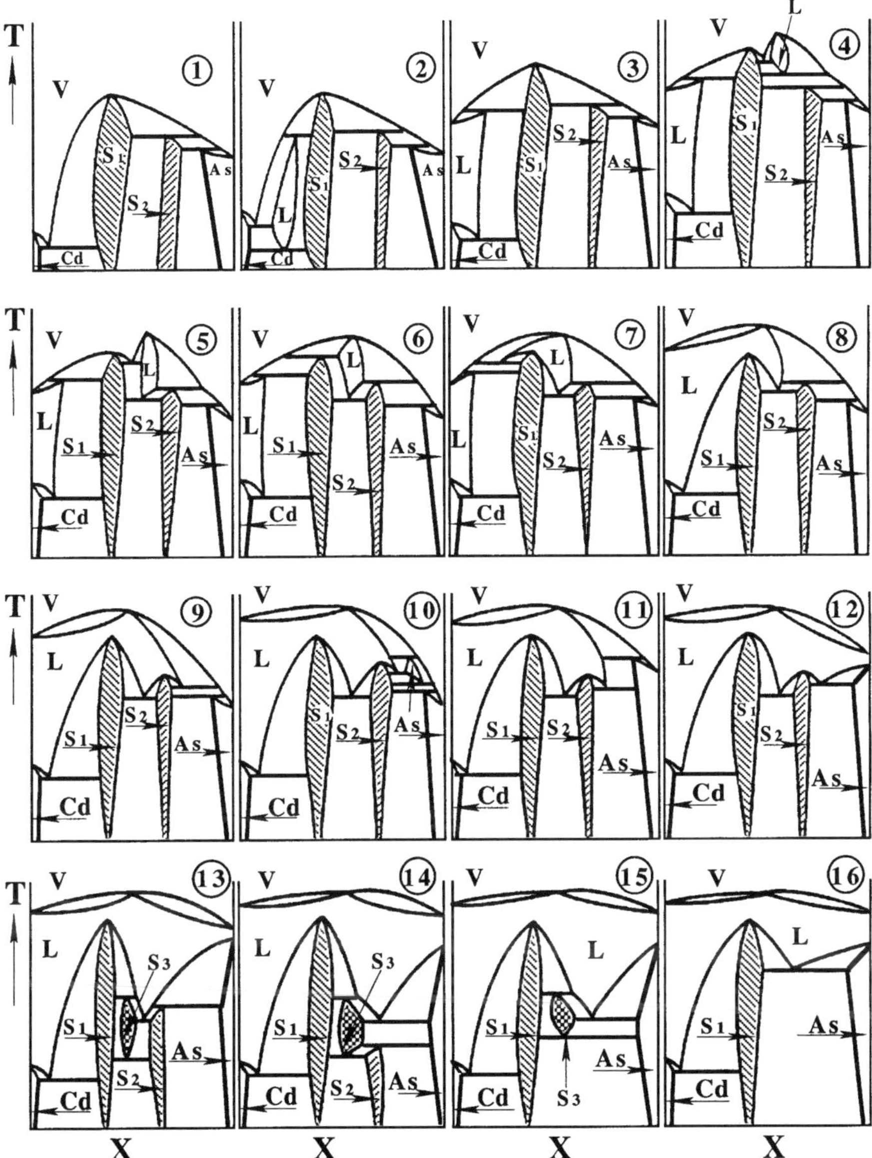

Fig. 105. Isobaric sections of the Cd–As diagram. 1-$P<P(E_0)$, 2-$P(E_0)$–$P(Cd)$, 3-$P(Cd)$–$P_{min}(S_1LV)$, 4-P_{cv}–$P(E_1)$, 5-$P(E_1)$–P_{cs}, 6-P_{cs}–$P_{cf}(S_1)$, 7-P_{cf}–$P_{max}(VLS_1)$, 8-$P_{max}(VLS_1)$–$P_{cf}(S_2)$, 9-$P_{cf}(S_2)$–$P(E_2)$, 10-$P(E_2)$–$P_{max}(S_2S_{As}V)$, 11-$P_{max}(S_2S_{As}V)$–$P(As)$, 12-$P(As)$–$P(E_4)$, 13-$P(E_4)$–$P(E_3)$, 14-$P(E_3)$–$P(E_6)$, 15-$P(E_6)$–$P(E_5)$, 16-$P>P(E_5)$

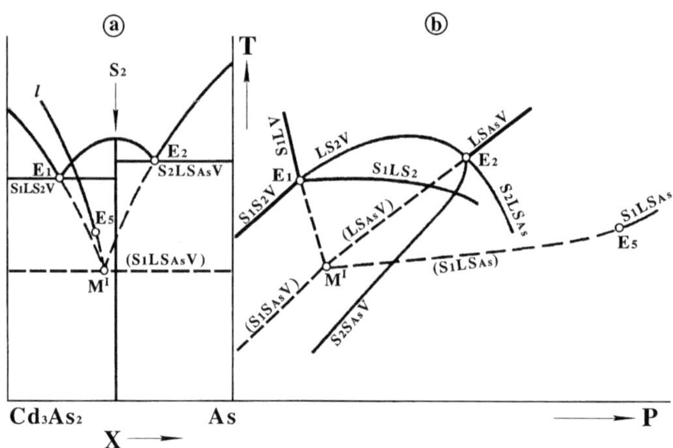

Fig. 106. *T–X* and *P–T* projections of the Type I Cd–As metastable diagram

The *P–T* projection of the LV equilibrium (Fig. 106b) is above (in temperature) and below (in pressure) the lines $LS_{As}V$ and S_1LV. The metastable field is $E_1M^IE_2$. In the *T–X* projection (Fig. 106a) this region is above the liquidus $S_1E_1M^IE_2S_{As}$ with the metastable part $E_1M^IE_2S_2$. When alloys in the composition interval $X(E_1)–X(M^I)$ (#1 in Table 16) are cooled from the LV equilibrium, in certain experimental conditions the melt may be supercooled below the temperature of the three-phase equilibrium LS_2V (Fig. 106b) down to the temperature corresponding to the E_1M^I line, at which S_1 may crystallize from this mixture. If the initial composition of the alloy falls within the $X(M^I)–X(E_2)$ interval (#2 in Table 16), then supercooling of the liquid may result in going down to the E_2M^I line, where metastable crystallization of arsenic will be observed. On further cooling, at temperature $T(M^I)$, the supercooled liquid should crystallize completely, and the metastable state $(S_1S_{As}V)$ may be preserved down to room temperature, passing along the M^IA line (Fig. 106b). These processes could be observed at low pressures [246, 247].

Supercooling the liquid is also possible at high pressures from the S_1L equilibrium, which in the *P–T* projection is above the three-phase equilibria S_1LV, S_1LS_2, and S_1LS_{As} (Fig. 106b). The metastable part of this region is between the lines S_1LS_2, E_1M^I, and M^IE_5 in the *P–T* projection, and the composition of the supercooled melt is determined by the metastable parts of the liquidus, M^IE_5, and M^IE_1 in the *T–X* projection. Supercooling the liquid from the narrow region of the composition, adjacent to $X(M^I)$ and corresponding to #3 in Table 16, may result in metastable crystallization of the $Cd_3As_2(s) + As(s)$ mixture along the M^IE_5 line, instead of the equilibrium crystallization of $CdAs_2(s)$ in the S_1LS_2 equilibrium.

Crystallization of the mixture (Cd_3As_2+As) instead of $CdAs_2(s)$ from the vapor is also possible, by a metastable decrease of the pressure in the $S_{As}V$ equilibrium (Table 16, #4). In the *P–T* projection, this equilibrium fills up the space between the $LS_{As}V$, $S_2S_{As}V$ equilibria and the sublimation curve of pure arsenic. As a result, the metastable state $(S_1S_{As}V)$ may be formed instead of the $S_2S_{As}V$ equilibrium (Fig. 106b).

The metastable state, Type I, was observed in all of the experimental studies of this phenomenon in the Cd–As system. In [260], crystallization of samples with the composition 66.58 at.% As and 66.72 at.% As resulted in three-phase metastable lines $(LS_{As}V)$ and $(S_1S_{As}V)$ and $T(M^I) = 788$ K. As mentioned earlier, other temperatures for point M^I were reported: 799, 803, and 781 K [246–248, 252]. This spread in temperature is quite understandable because kinetic specifics of the experiments could result in crystallization of the melt even before reaching the $T(M^I)$ temperature.

Table 16. Three-phase metastable states in the Cd–As system

#	ITE[a]	Initial composition	Process	NTE[b]	MCP[c]	TMS[d]	Labeling, Fig.106–108
				Type I			
1	LV	$X(E_1)–X(M^I)$	Cooling	LS_2V	(S_1)	(S_1LV)	E_1M^I
2	LV	$X(M^I)–X(E_2)$	Cooling	S_2LV,LS_2V	(S_{As})	$(LS_{As}V)$	M^IE_2
3	S_1L	$X(E_5)–X(M^I)$	Cooling	S_1LS_2	(S_{As})	(S_1LS_{As})	M^IE_5
4	$S_{As}V$	~100 at.% As	Pressure decrease	$S_2S_{As}V$	(S_1)	$(S_1S_{As}V)$	AsM^I
				Type II			
5	S_1L	$X(M_1^{II})–X(E_4)$	Cooling	S_1LS_2	(S_3)	(S_1S_3L)	$M_1^{II}E_4$
6	LS_{As}	$X(E_3)–X(M_2^{II})$	Cooling	S_2LS_{As}	(S_3)	(S_3LS_{As})	$E_3M_2^{II}$
7	LV	$X(E_1)–X(M_1^{II})$	Cooling	LS_2V	(S_1)	(S_1LV)	$E_1M_1^{II}$
8	LV	$X(M_2^{II})–X(E_2)$	Cooling	LS_2V,S_2LV	(S_{As})	$(LS_{As}V)$	$M_2^{II}E_2$
9	LV	$X(M_1^{II})–X(M_2^{II})$	Cooling	LS_2V,S_2LV	(S_3)	(S_3LV)	$M_1^{II}M_2^{II}$
				Type III			
10	LS_{As}	$X(M_2^{III})–X(M_3^{III})$	Cooling	S_2LS_{As}	(S_4)	(LS_4S_{As})	$M_2^{III}M_3^{III}$
11	S_1L	$X(M_1^{III})–X(M_3^{III})$	Cooling	S_1LS_2	(S_4)	(S_1LS_4)	$M_1^{III}M_3^{III}$
12	LV	$X(M_2^{III})–X(M_1^{III})$	Cooling	LS_2V,S_2LV	(S_4)	(LS_4V)	$M_1^{III}M_2^{III}$

[a] ITE: initial two-phase equilibrium
[b] NTE: non-realized three-phase equilibrium
[c] MCP: metastable crystallizing phase
[d] TMS: three-phase metastable state

Metastable states, Type II. At high pressures, phase formation in the Cd–As system depends on the thermodynamic properties of CdAs(s). Kinetic specifics of the experiment [252] could lead to crystallization of CdAs(s) and arsenic instead of $CdAs_2$(s). Supercooling the liquid from the S_1L region (between the three-phase curves S_1LV, S_1LS_2, S_1S_3L and S_1LS_{As} in the P–T projection, Fig. 107b) in the

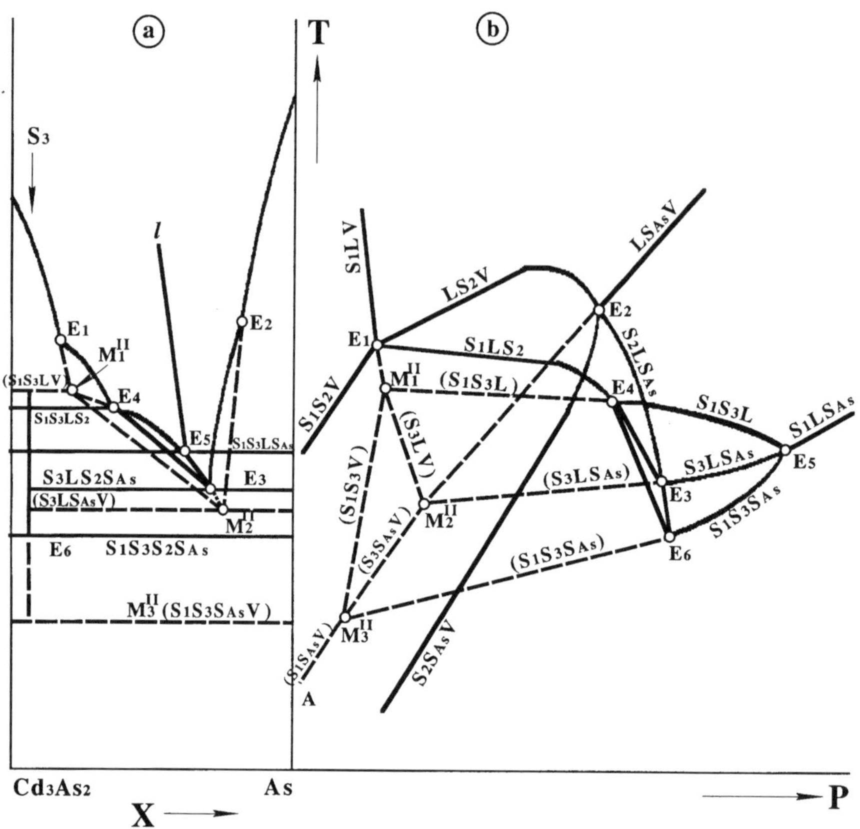

Fig. 107. *T–X* and *P–T* projections of the Type II Cd–As metastable diagram

composition interval $X(M_1^{II})$–$X(E_4)$ (Table 16, #5) leads to crystallization of S_3 at temperatures determined by the metastable line (S_1S_3L). The composition of the melt in this case varies along the metastable liquidus $M_1^{II}E_4$ (Fig. 107a).

CdAs(s) can also be formed in the composition interval #6 in Table 16 from the LS_{As} equilibrium   (the region between $LS_{As}V$, S_2LS_{As}, S_3LS_{As}, and S_1LS_{As} in Fig. 107b) along the (S_3LS_{As}) curve down to the four-phase metastable state M_2^{II}. In this case, the composition of the melt moves along the metastable liquidus $E_3M_2^{II}$ (Fig. 107a) down to the $(S_3LS_{As}V)$ horizontal. The metastable solidus S_3, shown in Fig. 107a by the dashed vertical line, is superimposed over the equilibrium solidus and lies between (S_1S_3LV) and $(S_1S_3S_{As}V)$.

When alloys with composition #1 and #2, Table 16, are rapidly cooled from the two-phase equilibrium LV, CdAs(s) can be formed from the supercooled melt (#9 in Table 16), along with $Cd_3As_2(s)$ and As(s) (#7 and #8). The metastable state (S_3LV) is the result of this process. But unlike the metastable diagram, Type I, in this case

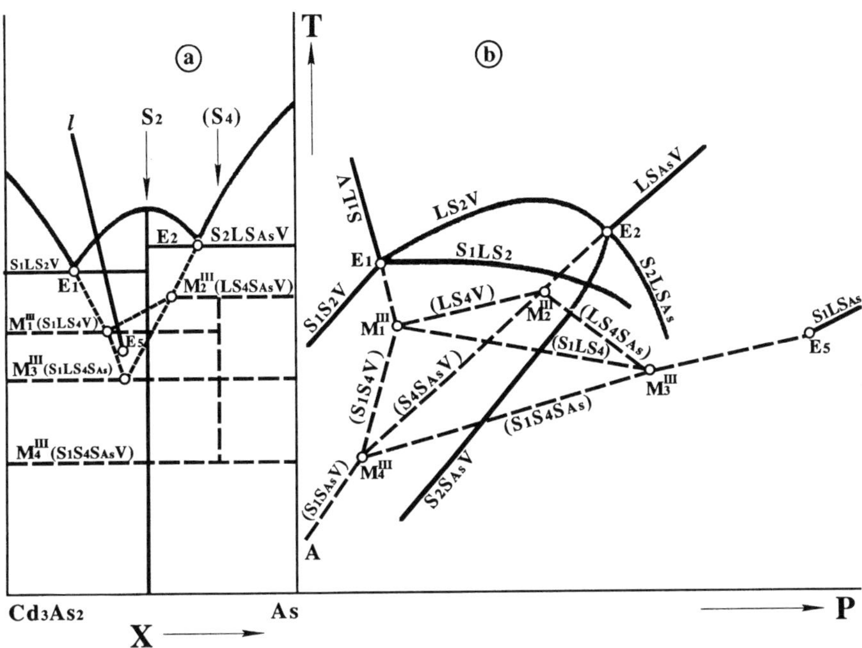

Fig. 108. *T–X* and *P–T* projections of the Type III Cd–As metastable diagram

there is no vapor in equilibrium with condensed phases, and phase transformations are determined by decomposition of $CdAs_2(s)$ into $CdAs(s)$ and crystalline arsenic.

The metastability of type II is associated with the high pressure phase $CdAs(s)$ [252] and therefore could not be observed at ambient pressures [247–249].

Metastable states, Type III. Compounds $CdP_4(s)$ and $ZnP_4(s)$ were reported in phosphide II–V systems [261,262]. On the other hand, detailed studies of the Cd–As system [255,260] did not reveal a similar compound. It could be attributed to thermodynamic instability of this phase relative to other cadmium arsenides. Nevertheless, the kinetic stability of $CdAs_4(s)$ (labeled S_4 in Fig. 108) could be sufficient for this phase to form from the melt or even from the vapors. Possible routes of metastable crystallization of $CdAs_4(s)$ are given in Table 16 (#10–12). Figure 108 shows the *P–T* and *T–X* projections of the corresponding metastable volumes that are contained within the lines of the metastable coexistence of the three phases. In Fig. 108, the single-phase metastable region of existence of $CdAs_4(s)$ is projected onto the *P–T* plane as the field $M_1^{III}M_2^{III}M_3^{III}M_4^{III}$, and in the *T–X* projection, it is the dashed vertical line S_4 between the temperatures $T(M_4^{III})$ and $T(M_2^{III})$. These were apparently the conditions of the experiment [249], where $CdAs_4(s)$ was isolated from the mixture with $Cd_3As_2(s)$ and was characterized by X-ray diffraction.

Thus, three different routes of metastable crystallization of alloys are possible in the Cd–As system. Depending on specific kinetic conditions of the experiment, it is obvious that various mixed types of crystallization could also be observed, which makes interpretation of the experimental results of equilibrium in this system exceptionally difficult.

Vapor pressure scanning of non-stoichiometry in Cd$_3$As$_2$. To study the deviation from stoichiometry in Cd$_3$As$_2$, 36 vapor pressure curves were obtained and analyzed in the two-phase equilibrium of Cd$_3$As$_2$(s) with the vapor [263]. The total data set comprised 654 experimental vapor pressure points.

The first step was to calculate P–T–X_S and P–T–X_V coordinates for every experimental point (X_S and X_V are the compositions of the phases in at.% As). This problem consists of solving a system of three equations with three unknown partial pressures because cadmium does not form gaseous arsenides, and the predominant vapor phase species are Cd(g), As$_2$(g), and As$_4$(g). The partial pressures are related by the equations of the total vapor pressure and the equilibrium constant of the vapor phase reaction As$_4$(g) = 2As$_2$(g). The former was measured experimentally, and the latter was calculated from the tabulated thermodynamic properties of As$_4$(g) and As$_2$(g) [264, 265]. The third equation was derived from the Gibbs energy ΔG of Cd$_3$As$_2$(s), which relates the partial pressures of Cd(g) and As$_4$(g). Estimates for different substances (see, for example [51]) showed that if the non-stoichiometric region of a solid ΔX_S does not exceed 1 at.%, the change in ΔG at T=const for this solid is expected to be within the confidence limits of the experimental measurement of ΔG. Hence, if $\Delta X_S < 1$ at.%, it may be taken that ΔG=const at T=const. Because the saturated vapor pressures for pure cadmium and arsenic are known [141, 246], this approximation, applied to the non-stoichiometric region of Cd$_3$As$_2$(s), determines the product of partial pressures $P(\text{Cd})P(\text{As}_4)^{\gamma/4}$ at T=const, where $\gamma = X_S/(1-X_S)$. Thus, if this product is measured for a specific composition within the non-stoichiometric region, it can be used for the whole region.

Univariant equilibria of α-Cd$_3$As$_2$ and β-Cd$_3$As$_2$ with the vapor were studied by Nipan et al. [266], and temperature dependences of the equilibrium constants for the sublimation reactions of both polymorphic modifications were reported. These results were used as the third equation to determine the partial pressures of Cd(g), As$_2$(g), and As$_4$(g) at every experimental point. From the partial pressures, the composition of the vapor X_V was calculated along with the composition of the conjugated solid X_S. For the latter, the evaporated masses of both components were subtracted from the initial masses. As a result of these calculations, every experimental (P,T) point in the solid–vapor equilibrium gave a pair of scanning points, (P,T,X_S) and (P,T,X_V), on the solidus and vaporus surfaces. The entire data file of experimental points treated in this way resulted in two sets of scanning points, $\{P,T,X_S\}$ and $\{P,T,X_V\}$, with one-to-one correspondence, which determine the arrangement of the solidus and vaporus surfaces in the P–T–X phase space.

For a quantitative description of the crystalline phase, analytical presentation of the solidus surface is necessary. For this purpose, P–T and T–X projections of every vapor pressure curve were presented in polynomial form $\log P_i = \sum a_i T^i$, $X_S = \sum b_j T^j$, $T = \sum c_k X_S^k$. Two results were derived from this treatment: dependence of the composition of the solid phase Cd_3As_2 on temperature and pressure and the thermodynamic properties of this non-stoichiometric phase.

P–T–X region of existence of Cd_3As_2. Experiments were done in the temperature range of existence of two modifications, α-Cd_3As_2 and β-Cd_3As_2. DSC measurements showed that the α–β transition in Cd_3As_2(s) is the first-order phase-transition whose temperature maximum is at 870.15 ± 0.2 K and invariant points at $T(VL\beta\alpha) = 868.25 \pm 0.1$ K and $T(\alpha\beta S_2 V) = 868.55 \pm 0.4$ K. Extrapolation to the phase-transition temperature of the $X_S = f(T)$ functions derived from the experimental vapor pressure curves, which crossed the region of coexistence of the two polymorphic forms, showed (Table 17) that the compositions of α- and β-forms at the invariant points are different. In the VL$\beta\alpha$ equilibrium, $X_\alpha > X_\beta$ (in at.% As), whereas in $\alpha\beta S_2 V$, the compositional sequence was the opposite, $X_\beta > X_\alpha$. Hence, the sequence in composition of α- and β-Cd_3As_2 in equilibrium with the vapor is changed, which proves that $\alpha \rightarrow \beta$ is a congruent phase-transition.

Geometrical analysis of the phase equilibria in the phase-transition region showed that the congruent phase-transition curve was on the Cd-side of the composition corresponding to the temperature maximum in α–β (Fig. 109, inset). Accord-

Table 17. Maximum non-stoichiometry of $Cd_3As_{2\pm\delta}$, X_S, and composition of the conjugated vapor X_V

T (K)	X_S(at.% As)	X_V(at.% As)	T (K)	X_S(at.% As)	X_V(at.% As)
	Equilibrium VLS$_1$			Equilibrium S$_1$S$_2$V (T<883 K) and S$_1$LV	
748.15	39.9869±0.001	1.3	791.85	39.9903±0.001	93.9±0.8
808.15	39.9797±0.0008	0.06	811.65	39.9944±0.002	94.9±2.5
823.15	39.979	—	830.55	39.9974±0.003	95.5±0.2
838.15	39.9792±0.001	—	853.15	39.9980±0.001	96.5±1.5
858.15	39.9818±0.001	—	869.15[a](α)	39.9982±0.0015	79.3±1.5
869.15[a](α)	39.9822±0.008	0.6	869.15[a](β)	40.0001±0.0015	75.0±1.5
869.15[a](β)	39.9800±0.0008	—	893.15	40.002±0.001	96.7±1.0
882.95	39.9803±0.0015	—	923.15	40.001±0.001	76.8±3.5
929.05	39.9798±0.0007	—	940.85	39.9999±0.0015	86.7±0.4
953.15	39.979±0.001	—	953.15	39.998±0.001	77.7±1.0

[a] α–β phase-transition

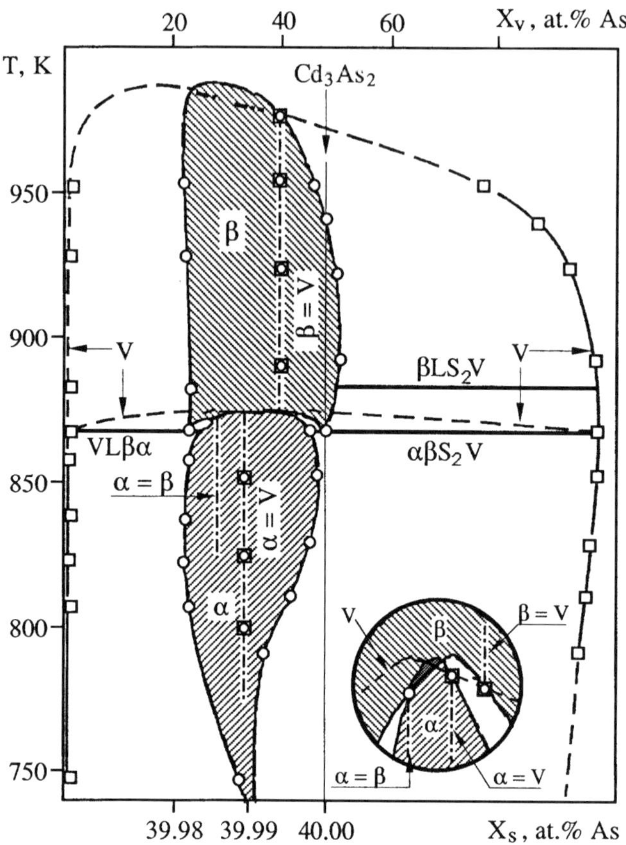

Fig. 109. T–X projection of the non-stoichiometry range of Cd$_3$As$_2$

ing to [267], the phase-transition temperature decreases with rising pressure, as reflected in Fig. 109.

To determine the maximum Cd and As non-stoichiometry in Cd$_3$As$_2$, the following system of equations was solved:

$$P_i = f_1(T),$$
$$Pi = f_2(T),$$
$$X_S = f_3(T).$$

Here i is Cd or As, functions f_2 and f_3 correspond to the bivariant solid–vapor equilibrium and f_1 corresponds to the univariant equilibria VLS$_1$ (for Cd non-stoichiometry), S$_1$LV, or S$_1$S$_2$V (for As non-stoichiometry). Here, the sequence of phases is given in the order of increasing As content in the phases. The form of the function f_1 was determined by the least-squares method [268].

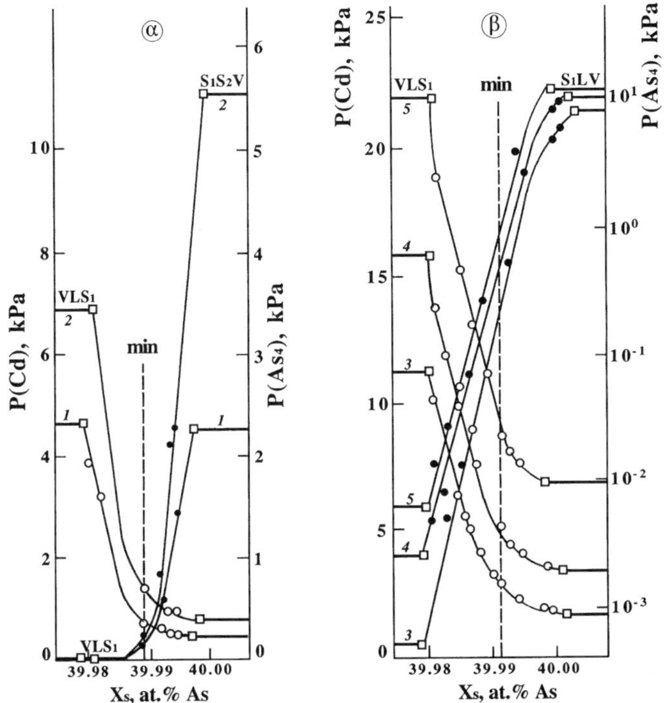

Fig. 110. Isotherms of Cd and As$_4$ partial pressures. T(K) = 823(*1*), 853(*2*), 893(*3*), 923(*4*), 953(*5*)

The calculations for the maximum non-stoichiometry of Cd$_3$As$_2$ are shown in Table 17. The *T–X* projection of the single-phase regions for α-Cd$_3$As$_2$ and β-Cd$_3$As$_2$ are shown in Fig. 109 together with the vaporus curve. It can be seen from Table 17 that the non-stoichiometry range for α-Cd$_3$As$_2$ is ~0.02 at.% and for β-Cd$_3$As$_2$, it is ~0.03 at.%. The existence regions of both forms are on the Cd-side of the stoichiometric plane X = 40 at.% As, so that α-Cd$_3$As$_2$ is completely outside of it, whereas β-Cd$_3$As$_2$ does not include the stoichiometric composition at T > 940 K. Below T = ~750 K the single-phase region is too small to be reliably measured by the vapor pressure scanning method (Fig. 109).

The spatial arrangement of the single-phase volumes for α-Cd$_3$As$_2$ and β-Cd$_3$As$_2$ in equilibrium with vapor was studied by solving the three previous equations. In this way, isotherms of partial pressures were obtained (Fig. 110), from which isothermal sections of the *P–T–X* phase diagram were constructed (Fig. 111). It can be seen in Fig. 111 that the two-phase equilibria αV and βV comprise a flat vapor pressure minimum that corresponds to the congruent sublimation of α-Cd$_3$As$_2$ and β-Cd$_3$As$_2$. Intersections of the $P_i(X_S)$ curves for Cd(g) and As$_4$(g) isotherms (Fig. 110) with these partial pressures in congruent sublimation [266] gave the congruent sublimation

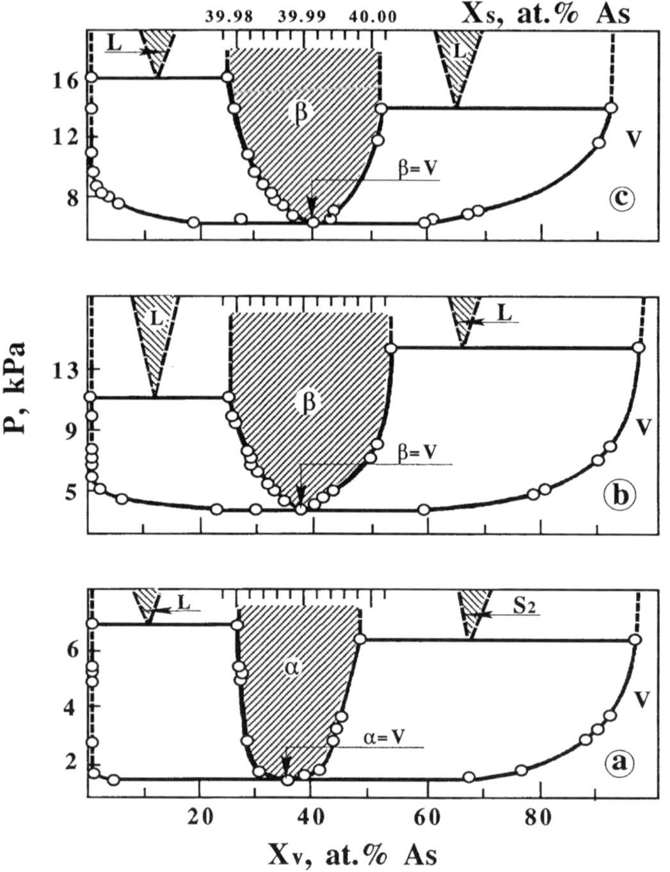

Fig. 111. *P–X* sections of the range of existence of Cd_3As_2. T(K) = 853(**a**), 893(**b**), 923(**c**)

composition X^{cs} at 823 and 853 K (for α-Cd_3As_2), and at 893, 923, and 953 K (for β-Cd_3As_2). It is seen in Fig. 110 that X^{cs} does not depend on temperature. For α-Cd_3As_2, X^{cs} = 39.989 at.% As, and for β-Cd_3As_2, it is 39.991 at.% As (Fig. 109).

Thermodynamic properties of Cd₃As₂. To compute partial thermodynamic functions, isopleths (X=const) of Cd(g) and As_4(g) partial pressures were calculated. For this purpose, equations $P_i = f_1(T)$ and $P_i = f_2(T)$ were combined with $T = f_4(X_S)$ and then temperature dependences of P(Cd) and P(As₄) were calculated for fixed X_S=const in the standard form $\log P_i = A - B/T$ (Table 18). To calculate activities, tabulated values for saturated vapor pressures of pure cadmium and arsenic were used [141, 246, 264]. In Table 18 the results are given separately for β-Cd_3As_2 in equilibrium with sub-stoichiometric vapor $X_V < 40$ at.% As (VS_1 equilibrium) and for $X_V > 40$ at.% As (S_1V equilibrium). Standard thermodynamic procedure was used to calculate the enthalpy and entropy.

Table 18. Partial pressures P_i(Pa), activities a_i, partial enthalpy ΔH^M(kJ/mole), and entropy ΔS^M (J/mole×K) for $Cd_3As_{2\pm\delta}$

X_S(at. %As)	Cadmium						Arsenic					
	$\log P = A - B/T$		$\log a = A - B/T$		ΔH^M	ΔS^M	$\log P = A - B/T$		$\log a = A - B/T$		ΔH^M	ΔS^M
	A	B	A	B			A	B	A	B		
	Equilibrium VS$_1$											
39.980	8.4	3909	−2.56	−1899	36.3	49.0	24.9	22636	2.77	3710	−71.0	−53.1
39.981	8.8	4349	−2.13	−1459	27.9	40.7	22.3	19996	2.13	3050	−58.4	−40.7
39.982	9.0	4648	−1.92	−1160	22.2	36.8	21.1	18202	1.82	2601	−49.8	−34.8
39.983	9.7	5209	−1.26	−599	11.5	24.2	17.2	14836	0.83	1760	−33.7	−15.9
39.984	9.9	5401	−1.08	−407	7.8	20.7	16.1	13684	0.56	1472	−28.2	−10.7
39.985	10.0	5633	−0.91	−175	3.3	17.3	15.0	12292	0.30	1124	−21.5	−5.7
39.986	10.2	5825	−0.72	18	−0.3	13.9	13.9	11140	0.02	836	−16.0	−0.4
39.987	10.2	5798	−0.76	−9.5	0.2	14.5	14.1	11302	0.07	876	−16.8	−1.4
39.988	11.3	6839	0.35	1032	−19.7	−6.6	7.5	5056	−1.58	−685	13.1	30.3
39.989	11.6	7168	0.71	1361	−26.0	−13.7	5.3	3082	−2.13	−1179	22.6	40.9
39.990	11.8	7408	0.89	1600	−30.6	−17.0	4.2	1639	−2.40	−1539	29.5	45.9
	Equilibrium S$_1$V											
39.992	12.0	7682	1.03	1875	−35.9	−19.7	3.4	−180	−2.61	−1994	38.2	49.9
39.993	12.0	7710	1.08	1903	−36.4	−20.6	3.1	173	−2.68	−1906	36.5	51.2
39.994	12.0	7682	1.03	1875	−35.9	−19.8	3.4	−18	−2.61	−1954	37.4	50.0
39.999	10.7	6580	−0.25	772	−14.8	4.8	11.1	6610	−0.69	−297	5.7	13.1
40.000	10.7	6611	−0.23	804	−15.4	4.4	11.0	6424	−0.72	−343	6.6	13.7

Thus, two main quantitative results were obtained for Cd_3As_2: the spatial arrangement of the single-phase volumes for α-Cd_3As_2 and β-Cd_3As_2 and the partial thermodynamic functions of the components. To do this, systems of algebraic equations were solved. These equations were constructed from polynomial approximations of the experimental vapor pressure results. In this way, two sets of scanning points with one-to-one correspondence were obtained on solidus and vaporus surfaces, from which these surfaces were reconstructed in the P–T–X phase space.

Because of a narrow range of single-phase existence of the solid phases (Table 17), it is very important to evaluate the confidence interval for the compositions X_S and X_V. For this purpose, the error accumulation law was applied because all experimental errors, as well as the uncertainties in the thermodynamic functions of $As_4(g)$, $As_2(g)$, and $Cd_3As_2(s)$, were known. These estimates were made for the entire P–T–X region investigated. It was shown that the main source of the errors δX_S and δX_V was the uncertainty in the measured vapor pressure because of the tendency of the Cd–As alloys to form metastable states. The estimates given in Table 17 include this phenomenon.

A certain measure of confidence in the results cited in Table 17 was added by comparing them with calculations at the intersection points of different pairs of vapor pressure curves. A total of 14 intersection points were obtained in the VS_l equilibrium and seven intersections in S_lV. The X_S values were then calculated at these intersections from Eq. (31) and compared with those computed by the polynomial approximation procedure. The reproducibility of X_S was within 10^{-4}–10^{-3} at.%.

Vapor pressure scanning of non-stoichiometry of CdAs₂. The procedure for vapor pressure scanning of $CdAs_2$ solidus [269] was similar to that described for Cd_3As_2 both experimentally and in treatment of the experimental results. The compositions of the conjugated solid X_S and vapor X_V were derived from the vapor pressure in the two-phase equilibrium $S(CdAs_2)V$. Because sublimation of $CdAs_2(s)$ results in a two-component vapor, Eq. (25) for this case can be written as follows:

$$X_S \text{ (at.\% As)} = \left[N(As) - n(As)\right] / \left\{ \left[N(Cd) + N(As)\right] - \left[n(Cd) + n(As)\right] \right\} \times 100\%.$$

The composition of the conjugated vapor X_V is given by

$$X_V \text{ (at.\% As)} = n(As) / \left[n(Cd) + n(As)\right] \times 100\%.$$

In the pressure range where Dalton's law applies, the evaporated quantities of the elements $n(i)$ can be calculated from their partial pressures:

$$n(Cd) = P(Cd)v/RT,$$
$$n(As) = \left[2P(As_2) + 4P(As_4)\right]v/RT.$$

These three partial pressures were calculated in the same way as those for Cd_3As_2 by solving the system of three equations at every temperature:

$$P = P(Cd) + P(As_2) + P(As_4)$$
$$K_P = P^2(As_2)/P(As_4)$$
$$\Delta G = RT\ln\left[a(Cd)a^\gamma(As)\right],$$

where P is the total vapor pressure, K_P is the equilibrium constant, ΔG, the Gibbs energy of $CdAs_2(s)$, and $\gamma = X_S/(1-X_S)$. Subsequently $T–X$ and $P(i)–T$ projections of every vapor pressure curve were derived in an analytical form, and the isopleths (X=const) of partial pressures were deduced from them. The results are presented in Table 19.

Table 19. Partial pressures P_i(Pa), activities a_i, partial enthalpy ΔH^M(kJ/mole), and entropy ΔS^M (J/mole×K) for $CdAs_{2-\delta}$

X_S(at. %As)	Cadmium						Arsenic					
	$\log P = A - B/T$		$\log a = A - B/T$		$-\Delta H^M$	ΔS^M	$\log P = A - B/T$		$\log a = A - B/T$		$-\Delta H^M$	$-\Delta S^M$
	A	B	A	B			A	B	A	B		
66.6605	9.9	6100	−1.04	300	5.9	19.9	15.2	9400	0.34	400	7.8	6.5
66.6615	10.2	6400	−0.75	650	12.2	14.4	14.6	8800	0.19	250	4.7	3.7
66.6625	9.8	6100	−1.16	350	6.2	22.2	15.4	9400	0.40	400	7.8	7.8
66.6635	10.1	6400	−0.86	650	12.2	16.5	14.8	8800	0.25	250	4.8	4.8
66.6645	9.8	6200	−1.16	400	8.1	22.2	15.5	9300	0.41	350	7.0	7.9
66.6655	9.9	6400	−0.98	550	11.0	18.8	15.1	8900	0.32	300	5.5	6.1

To obtain isotherms of the partial pressures, T–X and $P(i)$–T projections of the vapor pressure curves were derived in the polynomial form $P(i) = f(T)$ and $X_S = \varphi(T)$. As an example, two isotherms of $P(Cd)$ and $P(As_4)$ are shown in Fig. 112 for temperatures 800 and 820 K. Subsequent treatment of the experimental data resulted in determining the P–T–X spatial arrangement of the solidus and vaporus surfaces in the two-phase equilibrium $S(CdAs_2)V$. Boundary compositions of the solid and conjugated compositions of the vapor are presented in Table 20.

Table 20. Maximum non-stoichiometry of $CdAs_{2-\delta}$, X_S, and composition of the conjugated vapor X_V

T (K)	X_S(at.% As)	X_V(at.% As)	T (K)	X_S(at.% As)	X_V(at.% As)
Cadmium non-stoichiometry			Arsenic non-stoichiometry[c]		
809.00[a]	66.659±0.0006	96.9±1.9	701.45	66.662±0.0006	99.7±3.60
872.35[a]	66.658±0.0007	97.3±0.3	774.7	66.667±0.0007	99.8±0.40
885.05[b]	66.657±0.0009	99.5±0.1	806.65	66.666±0.0006	99.8±0.20
891.25[b]	66.657±0.0006	99.3±0.2	819.65	66.666±0.0007	99.8±0.16
896.85[b]	66.658±0.0010	99.8±0.1	853.25	66.662±0.0010	99.8±0.06
900.5[d]	66.660±0.0010	99.6±0.1	862.65	66.663±0.0010	99.8±0.06
900.5[e]	66.607	100	880.65	66.664±0.0011	99.9±0.05

[a] Equilibrium $S(Cd_3As_2)S(CdAs_2)V$
[b] Equilibrium $LS(CdAs_2)V$
[c] Equilibrium $S(CdAs_2)S(As)V$
[d] $T_{max}(CdAs_{2-\delta})$
[e] Calculations in the approximation $P(Cd) = 0$

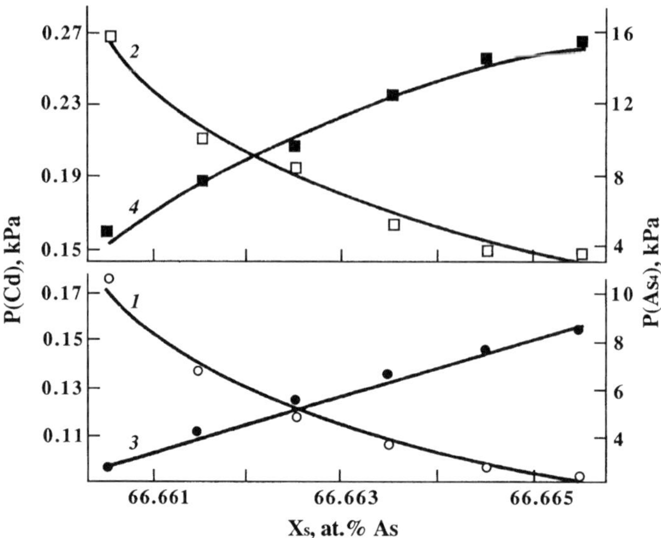

Fig. 112. Isotherms of Cd (*1,2*) and As_4 (*3,4*) partial pressures at 800 K (*1,3*) and 820 K (*2,4*) in the non-stoichiometry range of $CdAs_2$

In Fig. 113, the *T–X* projection of the $CdAs_2$ solidus is presented together with the conjugated vaporus curve. The temperatures on the solidus surface were found as the intersections between the vapor pressure curves in the two-phase equilibrium $S(CdAs_2)V$ and the three-phase equilibrium $S(Cd_3As_2)S(CdAs_2)V$ or $S(CdAs_2)S(As)V$. The corresponding compositions X_S and X_V were calculated from the polynomial approximations $X_S = \psi(T)$ of the experimental vapor pressure curves. It follows from Table 20 and Fig. 113 that the maximum range of existence for $CdAs_2$ is ~0.01 at.%, and the stoichiometric plane only touches the single-phase volume. It means that the stoichiometric composition $CdAs_2$ is a two-phase mixture of $CdAs_{2-\delta}$ and crystalline arsenic. The vapor phase in the $S(Cd_3As_2)S(CdAs_2)V$ equilibrium contains ~97 at.% As (Table 20). Above the melting point of the eutectic, the vapor is enriched in arsenic up to $X_V > 99$ at.%. At the maximum melting point of the $CdAs_{2-\delta}$ phase, the composition of the vapor is $X_V = 99.6$ at.% As, and in the $S(CdAs_2)S(As)V$ equilibrium, the vapor is almost pure arsenic.

The shape of the isotherms of the partial pressures (Fig. 112) shows that no ordering of the defects is observed in the homogenous region of $CdAs_{2-\delta}$. Analysis [269] of the compositional dependences of cadmium and arsenic activities in the approximation of statistically distributed neutral non-interacting [270] or interacting defects [271] showed that neither of these models adequately described the experimental results.

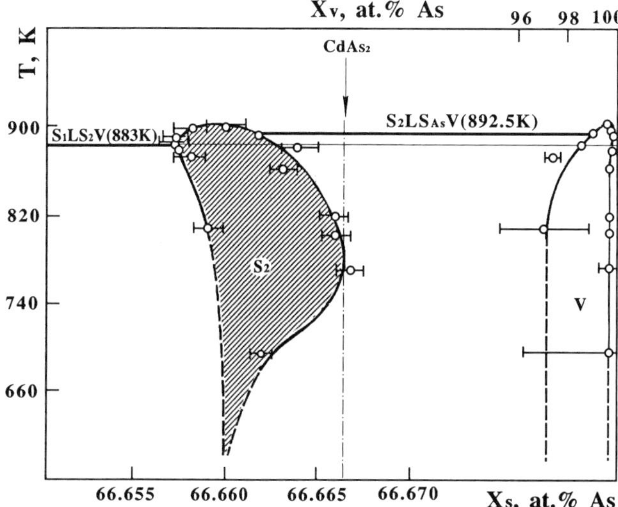

Fig. 113. T–X projection of the non-stoichiometry range of $CdAs_2$ and vaporus curve in three-phase equilibria S_1S_2V, LS_2V, S_2LV, $S_2S_{As}V$

Because the non-stoichiometry range of $CdAs_{2-\delta}$ is only about 0.01 at.%, special attention was given to the errors associated with the composition of the solid [269]. These errors originate from two groups of factors: experimental errors and validity of the ΔG_T = const approximation. Calculations based on the error accumulation law showed that the confidence interval for X_S was strongly dependent on the experimental conditions and the (P,T) region of the experiment. The measurements [269] were made in a wide interval of the m/v ratios (mass of the sample per volume), 0.05 to 1.1 g/ml, with initial composition 66.7 to 68.1 at.% As. Typical confidence intervals for the $CdAs_{2-\delta}$ composition are shown in Table 20. The principal source of the uncertainties δX_S was the errors in determining the initial masses of the components. The errors in measuring the volume, vapor pressure and temperature influenced δX_S essentially less (one to two orders of magnitude). The uncertainties in the thermodynamic functions of $CdAs_2(s)$ and vapor phase species had still smaller effect on δX_S.

The confidence interval δX_V for the vapor composition also depends on the $(P–T–X)$ interval of measurement (Table 20). Although the precision in X_V is essentially less than that for X_S, it proved high enough to determine the coordinates of the vaporus curve in the three-phase equilibria of $CdAs_2$ with $Cd_3As_2(s)$ liquid, and crystalline arsenic (Fig. 113). It should be stressed that, in spite of the low concentration of Cd in the vapors (1–2 %), it cannot be ignored. Calculations showed that if Cd(g) is disregarded, the resulting X_S values could shift as much as 0.05 at.% (Table 20, the last row), which for $CdAs_2$ is more than the entire single-phase range of existence.

3.2 Oxide systems

Oxides are probably the most long-standing inorganic materials that mankind has used. Contemporary applications of oxides cover almost all of scientific and technological activities, from construction to microelectronics, from refractory and ceramic applications to high-temperature superconductors. One of the important fields of oxide application is high-temperature technology. For example, yttria-stabilized zirconia (YSZ) is used as high-temperature ceramics in production of crucibles, tubes, etc. It is also one of the best protective refractory coatings. YSZ is widely known as a high-temperature solid electrolyte with oxygen conductivity that has potential application in solid oxide fuel cells. Advanced oxide materials, such as $BaWO_4$, $BaZrO_3$, $BaTiO_3$, play an important role in modern electronics in a wide range of applications: ferroelectrics, dielectrics, non-linear materials, substrates, etc. (see, for example, [272–275]. Same as for other inorganic materials, the scientific basis for controlled synthesis of oxides is phase equilibrium in the corresponding systems. This is the reason why phase equilibrium in oxide systems was extensively studied, and many reference books have been published on phase diagrams of binary, ternary and multinary oxide systems [276–291]. But because the bulk properties of a great number of oxides are not very sensitive to oxygen non-stoichiometry, the majority of phase equilibrium studies were confined to condensed phase *T–X* diagrams constructed from experiments conducted in open air, i.e., at constant oxygen pressure. In this section, some systems will be discussed, for which *P–T–X* data are also available.

3.2.1 High-temperature oxides

BaO–WO₃ system. Three solid-state compounds were identified by DTA and XRD methods in the $BaO–WO_3$ system: $BaWO_4$ (in subsequent discussion referred to as S_1), Ba_2WO_5 (S_2), and Ba_3WO_6 (S_3) [292]. The corresponding congruent melting temperatures for them are 1748, 1923, and 2063 K. Because melting, sublimation, and vaporization of BaO (labeled S^I) and WO_3 (S^{II}) proved to be congruent processes [293, 294], the $BaO–WO_3$ system may be treated as a quasi-binary section of the ternary system Ba–W–O. Mass spectrometric study of this system [295] showed that the main vapor phase species were $BaO(g)$, $Ba_2WO_5(g)$, $BaWO_4(g)$, $(BaWO_4)_2(g)$, $BaW_2O_7(g)$, and $(WO_3)_n$ where n = 2, 3, and 4. The *P–T–X* phase equilibrium in $BaO–WO_3$ is presented in *P–T* and *T–X* projections in Fig. 114 and in isothermal sections in Fig. 115. The total vapor pressure in *P–T* projection and the vaporus curve in the *T–X* projection were derived from mass spectrometric measurements of the partial pressures of vapor phase species, thermodynamic properties of barium–tungsten oxides, and thermodynamic correlations deduced for this system [295]. *P-T-X* coordinates for the invariant equilibria are presented in Table 21.

Table 21. P–T–X coordinates of invariant equilibria in the BaO–WO$_3$ system

	Equilibrium [a]	T (K)	P (Pa)	X_L	X_V
E$_1$	S^{I}VLS$_3$	1863	6.4	0.10	0.01
E$_2$	VS$_3$LS$_2$	1913	5.73	0.32	0.23
E$_3$	VS$_2$LS$_1$	1593	5.9×10^{-3}	0.42	0.32
E$_4$	S$_1$LVS^{II}	1208	8.4×10^{-4}	0.75	1.00

[a] The sequence of phases in each equilibrium corresponds to the increase of WO$_3$

To calculate the total vapor pressure P and the analytical composition of the vapor X_V in all three-phase equilibria, solid–liquid–vapor, it was necessary to know the partial pressures in the saturated vapor above the liquidus of fixed composition X_L. In the quasi-binary BaO–WO$_3$ system, this two-phase equilibrium is bivariant. Nevertheless, to calculate all of the partial pressures for an isopleth (X_L=const) three independent parameters had to be fixed because the analytical relation between X_L and X_V was unknown. In mass spectrometric experiment it is convenient to take the temperature and two arbitrary partial pressures as independent variables. If these two are measured, then the rest are readily calculated from the temperature dependence of the equilibrium constants for vapor phase reactions, studied in the mass spectrometric experiment. Partial pressures in equilibrium with the liquid X_L obtained in this way at a given temperature may then be recalculated to the liquidus temperature T_L, which corresponds to X_L, and at this point the total vapor pressure P and the composition of the vapor X_V can be found. It is clear that a sequence of such points over the whole range of X_L is indeed the vaporus curve in the T–X projection and a system of three-phase equilibrium curves in P–T.

In Table 22, calculations of the total vapor pressure and the vapor composition are presented for all three-phase equilibria, solid–liquid–vapor, in the BaO–WO$_3$ system for compositions $X = 0.45$ to 0.80. Also in Table 22, partial pressures are given as a function of the temperature for vapor phase species taken as independent. The resulting P–T and T–X projections for BaO–WO$_3$ are presented in Fig. 114.

In all of the phase reactions, the mole fractions X of BaO and WO$_3$ are so related that $X(\text{BaO}) + X(\text{WO}_3) = 1$. This is the consequence of the congruent sublimation of BaO(s) and WO$_3$(s), and it means that neither BaO nor WO$_3$ is accumulated in the condensed phase; this is the proof of the quasi-binary behavior of the BaO–WO$_3$ system.

Because the condensed phase univariant equilibria are projected on the P–T plane as almost vertical lines originating from the corresponding eutectic points, they are shown only on the insets of the near-eutectic areas. The phases in all of the univariant equilibria on the P–T projection are labeled in a sequence according to the increase of WO$_3$. Three-phase equilibria S^{I}VS$_3$ and LVS^{II} almost coincide with the corresponding sublimation curves of BaO (S^I) and WO$_3$ (S^{II}), which means that

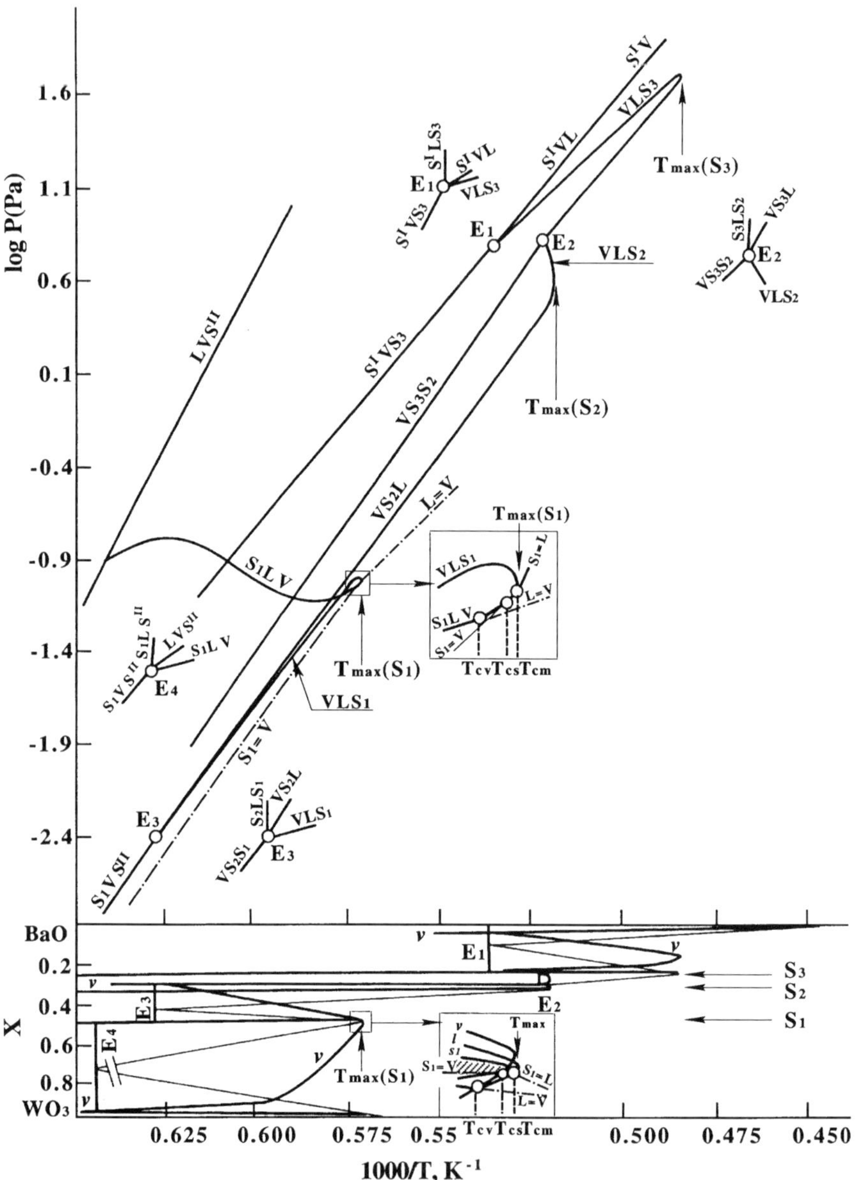

Fig. 114. *P–T* and *T–X* projections of the BaO–WO₃ diagram. S₁-BaWO₄, S₂-Ba₂WO₅, S₃-Ba₃WO₆, S¹-BaO, S¹¹-WO₃

mutual solid-state solubility of BaO and WO_3 is outside the limits of the mass spectrometric experiment.

The results in Table 22 show that the vapor pressures in LVS^{II} and the low temperature portion of S_1LV are much higher than those for the rest of the equilibria. Because of that, these two curves are given on an arbitrary scale in Fig. 114. Also schematic is the vaporus curve (T–X projection) in $S^I VS_3$ and LVS^{II} because according to Table 21, the vaporus in these equilibria is almost pure BaO or WO_3.

Two vapor pressure extrema (minimum and maximum) were observed in the three-phase equilibrium S_1LV. Because of that, special attention was given to this equilibrium. Four isopleths were examined (Table 22). The liquidus temperatures were taken both from DTA [292] (for X = 0.55, 0.67, and 0.70) and mass spectrometry (X = 0.63) [295]. In the latter case, the liquidus point is registered as a break on the evaporation isotherm, which corresponds to the phase transformation $LV \rightarrow S_1LV$.

On the P–T projection (Fig. 114), the S_1LV curve intersects four other three-phase lines. It is clear, however, that no intersections exist in the P–T–X phase space

Table 22. Independent partial pressures as a function of temperature $\log P$ = A–B/T and coordinates of three-phase equilibrium points

X_L	Equilibrium	Species	A	B	T(K)	P(Pa)	X_V
0.45	VLS_1	$BaWO_4$	11.96	22500	1691	0.045	0.43
		BaO	11.76	23000			
0.50	LV	$BaWO_4$	11.97	22500	1748	0.11	0.49
		BaW_2O_7	13.32	29000			
0.50	$S_1 = V$	$BaWO_4$	15.13	28000			0.50
		$(BaWO_4)_2$	16.72	35500			
0.55	S_1LV	BaW_2O_7	13.32	24500	1673	0.087	0.71
		$(WO_3)_3$	10.97	22000			
0.63	S_1LV	BaW_2O_7	13.32	23000	1523	0.17	0.97
		$(WO_3)_3$	10.97	18000			
0.67	S_1LV	BaW_2O_7	13.32	23000	1420	0.063	0.99
		$(WO_3)_3$	10.92	17000			
0.70	S_1LV	BaW_2O_7	13.32	23000	1337	0.037	1.00
		$(WO_3)_3$	10.92	16500			
0.80	LVS^{II}	BaW_2O_7	13.32	23000	1413	1.2	1.00
		$(WO_3)_3$	10.97	15500			

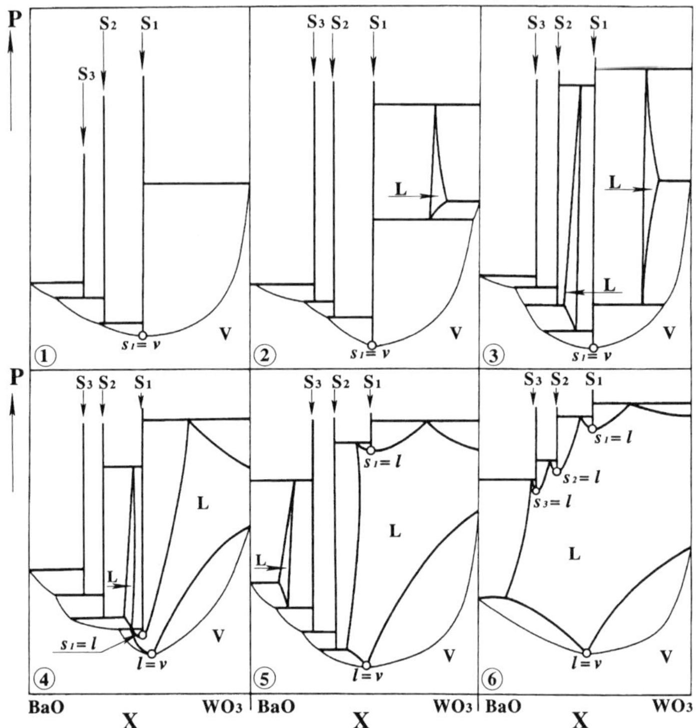

Fig. 115. Isothermal sections of the BaO–WO$_3$ diagram. *1* - $T < T(E_4)$, *2* - $T(E_4)<T<T(E_3)$, *3* - $T(E_3)<T<T_{max}(S_1)$, *4* - $T_{max}(S_1)$, *5* - $T(E_1)<T<T(E_2)$, *6* - $T_{max}(S_3)$

because different equilibria comprise either different phases or various compositions of the same phase. It is clearly seen, for example, in Fig. 115,3, which is the isotherm of the diagram at the intersection temperature of S$_1$LV and VS$_2$L: although the horizontal conodes for these two equilibria are at the same pressure, they are the tie-lines for different solid phases (S$_1$, S$_2$) and different compositions of the same phase (V and L). On the isothermal sections (Fig. 115), the composition of the vapor is taken from mass spectrometric experiments, and the condensed phase compositions correspond to the diagram reported by Kreidler [292]. The condensed three-phase equilibria are shown schematically.

BaWO$_4$(s) is of primary importance in the BaO–WO$_3$ system. Therefore, it is appropriate to examine the details of phase equilibria in the region of $X = 0.5$. Below the melting temperature of BaWO$_4$ (1748 K), a vapor pressure minimum is observed, which is a consequence of of the congruent sublimation S$_1$ = V. It is plainly seen in isotherms at $T < T_m(BaWO_4)$, Fig. 115,1–3, that BaWO$_4$(g) and (BaWO$_4$)$_2$(g) are the only vapor species observed at $T < 1748$ K. Hence, it may be assumed that the congruently subliming composition is $X^{cs} = 0.50$ and it does not change with the temperature. When $T > T_m(BaWO_4)$ the liquid–vapor equilibrium at $X = 0.50$ is no longer univariant; the vapor is enriched in BaO compared to the liquid (Table 22). This means that the congruent vaporizing composition of the melt is $X^{cv} > 0.50$ and probably is a function of the temperature. The intersection points of the azeotrpic curve $L = V$ with the isothermal planes are seen in Fig. 115,4–6 where it is evident that the isopleth $X = 0.50$ at $T > T_m(BaWO_4)$ corresponds to two-phase equilibrium LV where $X_L > X_V$.

Phase equilibria at the maximum melting point of $BaWO_4(s)$ are shown in Fig. 115,4 and in the insets of the $P–T$ and $T–X$ projections, Fig. 114. It is clearly seen that at T_{max} $BaWO_4(s)$ is in three-phase equilibrium with the BaO-rich melt and vapor, also enriched in BaO (VLS_1 conode in Fig. 115,4, and X_L and X_V compositions on the $T–X$ projection inset, Fig. 114). It is obvious that the crystal grown from VLS_1 is expected to be BaO-saturated. On the other hand, the crystal growth of $BaWO_4$ from the stoichiometric melt (the congruent melting curve $S_1 = L$) is expected to be very difficult because the process is to be conducted in the absence of vapors. In principal, $BaWO_4$ can also be grown from the congruent vapor ($S_1 = V$ equilibrium). Such a process would be extremely slow because of the very low vapor pressure of $BaWO_4(g)$. These considerations show that the crystal growth of stoichiometric $BaWO_4(s)$ is a formidable task.

BaO–ZrO₂ system. Condensed phase equilibria in the $BaO–ZrO_2$ system were studied by the XRD method [296,297]. Three barium zirconates were observed: Ba_2ZrO_4 (labeled S_1 in subsequent discussion), $Ba_3Zr_2O_7$ (S_2), and $BaZrO_3$ (S_3). The composition of the vapors was reported for the entire $BaO–ZrO_2$ compositional range from Knudsen cell mass spectrometry in the temperature interval 1650–2300 K [298, 299]. The partial pressures of the main vapor phase species are presented in Table 23.

Table 23. Partial pressures (in atm) of the vapor phase species in the $BaO–ZrO_2$ system

Equilibrium	$T(K)$	BaO	Ba_2O_2	Ba	O_2	O
$S^IS_1V^a$	1775	2.3×10^{-5}	1.5×10^{-7}	4.6×10^{-8}	6.0×10^{-9}	7.3×10^{-9}
S_SS_3V	1775	7.7×10^{-6}	1.7×10^{-8}	1.8×10^{-8}	4.3×10^{-9}	6.2×10^{-9}
$S_3S^{II}V^a$	2210	1.1×10^{-5}	4.2×10^{-10}	1.0×10^{-6}	1.3×10^{-8}	3.2×10^{-7}
$P(min)$	2210	1.1×10^{-5}	4.2×10^{-10}	5.6×10^{-7}	4.0×10^{-8}	5.6×10^{-7}

^{a} S^I is for $BaO(s)$, S^{II} is $ZrO_2(s)$

Mass spectrometric results together with the $T-X$ data for the $BaO–ZrO_2$ system and the general shape of $MO–ZrO_2$ phase diagrams [300,301] were the basis for constructing the $P–T–X$ phase diagram $BaO–ZrO_2$ [299]. It is shown in Fig. 116 as a succession of isothermal sections starting with $T_{max}(Ba_3Zr_2O_7)$ up to the maximum melting point of $BaZrO_3$. The sections are made at representative temperatures used in barium zirconate technology. All barium zirconates are shown in Fig. 116 as stoichiometric compounds because non-stoichiometry proved to be outside the experimental limits of mass spectrometric detection [299]. Three-phase equilibria vapor–BaO(s)–Ba_2ZrO_4(s), vapor–Ba_2ZrO_4(s)–$BaZrO_3$(s), and vapor–$BaZrO_3$(s)–ZrO_2(tetragonal) are shown in Fig. 116 based on the experimental data (Table 23). The compound $Ba_3Zr_2O_7$(s) at $T > 1650$ K exists only at high pressures. Fig. 116,2–4 shows that Ba_2ZrO_4 (S_1) can be crystallized from the melt in the presence of the vapors (three-phase horizontal conodes VLS_1) and from the vapors in the two-phase equilibrium VS_1. In the former case the liquid is enriched in BaO, and the composition of the vapor is ~100 mol.% BaO. Crystallization of $BaZrO_3$ (S_3) at the maximum melting temperature (Fig. 116,6, three-phase equilibrium VLS_3) also requires a BaO-rich melt and vapor $X_V = $ ~100 mol.% BaO. On the other hand, $BaZrO_3$ can also be crystallized at higher pressures from two-phase equilibria LS_3 or S_3L, where the melt is enriched either in BaO or ZrO_2, or from the three-phase equilibrium VS_3L, where the melt is ZrO_2-rich and the vapor is almost pure BaO (Fig. 116,5). Figure 116 furthermore shows that BaO is a poor stabilizer for cubic ZrO_2, which in this system is stable only at temperatures $T > 2553$ K: no single-phase volume is seen in Fig. 116,1–3 for ZrO_2(c).

Stabilization of cubic ZrO_2. ZrO_2 forms three polymorphs: monoclinic S_M, tetragonal S_T, and cuic S_C [280]. In ZrO_2 technology, it is important to avoid cracking the material as a result of the phase-transitions from the high-temperature cubic form down to low-temperature phases. Usually it is achieved by doping ZrO_2 with different oxides. The mechanism of such doping can be readily understood from the phase equilibrium point of view, if the corresponding quasi-binary systems are treated in terms of the concept of polymorphism described earlier in Section 1.

The equilibrium S_M–S_T is Type II (Fig. 1b) considering ZrO_2 as a quasi-component. The triple point temperatures are 1478 K for S_MS_TV and 2983 K for S_TLV at $P = 0.088$ atm [280]. The third polymorph, S_C, is in Type III (Fig. 1c) equilibrium with the S_T polymorph. The invariant point S_TS_CL appears at $T = 2973$ K and $P = 4000$ atm [280], i.e., pure cubic ZrO_2 is not involved in equilibrium with the vapor and exists only at high temperatures and pressures. Nevertheless, it appears that in many ZrO_2–M_mO_n systems cubic ZrO_2-based solid solutions are observed in tangible compositional ranges. And these solid solutions are of the highest applied interest because doping with a second component makes it possible to decrease considerably the low-temperature limit of existence for the cubic ZrO_2, or, as it is called, to stabilize the cubic ZrO_2 down to low temperatures.

If the α-form in Fig. 27 is considered tetragonal ZrO_2 and the γ-form is cubic ZrO_2, then the physico-chemical nature of stabilizing the high-temperature γ-ZrO_2 can be understood. Different stabilizing abilities of various oxides M_mO_n in relation to ZrO_2 can also be rationalized as well as possible reasons for differences in phase diagrams of the same system published by different authors. It can be seen in Fig. 27 that the low-temperature limit of existence of the γ-phase at ambient pressure is the invariant point N_3. Therefore, if an oxide M_mO_n (in Fig. 27 it corresponds to the

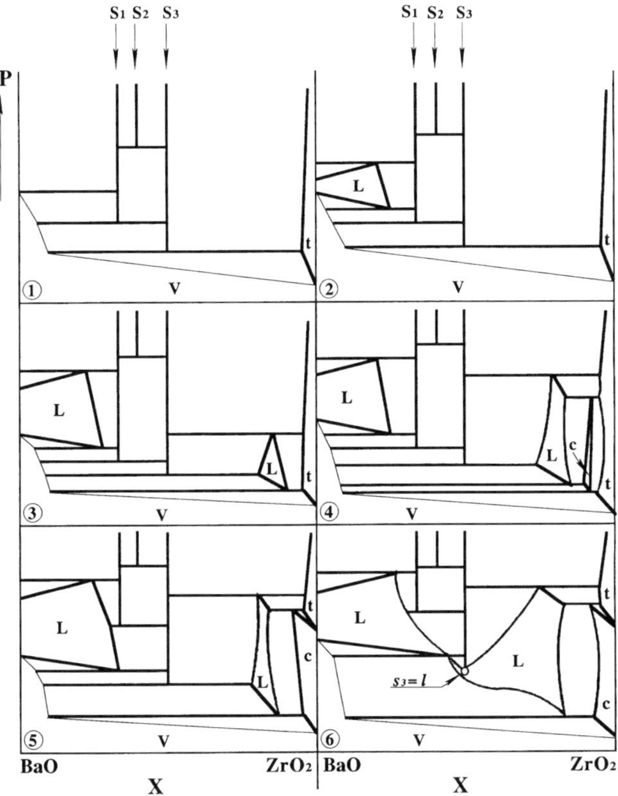

Fig. 116. Isothermal sections of the P–T–X diagram BaO–ZrO$_2$ at $T_{max}(S_2)$=2290K(1), 2290–2513 K (2), 2513–2553 K (3), 2553–2623 K (4), $T_f(S_1)$=2813 K (5), $T_{max}(S_3)$(6). S$_1$–Ba$_2$ZrO$_4$, S$_2$–Ba$_3$Zr$_2$O$_7$, S$_3$–BaZrO$_3$, t–ZrO$_2$(tetragonal), c–ZrO$_2$(cubic)

β-phase) is involved in the four-phase equilibrium αγVβ with γ-ZrO$_2$ at low temperature, then the solid solution γ prepared at an arbitrary temperature (high temperature, in particular, up to the melting point of γ at the corresponding pressure) could be cooled down to a low temperature without phase-transition to the α-polymorph. This would prevent mechanical cracking of the sample due to the structural changes accompanying the solid-state phase-transition. Y$_2$O$_3$ appears to be such an oxide. Experimental data [302] suggest that this quasi-binary system, ZrO$_2$–Y$_2$O$_3$, is actually the isobaric section, Fig. 27,III, above the three-phase equilibrium temperature for αγβ. The presence of the congruent sublimation curve γ = V in the ZrO$_2$–Y$_2$O$_3$ system [303] makes Y$_2$O$_3$ virtually an ideal stabilizer for γ-ZrO$_2$ because high-temperature exposure of such a protective coating or ceramic, even at low pressures (Fig. 27,I), would not change the composition of the solid due to partial sublimation. It should be noted that the congruent sublimation composition for ZrO_2–Y_2O_3 reported in [303] (18–20 mol.% Y_2O_3) is in agreement with more recent results (Fig. 117,a,b) for the compositional limit of the cubic solid solution (~15 mol.% Y_2O_3 [304,305]).

On the other hand, it can be seen in Fig. 117c,d that MgO (as well as other alkaline-earth metals) is a poorer stabilizer of cubic ZrO_2 because in this case the temperature for the cubic–tetragonal phase-transition is as high as 1573 K and below this temperature the cubic polymorph is metastable. Moreover, no congruent sublimation $\gamma = V$ was found in the ZrO_2–MgO system [303]. In addition, it is worthwhile to point out here that the condensed phase ZrO_2–MgO diagrams presented in [306,307] are essentially those of the isobars, Fig. 27,III and Fig. 27,IV without vaporus. Therefore, they probably just belong to different pressure intervals.

3.2.2 High-T_c superconductors

Ever since the discovery of high-T_c superconductivity in oxides, it was almost immediately realized that non-stoichiometry is a crucial factor in superconducting properties. Oxygen non-stoichiometry, it was proved, determines the superconducting properties of rare-earth 123 cuprates, cation non-stoichiometry is critical for Bi high-T_c superconductors, etc. That is why, from the very first steps, the high-T_c superconductor materials science relied heavily on phase equilibrium studies in the corresponding systems and great effort was invested in investigating phase diagrams. At present, two classes of materials are the leading candidates for technical application of high-T_c superconductivity, rare earth $LnBa_2Cu_3O_{7-\delta}$ and $Bi_2Sr_2Ca_2Cu_3O_{10}$ [308]. Not surprisingly, it is for these systems that the major part of the phase equilibrium data has been accumulated.

Rare-earth high-T_c superconductors. Figures 118 and 119 summarize the subsolidus phase equilibrium at 950°C in air for Ln–Ba–Cu–O systems according to Wong-Ng et al. [309–313] (Ln = Nd, Sm, Eu, Gd, Er), Kilbanow et al. [314] for Ln = La, and Roth et al. [315] for Ln = Y. The systems are treated as pseudo-ternaries $^1/_2Ln_2O_3$–BaO–CuO. It can be seen that phase equilibrium (the number of ternary oxides and the range of solid solutions) strongly depends on the size of the rare earth cation. In the La system, in addition to the 123 and 211 compounds ($LnBa_2Cu_3O_{6+z}$ and Ln_2BaCuO_5), common to all of the systems, 212, 336, and 415 compounds were also found.

The following specifics of phase equilibrium in these systems were noted by Wong-Ng et al. [309]. The superconductor compounds $LnBa_2Cu_3O_{6+z}$ exhibit deviation from cation stoichiometry for the first half of the lanthanide series to form solid solutions with the formula $Ln_{1+x}Ba_{2-x}Cu_3O_{6+z}$. The range of this non-stoichiometry varies with the ionic radius of Ln. The tendency is observed for the tie-lines between Ln_2BaCuO_5, CuO, $Ln_{1+x}Ba_{2-x}Cu_3O_{6+z}$ and Ln_2CuO_4 (or $Ln_2Cu_2O_5$). For smaller rare earth elements (Ln = Eu and beyond), the tie-line connection switches to link Ln_2BaCuO_5 and CuO compounds, whereas for Ln = Dy and beyond, the binary oxide $Ln_2Cu_2O_5$ appears in the phase diagram instead of Ln_2CuO_4. For larger lantha-

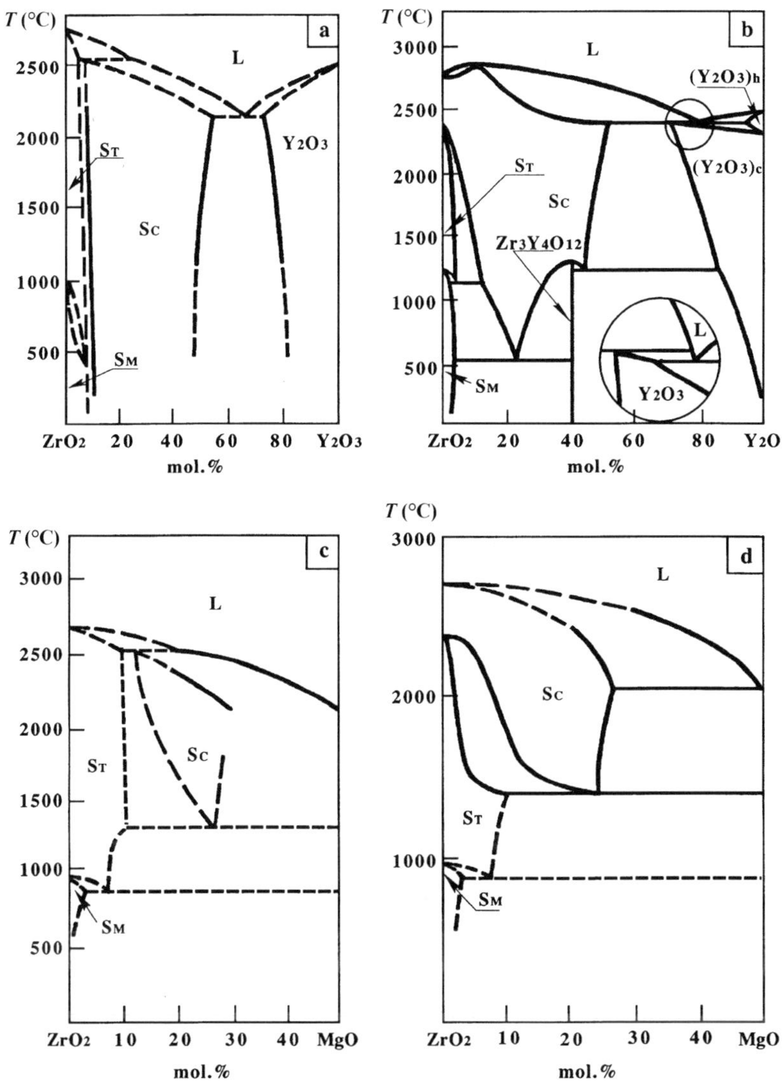

Fig. 117. Condensed phase diagrams ZrO_2–Y_2O_3 (**a**–[302], **b**–[305]) and ZrO_2–MgO (**c**–[306], **d**–[307])·

nide elements (La and Nd), the "brown phase" Ln_2BaCuO_5 is formed instead of the "green phase" that has the same 211 stoichiometry found for Ln = Sm, Eu, Gd, Dy, Ho, Y, Er, Tm, Yb, and Lu [309]. The structures of these phases were proved to be different.

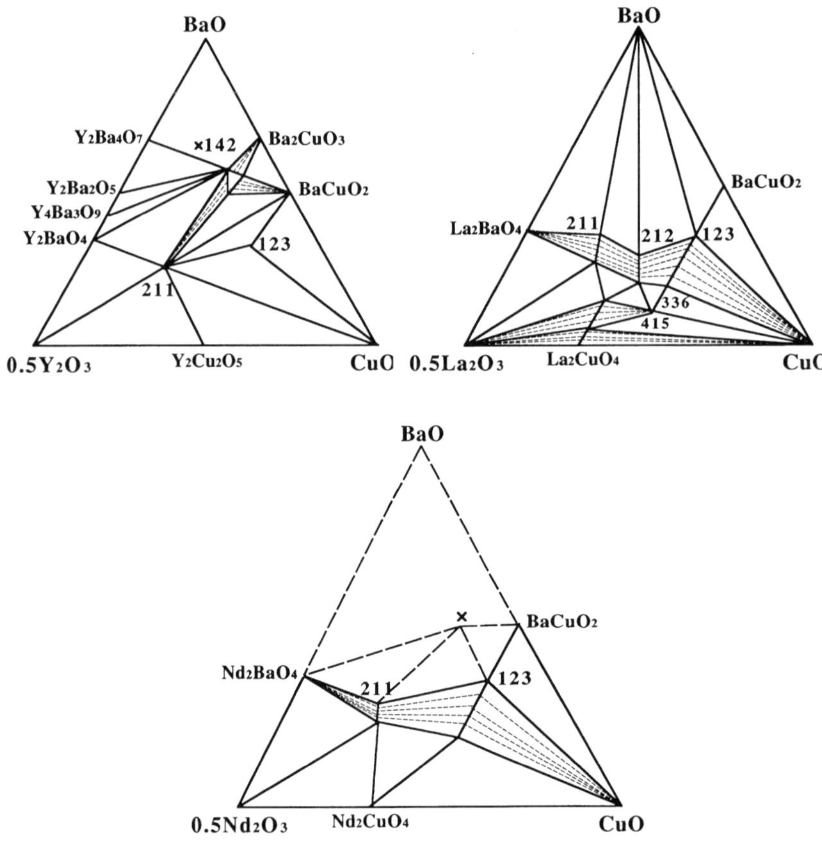

Fig. 118. Phase diagrams of 0.5 R_2O_3–BaO–CuO, R = Y, La, Nd

The thermal stability of the $LnBa_2Cu_3O_{6+z}$ phase also depends on the ionic radius of Ln^{3+}: the highest decomposition temperatures were registered for the larger lanthanides La^{3+} and Nd^{3+}. The size of the rare earth cation was found to be instrumental in the temperature of the phase-transition between orthorhombic (superconductor) and tetragonal (insulator) phases: the larger the ionic radius of Ln^{3+}, the lower the phase-transition temperature. It means that smaller rare earth cations stabilize the orthorhombic phase at higher temperatures (and lower oxygen content). The cation non-stoichiometry of $Ln_{1+x}Ba_{2-x}Cu_3O_{6+z}$ also influences the phase-transition temperature. For Ln = Nd [309], the phase transformation was registered at 550–570°C for the exact stoichiometry and went up to 950°C for x = 0.2–0.3. The non-stoichiometry limit in $Nd_{1+x}Ba_{2-x}Cu_3O_{6+z}$ was shown to be x ≤ 0.7, whereas for the 211 solid solution $Nd_{4-2x}Ba_{2+2x}Cu_{2-x}O_{10}$, the range of non-stoichiometry is much smaller, 0 ≤ x ≤ 0.1. The analogous non-stoichiometry range for the La 211 solid

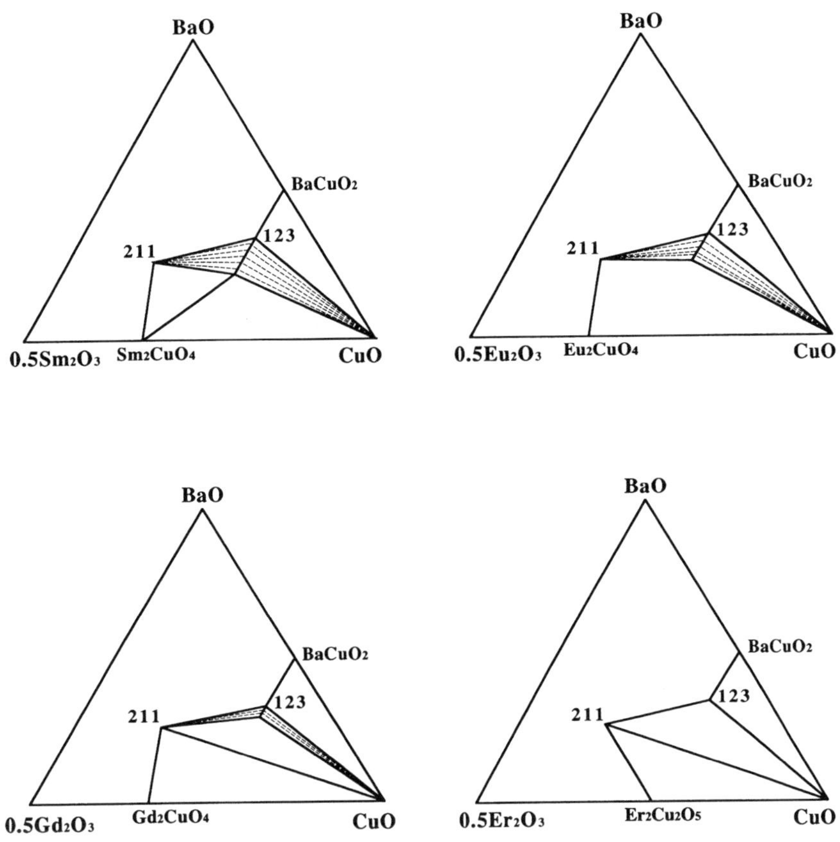

Fig. 119. Phase diagrams of 0.5 R_2O_3–BaO–CuO, R = Sm, Eu, Gd, Er

solution was $0.15 \leq x \leq 0.25$ due to a closer match between the ionic radii of La^{3+} and Ba^{2+}.

P–T–X phase equilibrium data for the Y–Ba–Cu–O system were compiled by Hauck [316] in the form of pseudobinary diagrams of $YBa_2Cu_3O_x$ (Fig. 120) and $YBa_2Cu_4O_x$ (Fig. 121). According to the Gibbs Phase Rule, the maximum number of phases in the invariant equilibria for this four-component system at constant pressure is equal to five, and the vertical lines describe three phase mixtures. The fields in Figs. 120 and 121 specify the regions of existence of phase mixtures within the corresponding temperature and oxygen pressure ranges (the pressures are given by the dashed lines). The composition of phases is listed in terms of the Y:Ba:Cu ratio.

According to Fig. 120, the range of equilibrium existence of the $YBa_2Cu_3O_x$ phase is within $P(O_2) < 10$ atm, $T < 1060°C$, and $6 < x < 6.5$. At higher temperature and pressure, this phase decomposes into the (211 + 023 + 247 + melt) mixture, whereas at $P(O_2) \leq 10^{-3}$ atm and $T < 932°C$, the decomposition products are (211 +

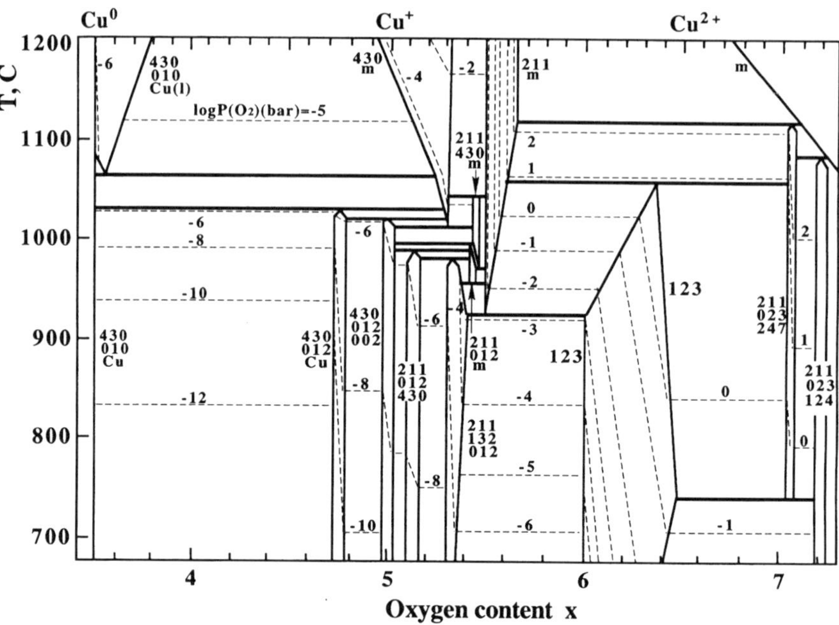

Fig. 120. Pseudo-binary $T\text{–}X$ phase diagram $YBa_2Cu_3O_x$. Phases are labeled according to Y:Ba:Cu ratios

$132 + 012$) at $x \cong 6$. It follows from this diagram that the superconducting $YBa_2Cu_3O_x$ (where $x > 6.5$) is a metastable phase and is expected to decompose during the use period of the corresponding devices. The $YBa_2Cu_4O_x$ phase is stable within a very narrow limit of the oxygen content $7.93 < x < 7.99$. Its melting temperature is above $1200°C$ at an oxygen pressure above 10^4 atm. However, this phase can also be obtained at ambient pressures, at temperatures below $850°C$ (Fig. 121).

Vapor pressure scanning of non-stoichiometry in $YBa_2Cu_3O_y$. $YBa_2Cu_3O_y$ has been studied in more detail compared to other high-temperature superconductor oxides. Oxygen non-stoichiometry was proved to be the main factor that determines the superconducting properties of this compound. Thus, the dependence of the oxygen index on temperature and vapor pressure, or the $P\text{–}T\text{–}X$ region of existence of the $YBa_2Cu_3O_y$ phase, is crucial for the technology of this material. Several thermogravimetric (TGA) studies of the oxygen non-stoichiometry of $YBa_2Cu_3O_y$ have been published [317–321]. The results of direct oxygen vapor pressure measurement in a closed volume were reported by Guskov et al. [322] at temperatures of 673 to 1173 K and oxygen pressures of 1 to 760 mmHg, which span the oxygen non-stoichiometry range of 6.2 to 6.97. Vapor pressure as a function of the temperature was measured for 16 samples of $YBa_2Cu_3O_y$ with different initial oxygen content. The results are presented in Fig. 122. Numerical values of the vapor pressures (an experimental data set comprising 611 points), as well as

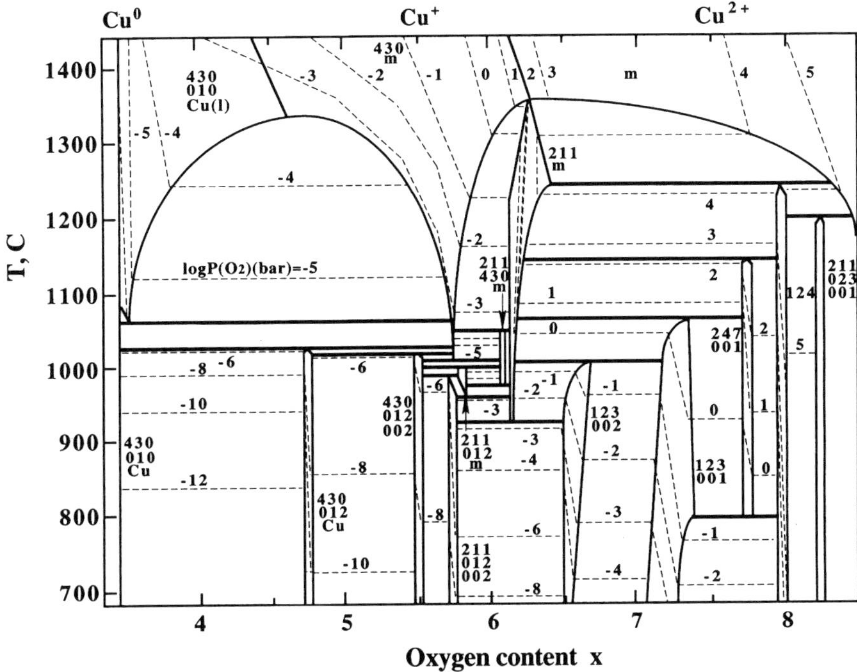

Fig. 121. Pseudo-binary T–X phase diagram $YBa_2Cu_4O_x$. Phases are labeled according to Y:Ba:Cu ratios

the volumes of the vapor, the initial mass of the samples, and iodometric oxygen indices, have been reported in [323].

In the temperature range under consideration, the predominant vapor species in equilibrium with $YBa_2Cu_3O_y$ is O_2. The partial pressures of other species are too small to influence the numerical results and can well be neglected in the calculations. It has also been shown [317–321, 324] that in this P–T–X range, the only condensed phase is $YBa_2Cu_3O_y$ which loses oxygen when heated according to the phase reaction

$$YBa_2Cu_3O_y \text{ (s)} \Rightarrow YBa_2Cu_3O_{y-\delta} \text{ (s)} + \tfrac{1}{2}\delta\, O_2 \text{ (g)}.$$

Consequently, at a fixed cation ratio, the system can be considered quasi-binary; one quasi-component is the sum of the metals and the other oxygen. The nonstoichiometry of the crystalline phase in this system can be probed by vapor pressure scanning in a way, similar to that, described in previous chapters for binary semiconductors. It is based on an explicit argument that, when heated in a closed volume, a part of the condensed phase sublimes, and because in general the compositions of the vapor and crystal do not coincide, the solid gradually changes in

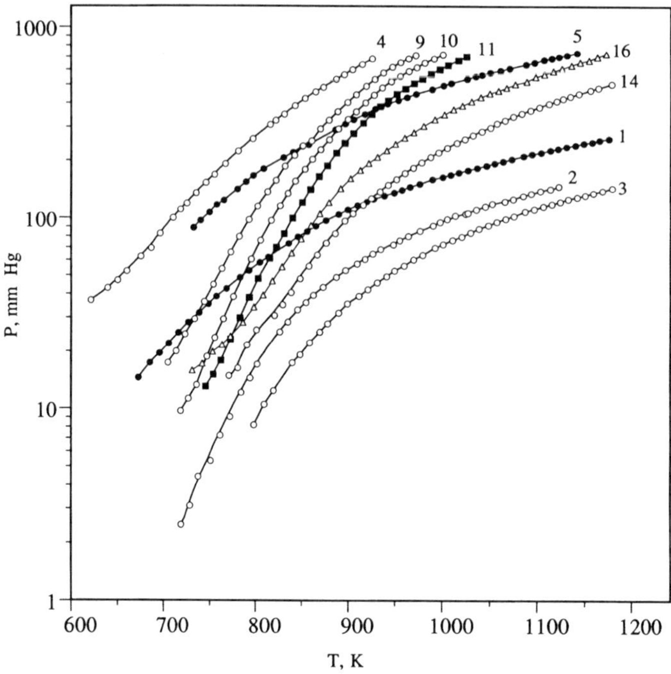

Fig. 122. Vapor pressure curves for different $YBa_2Cu_3O_y$ samples. Details are given in Table 24

composition when heated, according to Eq. (25). Assuming that the vapor in equilibrium with $YBa_2Cu_3O_y$ is an ideal O_2 gas, Eq. (25) can be rewritten as

$$X_S = (N_O - 2Pv/RT)/[(N_M + N_O) - 2Pv/RT] \times 100\%,\tag{52}$$

where N_M is the total mole number of metallic elements in the sample; N_O is the initial oxygen content in the sample; v is the volume of the vapor phase, and R is the gas constant.

Oxygen non-stoichiometry in oxides is usually described in terms of the oxygen index. In the vapor pressure experiment, the oxygen index y at every (P,T) point is associated with the initial oxygen index y_0 :

$$y = y_0 - 2Pv(a + by_0)/mRT,\tag{53}$$

where a is the total molar mass of the metals (Y + 2Ba + 3Cu), b is the atomic mass of oxygen, and m is the initial mass of the sample.

It has been shown in earlier chapters that in a binary system the composition of the solid at an intersection of two vapor pressure curves, determined by Eq. (31), is independent of the composition of the conjugated vapor. It depends solely on the initial composition of the two samples (1 and 2) and the volumes of the vapor in these two experiments. In terms of the oxygen index Eq. (31) can be rearranged to the following form:

$$y_{12} = [m_1 v_2 y_{01}/(a + b y_{01}) - m_2 v_1 y_{02}/(a + b y_{02})]/[m_1 v_2/(a + b y_{01}) - m_2 v_1/(a + b y_{02})] \qquad (54)$$

Here, y_{12} is the oxygen index of the $YBa_2Cu_3O_y$ phase at the intersection of two $P(T)$ curves for samples with initial indices y_{01} and y_{02}. Equations (53 and 54) can be used in two ways: either to determine the composition of the condensed phase at a high temperature (the intersection) if the initial compositions are known, or to determine these initial compositions from the known X_S or y_{12}. The first approach results in a number of scanning (P,T,X_S) points on the solidus, which are subsequently used to reconstruct the solidus surface in the $P–T–X$ phase space.

Analysis of the experimental data showed that iodometric determination of the oxygen index is not accurate enough for quantitative interpretation of the vapor pressure results. Therefore, the experimental coordinates of the intersections were used to determine the initial oxygen indices. They were calculated via an iteration procedure arranged so as to find the minimum of the sum of squares of residuals, $\min(\Sigma d^2)$, between oxygen indices calculated from Eqs. (53 and 54) at the total of n intersection points:

$$d^2 = (y_i - y_{ij})^2 + (y_j - y_{ij})^2 + (y_i - y_j)^2. \qquad (55)$$

Here y_i and y_j were calculated from Eq. (53) and y_{ij} from Eq. (54).

Two approaches were used for this purpose. At first, one of the iodometric indices, y_{01}, was used to calculate y_0's for all of the other samples from Eqs. (53 and 54); then the second, y_{02}, was used similarly; y_{03}, ... , y_{0n}. This first iteration cycle led to the mean values of all y_0's, which were used for a second cycle, and so on, until two conditions were met: (a) the minimum value of Σd^2 was reached, and (b) the difference in y_i's, calculated from Eq. (53) and Eq. (54), was of the order of magnitude of 10^{-3}. The second approach was to solve an overdetermined system of equations (53,54) and $y_{0i} = y_{iod}$, where y_{iod} is a set of all of the iodometric indices. These two approaches led to similar results cited in Table 24 as y_I and y_{II}. In subsequent treatment, y_I values were used because for those, the sum of squares of residuals was somewhat lower.

To find the (P,T) co-ordinates of the intersections, polynomial fits $P = \Sigma a_i T^i$ of the vapor pressure curves were used separately for the tetragonal and orthorhombic phases. The phase-transition showed on the vapor pressure curves as breaking points in the $\log P = f(1/T)$ representation (dashed line in Fig. 123).

An independent check of the initial composition of the samples can be made if the composition of the condensed phase at the intersection is written in weight per cent of the non-volatile component M:

$$X_S(\text{wt.\% M}) = [G_M/(G - g)] \times 100\%. \qquad (56)$$

Table 24. Iodometric results (y_{iod}) and vapor pressure determination of oxygen indices in the $YBa_2Cu_3O_y$ samples[a]

Exp. #	m(g)	V(mL)	y_{iod}	y_I	δy_I	y_{II}
1	0.93666	135.7	6.97	6.934	0.003	6.932
2	0.93060	135.7	6.70	6.673	0.006	6.667
3	0.92847	135.6	6.60	6.589	0.007	6.573
4	1.96508	33.6	6.97	6.934	0.003	6.932
5	0.84895	3.69	6.97	6.977	0.004	6.978
6	0.85643	48.0	6.74	6.698	0.005	6.697
7	0.85561	48.0	6.69	6.665	0.005	6.659
8	0.85268	48.0	6.56	6.526	0.005	6.523
9	3.18785	47.3	6.78	6.809	0.004	6.809
10	3.18472	47.3	6.74	6.765	0.004	6.764
11	3.18126	47.3	6.70	6.720	0.004	6.721
12	1.38649	34.7	—	6.819	0.004	6.819
13	1.38363	34.7	—	6.730	0.004	6.731
14	0.95111	35.4	6.60	6.633	0.006	6.631
15	1.19842	30.7	6.86	6.866	0.004	6.868
16	1.19249	30.6	6.65	6.683	0.006	6.670
Sum of squares of residuals at the intersections			0.20	3.25×10^{-3}		4.01×10^{-3}

[a] Two calculation procedures (y_I and y_{II}) are described in the text

Here G_M is the mass of the non-volatile component in the sample, G is the total mass of the sample, and g is the mass of the vapor phase, which in the case of pure oxygen is $g = 32Pv/RT$. Thus, at the intersection (P,T) of the two vapor pressure curves (1 and 2), where the composition of the condensed phase should be the same for both curves, the mass ratio of the non-volatile component in the two samples is

$$G_{M1}/G_{M2} = (G_1 - g_1)/(G_2 - g_2). \tag{57}$$

In the case under consideration, G_{Mi} is the total mass of the metallic elements in the sample i, irrespective of the actual ratio between Y, Ba, and Cu in it. On the other hand, G_{Mi} can be calculated for each sample i from the total mass of the sample G_i and the oxygen index y_{0i}:

$$G_{Mi} = aG_i/(a + by_{0i}). \tag{58}$$

Comparison of the results from Eq. (57) and Eq. (58) is a measure of confidence in the oxygen indices calculated from Eq. (54) and the initial assumption of the quasi-binary behavior of the system. The actual calculations showed that the G_{Mi}/G_{Mj} ratios were well within the limits of the experimental weighing errors of the balance used for preparing the material.

Table 25. Polynomial fits $X = \sum a_i(10^3/T)^i$ for $YBa_2Cu_3O_{7-x}$ at different oxygen pressures

P (atm)	a_0	a_1	a_2	a_3	T (K)
0.01	5.81881	−6.79265	2.01781	—	625–813
0.10	−5.97506	20.0794	−19.0581	5.58962	685–1050
0.21	−2.78437	12.4477	−13.2767	—	735–1250
0.50	1.93452	−1.43321	8.6021×10^{-3}	—	820–1190
1.00	1.75304	−1.20839	−0.10144	—	940–1190

Once all of the initial oxygen indices were found, each experimental run provided a set of (P,T,X_S) points, and all of the experimental database resulted in 611 scanning points, from which the solidus surface of the $YBa_2Cu_3O_y$ phase was reconstructed in the P–T–X phase space. The space arrangement of the solidus was determined in an analytical form. For this purpose, sections of the space model were made by three orthogonal planes: T = const, P = const, and X_S = const. To do this, the following polynomial fits of each vapor pressure curve i were made: $P_i(T)$, $T_i(P)$, $X_{Si}(T)$, $T_i(X_S)$, $P_i(X_S)$, and $X_{Si}(P)$. Isobaric sections were derived from a set of equations $\{X_{Si} = X_{Si}(P), T_i = T_i(P)\}$, solved for several pressures and presented in Table 25. Similarly, sets of equations $\{X_{Si} = X_{Si}(T), P_i = P_i(T)\}$ and $\{T_i = T_i(X_S), P_i = P_i(X_S)\}$ were solved for a number of temperatures and solid compositions to obtain isothermal and isoplethal (X_S = const) sections. The corresponding results are listed in Tables 26 and 27. The isopleths are shown separately in Fig. 123.

In Table 28, partial molar functions of oxygen, calculated from the vapor pressure data, are compared with the corresponding values derived from different thermogravimetric experiments. It can be seen in Table 28 that, although different TGA measurements do not coincide with each other, they tend to lead to higher thermodynamic functions compared to the values derived from the vapor pressure data. Two main reasons could be responsible for this discrepancy: uncertainties in the initial oxygen index and non-equilibrium conditions of the TGA measurements. The latter is expected to be the more probable, the lower are the temperature and the pressure [317]. This should lead to progressive underestimation of the vapor pressure with decreasing temperature and, as a consequence, to higher

Table 26. Isopleths of the oxygen pressure, $\log(P_{O_2}, \text{atm}) = A - B/T$, for $YBa_2Cu_3O_y$

y	A	B	T (K)
	Tetragonal phase		
6.30	6.58 ± 6.62	7900 ± 700	940-1205
6.35	6.01 ± 0.79	7100 ± 900	900-1175
6.40	6.33 ± 0.45	7100 ± 500	870-1130
6.45	6.59 ± 0.45	7200 ± 400	850-1075
6.50	6.91 ± 0.43	7200 ± 400	830-1040
	Orthorhombic phase		
6.65	8.14 ± 0.36	7700 ± 300	770-925
6.70	8.82 ± 0.52	8000 ± 400	760-910
6.75	9.30 ± 1.15	8200 ± 900	730-875
6.80	9.29 ± 0.59	7900 ± 500	700-845
6.85	9.89 ± 1.72	8000 ± 1300	680-805

slopes of the $P(T)$ functions resulting in overestimated enthalpy values. This is exactly what is seen in Table 28: equilibrium vapor pressure results [322] produced lower H_{O_2} values than those calculated from TGA. There might be one more reason for this discrepancy: the majority of the TGA $\log P(1/T)$ straight lines in [317–321], from which the H_{O2} and S_{O2} values were derived, were drawn through three to four (P,T) points (some of them even through two points),

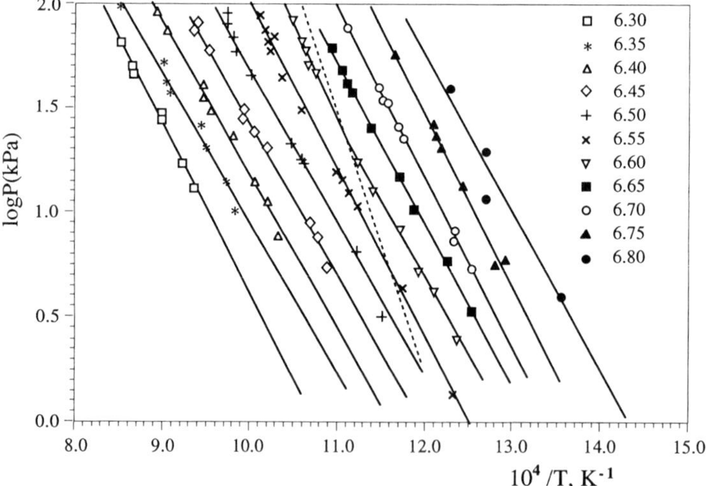

Fig. 123. Temperature dependences of the oxygen pressure for different oxygen indices in $YBa_2Cu_3O_y$. The dashed line represents the experimental tetragonal $\leftrightarrow$ orthorhombic phase-transition

Table 27. Isotherms of the oxygen pressure for the $YBa_2Cu_3O_{7-x}$ phase

Orthorhombic phase, $P(\text{atm}) = \sum a_i x^i$

T (K)	a_0	a_1	a_2	a_3	a_4	X
773	4.23600	−54.4900	267.892	−588.223	483.561	0.10–0.39
823	6.90307	−68.2045	257.467	−434.449	274.587	0.16–0.40
873	8.61699	−61.3831	159.423	−168.689	52.7924	0.24–0.44

Tetragonal phase, $P(\text{atm}) = \alpha \times x^{\beta}$

T(K)	$\alpha \times 10^2$	β	X
873	0.10364	−5.35623	0.44–0.55
923	0.24087	−5.61639	0.38–0.62
973	0.37233	−6.40011	0.41–0.68
1023	0.67373	−6.83363	0.49–0.73
1073	1.29981	−6.97083	0.53–0.76
1123	1.94813	−7.59700	0.59–0.80
1173	2.50969	−8.84175	0.66–0.80

Table 28. Partial molar oxygen enthalpies (H, kJ/mole) and entropies (S, J/mole×K) for $YBa_2Cu_3O_y$

y	[322]		[317]		[318]		[319]	[320]		[321]	
	$-S_{O_2}$	$-H_{O_2}$	$-S_{O2}$	$-H_{O_2}$	$-S_{O2}$	$-H_{O2}$	$-H_{O2}$	$-S_{O2}$	$-H_{O2}$	$-S_{O2}$	$-H_{O2}$
					Tetragonal phase						
6.30	126	152	122	155	144	173	169	121	153	117	148
6.40	121	137	140	162	148	166	160	132	157	128	148
6.50	132	138	160	170	171	176	164	131	144	136	146
					Orthorhombic phase						
6.60	154	149	170	169	196	188	166	157	159	159	157
6.70	169	154	179	165	211	190	172	171	162	169	157
6.80	178	151	188	162	220	184	173	—	—	187	162

whereas the vapor pressure data [322] were obtained from seven to thirteen (P,T) points, which is a statistically representative set of data. A tendency can be seen in Table 28 for H_{O2} to increase with the oxygen index approaching $y = 6$. This trend was also noted by Yamaguchi et al. [319], for both $y = 6$ and $y = 7$ extremes.

Comparison of different P–T–X results shows that at higher temperatures and pressures, the vapor pressure data coincide with TGA within the limits of different TGA results ($\Delta y = \pm 0.02$ and $\Delta T = \pm 10$ K [317]), whereas at $P = 0.01$ atm the disagreement in these values increases up to ± 0.08 and ± 40 K.

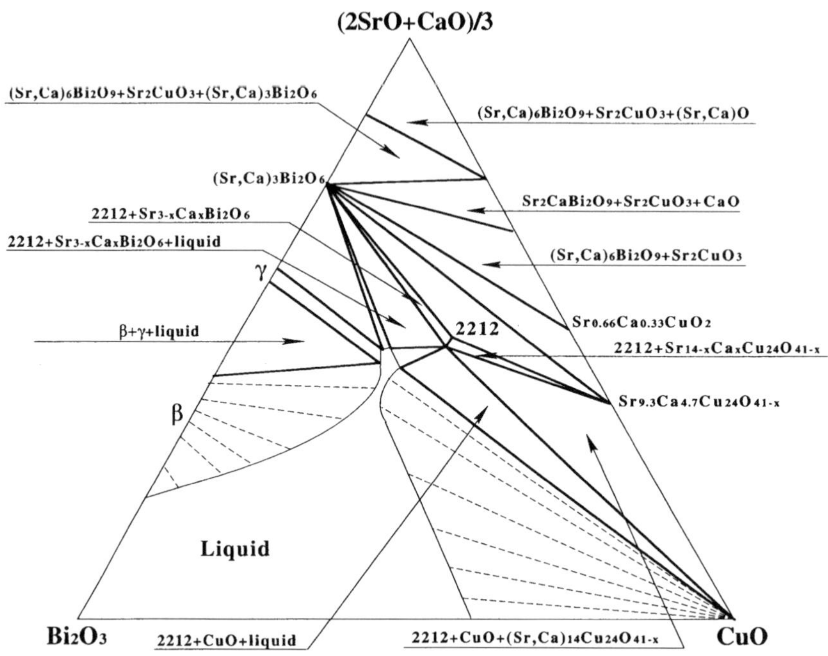

Fig. 124. Isothermal-isobaric section of the Bi_2O_3–SrO–CaO–CuO diagram at 850°C in air. Sr:Ca = 2:1

Bismuth high-T_c superconductors. Two high-T_c superconductors were discovered in the Bi–Sr–Ca–Cu–O system: the so-called 80 K phase $Bi_2Sr_2CaCu_2O_8$ (nicknamed 2212) and the 110 K phase $Bi_2Sr_2Ca_2Cu_3O_{10}$ (nicknamed 2223). A complete geometrical description of phase equilibrium in this five-component system requires six-dimensional space. Therefore, for practical considerations of phase equilibrium, sections of the phase diagram are to be made. An isobaric-isothermal section at T=const, P=const of the quasi-quaternary system Bi_2O_3–SrO–CaO–CuO is a concentration tetrahedron whose oxide quasi-components are vertices. The edges of this tetrahedron represent isobaric-isothermal sections of six quasi-binary systems, and the faces correspond to four quasi-ternary systems. Such tetrahedra were constructed for several temperatures at constant pressure $P(O_2)$ = 0.21 atm (in air) [325]. It turned out nevertheless, that this representation visualizes only relative positions of different individual phases in the tetrahedron. To describe phase compatibility, further simplification had to be made. For this purpose, solid solution (Sr,Ca)O of different compositions was considered a quasi-component. The resulting sections at 850°C in air for two (Sr,Ca)O compositions corresponding to Sr/Ca ratios in 2212 and 2223 compounds are presented in Figs. 124 and 125. The decomposition products of the 2212 and 2223 phases, as well as the phase states of the system along the entire compositional triangles, are quoted according to the review paper of Majewsky [325].

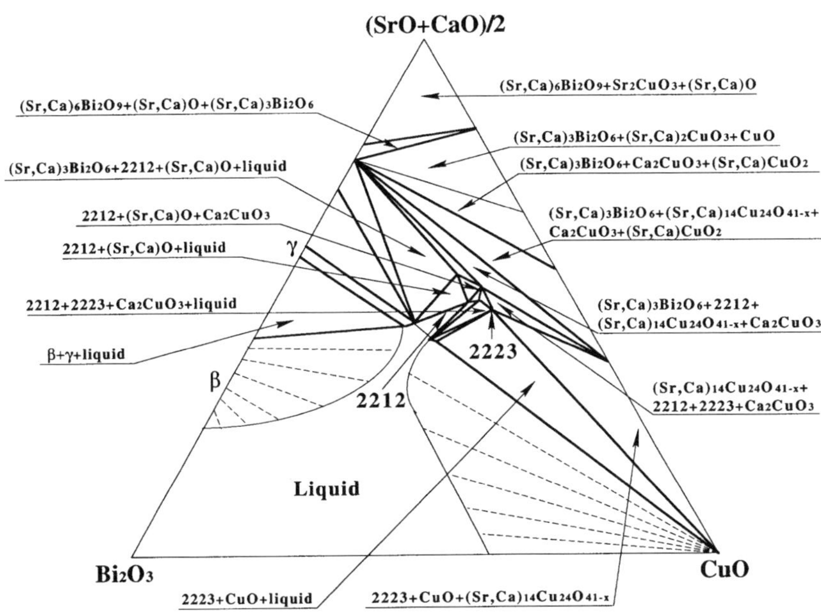

Fig. 125. Isothermal-isobaric section of the Bi_2O_3–SrO–CaO–CuO diagram at 850°C in air. Sr:Ca = 1:1

It has been shown that the 2212 phase is stable in a comparatively wide temperature range and coexists with a variety of other phases, whereas the temperature range of stability of the 2223 phase is essentially narrower. It follows from Fig. 125 that rather small deviations in composition could lead to failure in preparation of the 2223 material even in mixtures with other phases. The range of single-phase stability of 2212 depends on both temperature and cation non-stoichiometry. For example, according to Majewsky [325], $Bi_{2+x}Sr_2CaCu_2O_{8+\delta}$ is stable as a single phase in the temperature range 650 to 900°C with x = 0.05–0.3, and Ca content in the single-phase $Bi_{2.18}Sr_{3-y}Ca_yCu_2O_{8+\delta}$ varies from y = 0.6 to y = 1.8 at $T = 670$–890°C. It was argued that the exact stoichiometric composition is outside the single-phase volume of the 2212 phase because numerous attempts to prepare this material via a solid-state reaction route from the stoichiometric mixture of oxides resulted in multiphase samples [325]. On the other hand, sol-gel technology was successfully applied to prepare stoichiometric 2212 [326, 327]. In the 2223 phase, the cation non-stoichiometry is much narrower: the Sr/Ca ratio is within 1.9/2.1 to 2/2, and the range of the Bi stoichiometric index is 2–2.5.

Single-phase regions of 2212 and 2223 were significantly enhanced by partially substituting Bi with Pb. The limits of such doping also depend on temperature and cation non-stoichiometry. For the Pb-doped $(Bi,Pb)_{2+x}Sr_{4-y}Ca_yCu_3O_{10+\delta}$, the range of single-phase existence is y = 1.8–2.2 at temperatures 800 to 885°C. A serious prob-

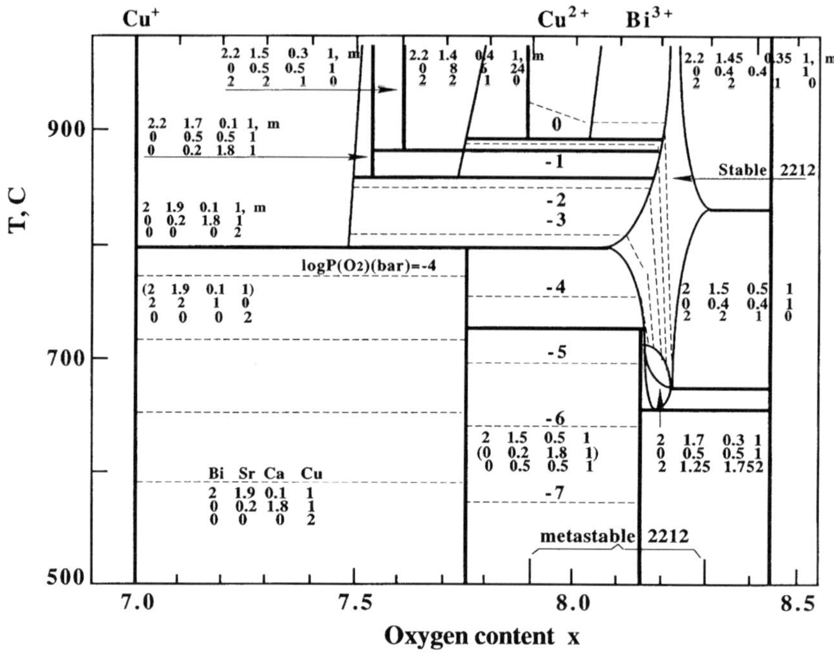

Fig. 126. Pseudobinary phase diagram of $Bi_2Sr_2CaCu_2O_x$

lem in preparing single-phase Pb-doped 2212 and 2223 compounds is the temperature dependence of the maximum Pb solubility in these materials because single-phase composition at high temperature may correspond to a multiphase mixture when cooled to room temperature.

Additional complications are encountered in preparing 2212 and 2223 wires and tapes when silver is used as a sheath. Althoug, it is believed, that Ag does not dissolve in either 2212 or 2223, it influences the phase equilibrium, both above the decomposition temperatures of 2212 (865°C) and 2223 (845°C) phases and at low temperatures, by forming the Ag–Pb–Cu–O eutectic at 650°C [325]. Silver can also affect the cation non-stoichiometry of 2212 and 2223.

P–T–X phase equilibrium in the Bi–Sr–Ca–Cu–O system was presented by Chernyaev et al. [328] as a pseudo-binary diagram M–O (Fig. 126), where M is the sum of cations in the ratio corresponding to the 2212 compound. The diagram was constructed for the range of the oxygen index in MO_x, x = 7–8.5. According to the phase rule, the vertical lines in this five-component system describe four phase equilibria, and six phases make up the invariant equilibria (horizontal lines). Oxygen partial pressures $P(O_2) = 10^{-7}$ to 1 atm are given in Fig. 126 by dashed lines, and cation content in phases comprising different phase states of the system are quoted according to the Bi:Sr:Ca:Cu ratio. For example, 2210 stands for $Bi_2Sr_2CaO_6$, 0002 is CuO, etc. It follows from Fig. 126 that the lower temperature limit of stability of the 2212 phase with exact cation stoichiometry is about 700°C. This data implies that the 2212 phase might melt congruently at temperature above 900°C at high oxygen pressure. Oxygen non-stoichiometry x depends on the cation ratio; for the exact 2212 composition x = 8.1–8.3.

Conclusion

Phase equilibrium represented geometrically as a pressure–temperature–composition phase diagram of a heterogeneous system is the thermodynamic basis for advanced materials science. A specific part of it, known as vapor pressure scanning of non-stoichiometry, is based on the idea of looking experimentally at the region of existence of the non-stoichiometric solid along the composition axis, where it appears as a vast P–T region, rather than along the pressure axis, where the solidus is often seen as a very narrow T–X field. In this approach the composition of the solid is determined in situ directly at a high temperature.

Phase diagrams comprising crystalline phases with small deviations from stoichiometry (less than 0.1 at.%) are sometimes called micro-diagrams. This implies that scrupulous experimental work and sophisticated methods are needed to study such systems.

Experimental thermodynamics is a very special sphere of scientific activity. Apart from considerable material investment, the researcher is expected to put in a tremendous amount of time and energy with only a slight chance of exciting scientific breakthroughs. Very often the only output of such enormous effort is just another (even though more accurate) entry in a reference database, to be subsequently used and, more often than not, not even quoted by scores of other scientists and engineers in their applied studies and while creating new technologies. Probably, that is why experimental thermodynamics has become progressively less popular among the scientific community, especially among young scientists, which is reflected in the ever diminishing number of publications in this field.

A significant part of thermodynamic studies, in particular high-precision–measurements of thermodynamic properties of substances and phase equilibrium investigations, was done during the period from the 1950s to the 1980s by the Russian thermodynamic schools of the Moscow and Leningrad State Universities, and the Kurnakov Institute of General and Inorganic Chemistry (Russian Academy of Sciences, Moscow). This vast experimental output was mainly published in different Russian journals and, as a result, remains only partially known to the Western scientific community, chiefly through a number of thermodynamic reference books and databases, such as IVTANTERMO [145]. Meanwhile, as can be seen in the present book, these results still remain sometimes the only source of information on phase equilibrium, in our case P–T–X diagrams for a number of oxide and semiconductor systems. The presentation and discussion of this data was one of the purposes of this book.

Josiah Willard Gibbs [7] defined the laws of thermodynamics as those that "express the approximate and probable behavior of systems of a great number of particles, or, more precisely, they express the laws of mechanics for such systems as they appear to beings who have not the fineness of perception to enable them to appreciate quantities of the order of magnitude of those which relate to single particles, and who cannot repeat their experiments often enough to obtain any but the most probable results". Nevertheless, statistical laws used for interpretation of specially designed experiments might prove to act as a "mathematical microscope", through which the properties of a system (even if not those of the individual constituent particles) can be examined to the smallest detail. In particular, if the system is a non-stoichiometric crystal, its homogeneity range can be scrutinized in a great many cases through the magnifying glass of vapor pressure scanning — a novel way of direct high-precision probing of non-stoichiometry in crystals.

References

1. Kurnakov NS. Selected works. Acad. Sci. (1960) Moscow (in Russian)
2. Wagner K, Schottky W. Z Phys Chem (1930) **B11** 163
3. Marsh JS. Principles of phase diagrams. McGraw-Hill (1935) New York
4. Prince A. Alloy phase equilibria. Elsevier (1966) Amsterdam
5. Glazov VM, Novikov II. Rus J Phys Chem (1974) **48** 1134 (in Russian)
6. Ricci J. The phase rule and heterogeneous equilibrium. Van Nostrand (1951) Toronto, New York, London
7. Gibbs JW. Thermodynamics and statistical mechanics. Nauka (1984) Moscow
8. Gerasimov YI (ed), Physical chemistry. Khimiya (1970) Moscow (in Russian)
9. Munster A. Chemical thermodynamics. Mir (1971) Moscow
10. Karapetyanz MK. Chemical thermodynamics. Khimiya (1975) Moscow (in Russian)
11. Storonkin AV. Thermodynamics of heterogeneous systems. Leningrad State University (1967) Leningrad (in Russian)
12. Denbigh K. The principles of chemical equilibrium. Cambridge University Press (1955) Cambridge
13. Swalin RA. Thermodynamics of solids. Wiley (1962) New York
14. Bazarov IP. Thermodynamics. Visshaya Shkola (1983) Moscow (in Russian)
15. Van der Vaals J, Kohnstamm P. Thermodynamics. ONTI (1936) Moscow
16. Wagner K. Thermodynamics of alloys. Metallurgizdat (1957) Moscow
17. Kurnakov NS. Introduction to physico-chemical analysis. Acad. Sci. USSR (1940) Moscow (in Russian)
18. Anosov VY, Pogodin SA. The basics of physico-chemical analysis. Acad. Sci. USSR (1947) Moscow, Leningrad (in Russian)
19. Gordon P. Principles of phase diagrams. McGraw-Hill (1968) New York
20. Kroger FA. The chemistry of imperfect crystals. North Holland (1964) Amsterdam
21. Zlomanov VP. $P–T–X$ diagrams of two-component systems. Moscow State University (1980) Moscow (in Russian)
22. Zlomanov VP, Novosyolova AV. $P–T–X$ phase diagrams of metal–chalcogen systems. Nauka (1987) Moscow (in Russian)
23. Schreinemakers FA. Invariant, univariant and bivariant equilibria. IL (1948) Moscow
24. Levinskii YV. P–T–X diagrams of two-component systems. Metallurgiya (1982) Moscow (in Russian)
25. Mertzlin RV. Heterogeneous equilibria. Saratov State University (1971) Saratov (in Russian)
26. Libowitz GG. Non-stoichiometry in chemical compounds. In Reiss H (ed) Progress in solid state chemistry. Pergamon Press (1965), London, vol 2, pp 216–264
27. Brebrick RF. Non-stoichiometry in binary semiconductor compounds. In Reiss H (ed) Progress in solid state chemistry. Pergamon Press (1966) London, vol 3, pp 213–263

28. Brebrick RF. Phase equilibria. In Hurle DTJ (ed) Handbook of crystal growth. North-Holland (1993) Amsterdam, London, New York, Tokyo, vol 1, pp 43-102

29. Eremin EN. Basis of chemical thermodynamics. Visshaya Shkola (1978) Moscow (in Russian)

30. Glazov VM. The basis of physical chemistry. Visshaya Shkola (1981) Moscow (in Russian)

31. Young DA. Phase diagrams of the elements. University of California Press (1991) Berkeley, Los Angeles, Oxford

32. Tonkov EY. Phase diagrams of compounds at high pressures. Nauka (1983) Moscow (in Russian)

33. Nipan GD, Greenberg JH, Lazarev VB. Polymorphism and metastable states in binary systems. In Glazov VM, Novosyolova AV (eds) Thermodynamics and semiconductor materials science (1992). Metallurgia, Moscow, pp 272–321 (in Russian)

34. Greenberg JH. Mater Sci Eng Rep (1996) **R16** 223

35. Greenberg JH, Lazarev VB, Guskov VN. Il Nuovo Cimento (1983) **2D** 1681

36. Nesmeyanov AN. Vapor pressure of the chemical elements. Gary R (ed), Elsevier (1963) Amsterdam, London, New York

37. Suvorov AV. Thermodynamic chemistry of the vapor state. Khimiya (1970) Leningrad (in Russian)

38. Greenberg JH. Vapor pressure investigation of multicomponent equilibria (1979). DSc Dissertation, Institute of General and Inorganic Chemistry, Moscow (in Russian)

39. Mascherpa G. Rev Chim Miner (1966) **3** 153

40. Kubaschewsky O, Alcock CB, Spencer PJ. Materials thermochemistry, 6th ed., Pergamon Press (1993) Oxford

41. Novosyolova AV, Pashinkin AS. Vapor pressure of volatile chalcogenides Nauka (1978) Moscow (in Russian)

42. Seith W, Krauss W. Z Elektrochem (1938) **44** 98

43. Herasimenko P. Acta Metall (1956) **4** 14

44. Jona F, Lever RF, Wandt HR. J Electrochem Soc (1964) **111** 413

45. Greenberg JH, Luzhnaya, Medvedeva ZS. Rus J Inorg Mater (1966) **2** 2130 (in Russian)

46. Thurmond CD, Frosh CJ. J Electrochem Soc (1964) **111** 184

47. Azhikina UA (1965) PhD Dissertation, Moscow State University (in Russian)

48. Brebrick RF, Liu H. High Temp Mater Sci (1996) **35** 215; J Phase Eq. (1996) **17** 495

49. Brebrick RF, Strauss AJ. J Chem Phys (1964) **41** 197; 3230

50. Brebrick RF. J Chem Phys (1964) **41** 1140; J Phys Chem Solids (1965) **26** 989

51. Yu TC, Brebrick RF. J Phase Eq (1992) **13** 476

52. Brebrick RF. J Chem Phys (1965) **43** 3846

53. Yu TC, Brebrick RF. Phase diagrams of Cd/Zn/Te/Se compounds. In Capper P (ed), Properties of narrow gap cadmium based compounds. INSPEC (1994)

54. Brebrick RF. J Chem Phys (1968) **48** 5741

55. Brebrick RF. J Electrochem Soc (1969) **116** 1274

56. Brebrick RF. J Electrochem Soc (1971) **118** 2014

57. Fang R, Brebrick RF. J Phys Chem Solids (1996) **57** 443

58. Richman D. RCA Rev (1963) **24** 596

59. Von Wartenberg H. Z Elektrochem (1913) **19** 482

60. Merten U. J Phys Chem (1959) **63** 443

61. Magomedov KA. Rus J Inorg Mater (1966) **2** 117 (in Russian)

62. Ferguson RR, Gabor T. J Electrochem Soc (1964) **111** 585

63. Seki H, Marijama K. Jap J Appl Phys (1967) **6** 785

64. Wehmeir FH. J Crystal Growth (1970) **6** 341

65. Gibart P, Goldstein L, Dormann JL, Guittard N. J Crystal Growth (1974) **24/25** 147

66. Taklaku K, Tosio K, Tetsue J. Rept Nat Res Inst Metals (1964) **7** 328

67. Silvestri VJ. J Electrochem Soc (1965) **112** 748

68. Rao DB, Dadape VV. J Phys Chem (1966) **70** 1349

69. Greenwood HC. Proc Royal Soc (1910) **83A** 483

70. Greenwood HC. Z Elektrochem (1912) **18** 319

71. Ruff O, Bergdahl G. Z Anorg Chem (1919) **106** 76

72. Ruff O, LeBoucher L. Z Anorg Chem (1934) **219** 376

73. Fischer J. Z Anorg Chem (1934) **219** 1

74. Hartman H, Schneider R. Z Anorg Chem (1929) **180** 275

75. Fischer W, Rahlfs O. Z Anorg Chem (1932) **205** 1

76. Bauer E, Brunner K. Helv Chim Acta (1934) **17** 958

77. Fischer J. Z Anorg Chem (1934) **219** 367

78. Novikov GI, Polyachonok OG. Rus J Inorg Chem (1961) **6** 1951(in Russian)

79. Novikov GI, Baev AK. Rus J Inorg Chem (1962) **7** 1349 (in Russian)

80. Smirnov NV, Detkov SP. Doklady USSR Acad Sci (1954) **98** 777 (in Russian)

81. Smirnov NV, Detkov SP. Rus J Phys Chem (1957) **31** 87 (in Russian)

82. Detkov SP. Rus J Phys Chem (1957) **31** 83 (in Russian)

83. Berg LG. Introduction to thermal analysis. Nauka (1969) Moscow (in Russian)

84. Berg LG, Rassonskaya IS. Proc I Conf on Thermal Analysis. Acad Sci USSR (1965) 13 (in Russian)

85. Berg LG, Rassonskaya IS, Buris EV. Trans Sector of Thermal Analysis (1956) **27** 239 (in Russian)

86. Biltz W, Meyer F. Z Anorg Chem (1928) **176** 23

87. Datz S, Smith WT, Taylor EH. J Chem Phys (1959) **63** 938

88. Hagenmark K, Blander M, Luchsinger EB. J Phys Chem (1966) **70** 276

89. Borshchevsky AS, Roenkov ND. Rus J Electron (1971) **14** 55 (in Russian)

90. Ugai YA, Pshestanchik VR, Anokhin VZ, Gukov OY. Rus J Inorg Mat (1974) **10** 1936 (in Russian)

91. Grigorowici R. Z Techn Phys (1939) **20** 102

92. Bodenstein M. Z Elektrochem (1916) **22** 327

93. Bodenstein M, Katayama M. Z Phys Chem (1909) **69** 26

94. Bodenstein M, Katayama M. Z Elektrochem (1909) **15** 244

95. Lazarev VB, Greenberg JH, Popovkin BA. Investigation of deviation from stoichiometry by vapor pressure measurement. In Kaldis E (ed) Current topics in materials science. North Holland (1978) Amsterdam, vol 1, pp 657–695

96. Greenberg JH, Lazarev VB. Vapor pressure investigation of P–T–X phase equilibria and non-stoichiometry in binary systems. In Kaldis E (ed) Current topics in materials science. Elsevier (1985) Amsterdam, vol 12, pp 119–204

97. Gaskov AM, Zlomanov VP, Novosyolova AV. Bull Moscow State University, Chem (1970) **1** 49 (in Russian)

98. Gaskov AM, Zlomanov VP, Sapozhnikov YA, Novosyolova AV. Bull Moscow State University, Chem (1968) **3** 48 (in Russian)

99. Karbanov SG, Zlomanov VP, Novosyolova AV. Bull Moscow State University, Chem (1970) **1** 51 (in Russian)

100. Ryazantzev AA, Popovkin BA, Novosyolova AV, Lyakhovitzkaya VA. Bull Moscow State University, Chem (1969) **3** 49 (in Russian)
101. Popovkin BA, Dolgikh VA, Trifonov VA, Belousov VI, Novosyolova AV. In Chemistry of gaseous inorganic compounds (Abstracts) Belorussian Acad Sci (1973) Minsk, p 153 (in Russian)
102. Valitova NR, Popovkin BA, Aslanov LA, Novosyolova AV. Rus J Inorg Mater (1973) **9** 796 (in Russian)
103. Novosyolova AV, Popovkin BA, Lyakhovitskaya VA, Trifonov VA. In Crystal growth and structure (Abstracts) Armenian Acad Sci (1972) Erevan, vol 1, p 45 (in Russian)
104. Trifonov VA, Dernovsky VI, Popovkin BA, Lyakhivitskaya VA, Belousov VI, Novosyolova AV. Rus J Phys Chem (1974) **48** 788 (in Russian)
105. Popovkin BA, Trifonov VA, Lyakhovitskaya VA, Novosyolova AV. Doklady Acad Sci USSR (1974) **215** 603 (in Russian)
106. Valitova NR, Alyoshin VA, Popovkin BA, Novosyolova AV. Rus J Inorg Mater (1976) **12** 225 (in Russian)
107. Greenberg JH. Doklady Acad Sci USSR (1975) **221** 117 (in Russian)
108. Greenberg JH, Lazarev VB, Epstein SI. Doklady Acad Sci USSR (1975) **224** 1345 (in Russian)
109. Greenberg JH, Lazarev VB. Rus J Phys Chem (1978) **52** 551 (in Russian)
110. Sedgwick TO. J Electrochem Soc (1965) **112** 496
111. Frank G. Ber Bunsen Ges Phys Chem (1965) **69** 119
112. Lever RF. J Electrochem Soc (1963) **110** 775
113. Schafer H, Nickl J. Z Anorg Allgem Chem (1953) **274** 250
114. Sedgwick TO, Aguke BJ. J Electrochem Soc (1966) **113** 54
115. Nitsche R, Richman D. Z Elektrochem (1962) **66** 709
116. Lyons VJ, Silvestri VJ. J Electrochem Soc (1962) **109** 963
117. Sandulova AV, Voronin VA, Prokhorov VA. Rus J Phys Chem (1871) **45** 2165 (in Russian)
118. Triboulet R, Rudolph P, Muller-Vogt G. In Kasper E, Parker EHC, Triboulet R, Rudolph P, Muller-Vogt G (eds) Proc 1995 E-MRS Spring Meeting. Elsevier Amsterdam
119. Parikh A, Pearson SD, Tran TK, Bichnell RN, Benz RG, Wagner BK, Schafer P, Summers CJ. J Crystal Growth (1996) **159** 1152
120. Karam NH, Sudharsanam R, Masrivito A, Sanfacon MM, Smith FT, Leonard M, El-Masry NA. J Electron Mater (1995) **24** 483
121. Palocz W, Szofran FR, Lehoczky SL. J Crystal Growth (1995) **148** 56
122. Rudolph P. Melt growth of II–VI compound single crystals. In Isshiki M (ed) Recent development of bulk crystal growth (1998) Sendai, pp 127–164
123. Isshiki M. Wide gap II–VI compounds for optoelectronic applications. Chapman & Hall (1992) London
124. Rudolph P, Schafer N, Fukuda T. Mater Sci & Eng (1995) **R15** 85
125. Okada H, Kawanaka T, Ohmoto S. J Crystal Growth (1996) **165** 31
126. Jordan AS, Zupp RR. J Electrochem Soc (1969) **116** 1264
127. Steininger J, Strauss AJ, Brebrick RF. J Electrochem Soc (1970) **117** 1305
128. Kulwicki BM. Phase equilibria of some compound semiconductors by DTA calorimetry. PhD Dissertation (1963) University of Michigan
129. Carides J, Fischer AG. Solid State Commun (1964) **2** 217

130. Greenberg JH, Guskov VN, Feltgen T, Fiederle M, Benz KW. P–T–X phase equilibrium and crystal growth of ZnTe from the vapor phase. E-MRS E-2000 Spring Meeting, Abstracts, P6, Strasbourg, France

131. Abrikosov NK, Bankina VF, Poretskaya LB, Skudnova EV, Chizhevskaya SN. Semiconductor chalcogenides. Nauka (1975) Moscow (in Russian)

132. Rau H. J Phys Chem Solids (1978) **39** 879

133. Calister WD, Varotto CF, Stevenson DA. Phys Stat Solidi (1970)**38** 45

134. Aven M, Prener JS. Physics and chemistry of II–VI compounds. North Holland (1967) Amsterdam

135. Albers W. Non-stoichiometry of inorganic solids. In Kaldis E (ed) Current topics in materials science. North Holland (1982) Amsterdam, vol 10, pp 191–247

136. Lorenz MR. J Phys Chem Solids (1962) **23** 939

137. Greenberg JH, Guskov VN, Lazarev VB. Mater Res Bull (1992) **27** 997

138. Greenberg JH, Guskov VN, Lazarev VB, Shebershneva OV. Mater Res Bull (1992) **27** 847

139. Greenberg JH, Guskov VN, Lazarev VB, Shebershneva OV. J Solid State Chem (1993) **102** 382

140. Greenberg JH. J Crystal Growth (1996) **161** 1

141. Hultgren R, Orr RL, Anderson PD, Kelley KK. Selected values of thermodynamic properties of metals and alloys. Wiley (1963) New York

142. Brooks LS. J Amer Chem Soc (1952) **74** 227

143. Glazov VM, Chizhevskaya SN, Glagoleva NN. Liquid semiconductors. Nauka (1967) Moscow (in Russian)

144. Goldfinger P, Jeunehomme M. Trans Faraday Soc (1962) **59** 2851

145. Thermodynamic data base IVTANTERMO. Acad Sci USSR

146. Smith FTJ. Metall Trans (1970) **1** 617

147. Berding MA. Appl Phys Lett (1999) **74** 552

148. Palocz W, Wiedemeir H. J Crystal Growth (1993) **129** 653

149. Laasch M, Kunz T, Eiche C, Fiederle M, Joerger W, Kloess G, Benz KW. J Crystal Growth (1997) **174** 696

150. Rudolph P. Fundamental studies on Bridgman growth of CdTe. In Progress in crystal growth and characterization. Elsevier (1994) Amsterdam, vol 29, pp 275-381

151. Rudolph P, Rinas U, Jacobs K. J Crystal Growth (1994) **138** 249

152. Raiskin E, Butler JE. IEEE Trans Nucl Sci (1988) **35** 81

153. Marbeuf A, Druilhe R, Triboulet R, Patriarche G. J Crystal Growth (1992) **117** 10

154. Onda T, Ito R. Jap J Appl Phys (1989) **28** 1544

155. Ruault MO, Kaitasov O, Triboulet R, Creston J, Gasgnier M. J Crystal Growth (1994) **143** 40

156. Brebrick RF. J Crystal Growth (1988) **86** 39

157. Medvedeva ZS, Greenberg JH, Boryakova VA. Rus J Inorg Chem (1968) **13** 1440 (in Russian)

158. Boryakova VA, Greenberg JH, Medvedeva ZS. Rus J Inorg Mater (1969) **5** 477 (in Russian)

159. Lieth RMA, Heijligers HIM, Heijden CWM. J Electrochem Soc (1966) **113** 798

160. Piacente V, Bardi G, Paopo VD, Ferro D. J Chem Thermodynam (1976) **8** 391

161. Dieleman J, Sanders FHN, Van Domelen JHJ. Philips J Res (1982) **37** 204

162. Suzuki H, Mori R. Jap J Appl Phys (1974) **13** 417

163. Uy OM, Muenow DW, Ficalora PJ, Margrave JL. Trans Faraday Soc (1968) **64** 2999

164. Alapini F, Flahaut J, Guittard M. J Solid State Chem (1979) **28** 309
165. Greenberg JH, Boryakova VA, Shevelkov VF. Rus J Inorg Mat (1972) **8** 2099 (in Russian)
166. Colin R, Drowart J. Trans Faraday Soc (1968) **64** 2611
167. Greenberg JH, Boryakova VA, Shevelkov VF. J Chem Thermodynam (1973) **5** 233
168. Burns RP, De Maria G, Drowart J, Inghram MG. J Chem Phys (1963) **38** 1035
169. Slavnova GK, Luzhnaya NP, Medvedeva ZS. Rus J Inorg Chem (1963) **8** 1199 (in Russian)
170. Illarionov VV, Lapina LM. Doklady Acad Sci USSR (1957) **114** 1021 (in Russian)
171. Rau H. Ber Bunsenges Phys Chem (1967) **71** 71
172. Berkowitz J, Chupka WA. J Chem Phys (1966) **45** 4289
173. Berkowitz J, Chupka WA. J Chem Phys (1968) **48** 5743
174. Abrikosov NK, Shelimova LE. Semiconductor materials $A^{IV}B^{VI}$. Nauka (1975) Moscow (in Russian)
175. Brebrick RF. J Chem Phys (1968) **49** 2584
176. Lu Tsun Hua, Pashinkin AS, Novosyolova AV. Doklady Acad Sci USSR (1963) **151** 1335 (in Russian)
177. Wiedemeir H, Siemers PA. Z Anorg Allgem Chem (1977) **431** 299
178. Karbanov SG, Karakhanova MI, Pashinkin AS. Rus J Inorg Mater (1971) **7** 1914 (in Russian)
179. Novosyolova AV, Zlomanov VP, Karbanov SG. Physico-chemical study of the germanium, tin and lead chalcogenides. Progress in Solid State Chemistry (1972) **7** 85
180. Lu Tsun Hua, Pashinkin AS, Novosyolova AV. Doklady Acad Sci USSR (1962) **146** 1092 (in Russian)
181. Karbanov SG, Zlomanov VP, Novosyolova AV. Bull Moscow State University, Chem (1968) **3** 96 (in Russian)
182. Karbanov SG, Statnova EA, Zlomanov VP, Novosyolova AV. Bull Moscow State University, Chem (1972) **5** 531 (in Russian)
183. Vinogradova GZ, Dembovsky SA, Sivkova NB. Rus J Inorg Chem (1968) **13** 2029 (in Russian)
184. Quenes P, Khodadat P. Compt Rend (1969) **268** 2294
185. Ross L, Bourgon M. Can J Chem (1969) **47** 2555
186. Quenes P, Khodadat P, Ceolin R. Bull Soc Chim France (1972) **1** 117
187. Chaussemy G, Fornazero J. Rev Phys Appl (1977) **12** 687
188. Wiedemeier H, Simers P. Z Anorg Allgem Chem (1975) **411** 90
189. Hansen M. Constitution of binary alloys. McGraw-Hill (1958) New York
190. Colin R, Drowart J. Phys Soc (1964) **68** 428
191. Chu D, Walser RM, Bene RW, Courtney TH. Appl Phys Lett (1974) **24** 479
192. Rau H. J Phys Chem Solids (1966) **27** 761
193. Colin R, Drowart J. J Chem Phys (1962) **37** 1120
194. Dumon A, Idchanot A, Gromb S. J Phys Chem Solids (1977) **38** 279
195. Kulyuhina EA, Zlomanov VP, Novosyolova AV. Rus J Inorg Mater (1977) **13** 237 (in Russian)
196. Colin R, Drowart J. Trans Faraday Soc (1964) **60** 673
197. Tsu R, Howard WE, Esaki L. Phys Rev (1968) **172** 779
198. Castanet R, Bergman C, Mathien J. Calphad (1979) **3** 205
199. Bloem J, Kroger FA. Z Phys Chem (1956) **7** 1

200. Novosyolova AV, Zlomanov VP. Thermodynamics and imperfections in lead chalcogenides. In Kaldis E (ed) Current topics in materials science. North Holland (1981) Amsterdam, vol 7, pp 643–710
201. Rau H. J Phys Chem (1981) **42** 257
202. Cubicciotti D. J Phys Chem (1963) **67** 118
203. Ohashi T, Kozuka Z, Morijama J. J Jap Inst Metals (1966) **30** 785
204. Brebrick RF. J Phys Chem Solids (1969) **30** 719
205. Brebrick RF, Smith FTJ. J Electrochem Soc (1971) **118** 991
206. Tenne R. Adv Mat (1995) **5** 965
207. Rothchild A, Cohen SR, Tenne R. Appl Phys Lett (1999) **75** 4025
208. Zhegalina VA, Arakelyan ZS, Kalinnikov VT, Greenberg JH. Rus J Inorg Chem (1980) **25** 2807 (in Russian)
209. Haraldsen H, Mehmed F. Z Anorg Chem (1938) **B239** 369
210. Babitzina AA, Chernitzina MA, Kalinnikov VT. Rus J Inorg Chem (1975) **20** 3357 (in Russian)
211. Chevreton M, Murat M, Eyraud C, Bertaut EF. J Phys (1963) **24** 443
212. Chevreton M, Dumout B. Compt Rend (1968) **267** 884
213. Wehmeier FH, Keve ET, Abrahams SC. Inorg Chem (1970) **9** 2125
214. Alikhanyan AS, Steblevsky Novosyolova AV, Kalinnikov VT, Lazarev VB, Greenberg JH. Rus J Inorg Mat (1977) **13** 1194 (in Russian)
215. Wenzl H, Oates WA, Mika K. Defect thermodynamics and phase diagrams in compound crystal growth processes. In Hurle DTJ (ed) Handbook of crystal growth. North-Holland (1993) Amsterdam, London, New York, Tokyo, vol 1, pp 105–188
216. Rudolph P, Jurisch M. J Crystal Growth (1999) **198/199** 325
217. Blom JM, Woodall JM. J Electron Mater (1988) **17** 391
218. Wenzl H, Mika K, Henkel D. Cryst Res Technol (1990) **25** 699
219. Hurle DTJ. J Phys Chem Solids (1979) **40** 613
220. Morozov AN, Bublik VT, Morozova OY. Cryst Res Technol (1986) **21** 749
221. Lazarev VB, Shevchenko VY, Greenberg JH, Soblev VV. Semiconductor compounds $A^{II}B^{V}$. Nauka (1978) Moscow (in Russian)
222. Kakishita K, Baba T, Suda T. Thin Solid Films (1998) **334** 25
223. Lazarev VB, Greenberg JH, Marenkin SF, Magomedgadgiev GG, Samiev SH. Rus J Inorg Mater (1978) **14** 1961 (in Russian)
224. Ugai YA, Gukov OJ, Domashevskaya EP, Ozerov LA. Rus J Inorg Mater (1968) **4** 147 (in Russian)
225. Berak J, Pruchnik Z. Z Rochniki Chem (1969) **43** 1141
226. Greenberg JH, Lazarev VB, Kozlov SE, Shevchenko VY. J Chem Thermodynam (1974) **6** 1005
227. Shoonmaker RC, Venkitaraman AR, Lee PK. J Phys Chem (1967) **71** 2666
228. Valov YA, Ushakova TN. Rus J Gen Chem (1969) **40** 524 (in Russian)
229. Alikhanyan AS, Steblevsky AV, Greenberg JH. Rus J Inorg Mater (1978) **14** 1966 (in Russian)
230. Berak J, Pruchnik Z. An Soc Chim Pol (1968) **42** 1403
231. Krebs H, Muller KH, Zurn G. Z Anorg Allgem Chem (1956) **B295** 15
232. Kovalyova IS, Dmitriev Novosyolova AV, Zhizheiko IA. Rus J Inorg Mater (1966) **2** 403 (in Russian)
233. Lazarev VB, Greenberg JH, Marenkin SF, Samiev SH. Rus J Inorg Mater (1979) **15** 1149 (in Russian)

234. Lazarev VB, Marenkin SF, Maksimova SI, Huseinov B, Shevchenko VY. Rus J Inorg Mater (1979) **15** 749 (in Russian)
235. Lazarev VB, Shevchenko VY, Marenkin SF. Rus J Inorg Mater (1979) **15** 1701 (in Russian)
236. Schoonmaker RC, Lemmerman KJ. J Chem Eng Data (1972) **17** 139
237. Lyons VJ. J Phys Chem (1959) **63** 1142
238. Munir ZA, Benavides ME, Meschi DJ. High Temp Sci (1974) **6** 73
239. Marenkin SF, Shevchenko VY, Steblevsky AV, Alikhanyan AS. Rus J Inorg Mater (1980) **16** 1757 (in Russian)
240. Lazarev VB, Guskov VN, Greenberg JH. Mater Res Bull (1981) **16** 1113
241. Greenberg JH, Guskov VN, Lazarev VB, Kotlyar AA. Mater Res Bull (1982) **17** 1329
242. Greenberg JH, Guskov VN, Lazarev VB. J Chem Thermodynam (1985) **17** 739
243. Guskov VN, Greenberg JH, Lazarev VB, Kotlyar AA. Rus J Inorg Mater (1987) **23** 1418 (in Russian)
244. Guskov VN, Greenberg JH, Lazarev VB. Doklady Acad Sci USSR (1987) **292** 651 (in Russian)
245. Glazov VM, Kasimova M. Rus J Electron Technol, ser 14, Materials (1968) **1** 66 (in Russian)
246. Stull DR, Sinke GC. Thermodynamic properties of the elements. Amer Chem Soc (1956) Washington, DC
247. Gukov OJ, Ugai YA, Pshestanchik VR, Goncharov EG, Pakhomova NV. Rus J Inorg Mater (1970) **6** 1926 (in Russian)
248. Ugai YA, Marshakova TA, Aleinikova KB. Rus J Inorg Mater (1985) **21** 749 (in Russian)
249. Pruchnik Z. Mater Sci (1977) **3** 121
250. Nipan GD, Lazarev VB, Greenberg JH. Rus J Inorg Chem (1982) **27** 1788 (in Russian)
251. Nipan GD, Greenberg JH, Lazarev VB. Mater Res Bull (1985) **20** 1115
252. Clark JB, Pistorius CWET. High Temp–High Pressure (1973) **5** 319
253. Clark JB, Range KJ. Z Naturforsch B: Anorg Chem (1975) **30** 688
254. Pistorius CWFT. High Temp–High Press (1975) **7** 441
255. Nipan GD, Greenberg JH, Lazarev VB. Rus J Inorg Mater (1987) **23** 1423 (in Russian)
256. Guskov VN, Lazarev VB, Greenberg JH. Rus J Inorg Mater (1983) **19** 532 (in Russian)
257. Demidenko AF, Koschenko VI. Rus J Inorg Mater (1977) **13** 214 (in Russian)
258. Danilenko GD, Demidenko AF. Rus J Inorg Mater (1977) **13** 1736 (in Russian)
259. Nipan GD, Greenberg JH, Lazarev VB. Rus J Inorg Mater (1987) **23** 1596 (in Russian)
260. Nipan GD, Greenberg JH, Lazarev VB. Vapor pressure investigation of the P–T–X phase equilibria in Cd–As. Experimental results. VINITI (1988) #3854, Moscow (in Russian)
261. Danilenko GN, Demidenko AF, Koschenko VI. Rus J Inorg Mater (1978) **14** 1973 (in Russian)
262. Marenkin SF, Samiev SK, Shevchenko VY. Rus J Inorg Mater (1978) **14** 1971 (in Russian)
263. Nipan GD, Greenberg JH, Lazarev VB, Zelvensky MY. Rus J Inorg Mater (1989) **25** 1947 (in Russian)
264. Rau H. J Chem Thermodynam (1975) **7** 27
265. Thermal constants of substances. Ed Glushko VP. VINITI (1968) Moscow, vol 3 (in Russian)

266. Nipan GD, Greenberg JH, Lazarev VB. Rus J Phys Chem (1989) **63** 325 (in Russian)

267. Pistorius CWFT. High Temp–High Pressure (1975) **7** 723

268. Nipan GD, Greenberg JH, Lazarev VB. Rus J Inorg Mater (1989) **25** 357 (in Russian)

269. Nipan GD, Greenberg JH, Zelvensky MY. Rus J Phys Chem (1989) **63** 1042 (in Russian)

270. Libowitz GG, Lightstone JB. J Phys Chem Solids (1967) **28** 1145

271. Libowitz GG, Lightstone JB. J Chem Phys (1969) **30** 1025

272. Krtil P, Yoshimura M. J Solid State Electron (1998) **2** 321

273. Ravez J, Broustera C, Simon A. J Mater Chem (1999) **9** 1609

274. Su B, Choy KL. J Mater Chem (1999) **9** 1629

275. Vitins A, Vitinz G, Kraspins J, Steins I, Zalite I, Lusis A. J Solid State Electron (1998) **2** 299

276. Bazuev GV, Shveikin GP. Complex oxides of *d*- and *f*-transition elements. Nauka (1985) Moscow (in Russian)

277. Fotiev AA, Trunov VK. Vanadates of the divalent metals. Nauka (1985) Moscow (in Russian)

278. Tretyakov YD. Solid state reactions. Khimiya (1978) Moscow (in Russian)

279. Tretyakov YD, Lepis K. Chemistry and technology of the solid state materials. Moscow State University Press (1985) Moscow (in Russian)

280. Toropov NA, Borzakovsky VP, Bondar IA, Udalov YP. Phase diagrams of the silicate systems. Nauka (1969) Leningrad (in Russian)

281. Sanders HY. Chem Eng News (1984) **62** 26

282. Tretyakov YD. Thermodynamics of ferrites. Khimiya (1967) Leningrad (in Russian)

283. Men AN, Vorobjov YP, Chufarov GI. Physico-chemical properties of the non-stoichiometric oxides. Khimiya (1973) Leningrad (in Russian)

284. Toropov NA, Bondar IA, Lazarev AN, Smolin IY. Silicates of the rare earth elements. Nauka (1971) Leningrad (in Russian)

285. Rubinchik YS. Compounds of the binary oxides of the rare earth elements. Nauka i Tekhnika (1974) Minsk (in Russian)

286. Kofstadt P. Deviation from stoichiometry, effusion and electrical conductivity in simple metal oxides. Mir (1975) Moscow

287. Chemistry of silicates and oxides. Ed Shultz MM. Nauka (1982) Leningrad (in Russian)

288. Kuzminov YS. Ferroelectric crystals for guided laser radiation. Nauka (1982) Moscow (in Russian)

289. Portnoy KI, Timofeeva NI. Rare earth oxides. Metallurgiya (1986) Moscow (in Russian)

290. Lazarev VB, Sobolev VV, Shapligin IS. Chemical and physical properties of metal oxides. Nauka (1983) Moscow (in Russian)

291. Lazarev VB, Krasov Guskov VN, Shapligin IS. Electrical conductivity of oxide systems and film structures. Nauka (1979) Moscow (in Russian)

292. Kreidler ER. J Am Ceram Soc (1972) **55** 514

293. Kazenas EK, Chizhikov DM. Vapor pressure and composition of the vapors of oxides. Nauka (1976) Moscow (in Russian)

294. Marushkin KN, Alikhanyan AS, Greenberg JH, Lazarev Lazarev VB, Gorgoraki VI. I Chem Thermodynam (1985) **17** 245

295. Alikhanyah AS, Marushkin KN, Greenberg JH, Gorgoraki VI. J Cem Thermodynam (1988) **20** 1035

296. Scholder R, Schwarz H./ Z Anorg Chem (1968) **362** 149
297. Appendio P, Rawonda L. Ann Chim (Rome) (1971) **61** 61
298. Odoj R, Hilpert K. Z Phys Chem (1976) **102** 191
299. Marushkin KN, Alikhanyan AS, Greenberg JH, Kurtasov OV, Gorgoraki VI. Rus J Inorg Chem (1989) **34** 902 (in Russian)
300. Shevchenko AV, Lopato LM, Gerasimyuk GI. Rus J Inorg Mater (1987) **23** 1495 (in Russian)
301. Noguchi T, Okubo T, Yonemochi O. J Am Ceram Soc (1969) **52** 178
302. Duwez PS, Brown FN, Odell F. J Electrochem Soc (1951) **98** 356
303. Belov AN, Semenov GA. Rus J Inorg Mater (1980) **16** 2200 (in Russian)
304. Chaim R, Heuer AH, Brandon DG. J Am Ceram Soc (1986) **69** 243
305. Degtyarev SA, Voronin GF. Rus J Phys Chem (1987) **61** 617 (in Russian)
306. Duwez PS, Odell F, Brown FN. J Am Ceram Soc (1952) **35** 107
307. Viechnicki D, Stubikan VS. J Am Ceram Soc (1965) **48** 292
308. Maple MB. J Magnetism & Magn Mater (1998) **177-181** 18
309. Wong-Ng W, Cook LP, Paretzkin B, Hill MD, Stalick JK. J Am Ceram Soc (1994) **77** 2354
310. Wong-Ng W, Paretzkin B, Fuller ER. J Solid State Chem (19990) **85** 117
311. Wong-Ng W, Cook LP, Chiang CK, Swartzendruber LJ, Bennett LH, Blendell JE, Minor D. J Mater Res (1988) **3** 832
312. Wong-Ng W, Kuchinski MA, McMurdie HF, Paretzkin B. Powder Diffr (1989) **4** 1
313. Wong-Ng W, Paretzkin B, Chiang CK, Fuller ER. Powder Diffraction (1990) **5** 26
314. Kilbanow D, Sujata K, Mason TO. J Am Ceram Soc (1988) **71** C-267
315. Roth RS, Rawn C, Whitler J, Beech F. Phase equilibria in the Ba-Y-Cu-O-CO$_2$ system. In Yan MF (ed) Ceramic superconductors. Am Ceram Soc (1988) Westerville, OH, p 13
316. Hauck J. Supercond Sci Technol (1996) **9** 1033
317. Lindemer TB, Hunley JF, Gates JE, Sutton AL, Brynestad J, Hubbard CR, Gallagher PK. J Amer Ceram Soc (1989) **72** 1775
318. Kishio K, Suzuki K, Hasegawa T, Yamamoto T, Kitazava K. J Solid State Chem (1989) **82** 192
319. Yamaguchi S, Terabe K, Saito A, Yahagi S, Iguchi Y. Jap J Appl Phys (1988) **27** L179
320. Strobel P, Capponi JJ, Marezio M. Solid State Commun (1987) **64** 513
321. Specht ED, Sparks CJ, Dhere AG, Brynestad J, Cavin OB, Kroeger DM, Oye MA. Phys Rev B Condensed Matter (1988) **37** 7426
322. Guskov VN, Tarasov IV, Lazarev VB, Greenberg JH. J Solid State Chem (1995) **119** 62
323. Tarasov IV, Guskov VN, Lazarev VB, Shebershneva OV, Kovba ML. Vapor pressure of YBa$_2$Cu$_3$O$_y$. Experimental Data. VINITI (1993) Moscow (in Russian)
324. Lazarev VB, Gavrichev KS, Gorbunov VE, Greenberg JH. Thermochimica Acta (1991) **174** 27
325. Majewski P. Supercond Sci Technol (1997) **10** 453
326. Horiuchi S, Hirasaka M, Tsutsumi M, Kosuda K, Szerer MY, Ben-Dor L. Microscopy Res & Technol (1995) **30** 258
327. Greenberg JH, Ben-Dor L, Selig H. J Thermal Anal (1997) **50** 105
328. Chernyaev S, Hauck J, Mozhaev A, Bickmann K, Altenburg H. Physica C (1995) **243** 139

Index

P

parameters of state 5
peritectoid 65
phase 6
 – diagram, types of 26
 – surfaces 23
 – processes 12
phase equilibrium
 – experimental methods 91
 – geometrical representation 8
phase rule, Gibbs 7
phosphides
 – cadmium 171
 – zinc 170
points
 – conjugated 11
polymorphism 15
 – non-parallel 43
 – of a compound 71
 – of components 40
 – parallel 40
 – types of 18

R

resultant condensed phase diagram 25
retrograde solubility 23

S

Schreinemakers rule 24
sections
 – isobaric 40
 – isothermal 25, 30
silicon–chalcogenides 144
single crystals
 – binary and ternary 105
solid solution
 – universality 3
 – composition of 36
solidus 22
solution
 – ideal 26
 – non-ideal 31
solution, solid
 – composition of the 36
space model of a binary system 23
species chemical 5

state
 – equation of 6
 – of the system 5
state function 5
stoichiometry 3
sublimation curve 9
superconductors
 – bismuth high-T_c 230
 – high-T_c 218
 – rare-earth 218
systems
 – binary 20
 – eutectic 35
 – indium–selenium 140
 – peritectic 38
 – state of 5
 – ternary 135
 – thermodynamic 5

T

Three-phase surface 23
tin
 – selenides 148
 – sulfides 148
 – tellurides 151
triple point 10, 32

U

universality of a solid solution 3

V

Van der Waals equation 22
vapor pressure
 – direct 96
 – measurement 91, 96
 – scanning of non-stoichiometry
    100–103
variance 7
 – effect of restrictions 7

Z

zinc
 – arsenides 172
 – phosphides 170
 – selenium 107
 – tellurium 108
 – zinc–sulfur 106

Springer Series in

MATERIALS SCIENCE

Editors: R. Hull   R. M. Osgood, Jr.   J. Parisi   H. Sakaki   A. Zunger

* The 2nd edition is available as a textbook with the title: *Laser Processing and Chemistry*

Printing (Computer to Film): Saladruck Berlin
Binding: Stürtz AG, Würzburg